The Duffing Equation and Its Applications in Science and Engineering

The Duffing Equation and Its Applications in Science and Engineering

By

Álvaro H. Salas

Cambridge Scholars Publishing

The Duffing Equation and Its Applications in Science and Engineering

By Álvaro H. Salas

This book first published 2026

Cambridge Scholars Publishing

Lady Stephenson Library, Newcastle upon Tyne, NE6 2PA, UK

British Library Cataloguing in Publication Data
A catalogue record for this book is available from the British Library

ISBN: 978-1-0364-6940-5
ISBN (Ebook): 978-1-0364-6941-2

Contents

List of Symbols **xiv**

1 Introduction **1**

1.1 Nonlinear dynamics as a unifying framework 1

1.2 The Duffing oscillator: origin and fundamental properties 1

1.3 Geometric interpretation and phase–space structure 2

1.4 Analytical and numerical challenges 2

1.5 Fractional and generalized Duffing–type models 5

1.6 Organization of the book 6

2 Preliminaries **10**

2.1 Jacobian Elliptic Functions 11

2.2 Approximations for the function cn 13

2.2.1 Approximation for the function sn 14

2.3 Approximation for the function $\frac{d}{dt}\text{cn}\ (t) = -\text{sn}(t)\text{dn}(t)$. 15

2.4 Approximation for the function dn 15

2.5 Applications 15

2.5.1 Differentiation with variable modulus 15

2.5.2 Integration 16

2.5.3 Differential equations 16

2.5.4 Laplace transforms 17

2.5.5 Generalized elliptic functions 17

2.6 Summary 17

3 The conservative Duffing oscillator equation **19**

3.1 The cubic-quintic Duffing oscillator. 26

3.1.1 Mechanical interpretation and energy integral 27

3.1.2 Equilibria and their stability in terms of p, q, r 27

3.1.3 Global boundedness vs blow-up: when is the motion confined? 29

3.1.4 Phase portraits and bifurcations: parameter cases 29

3.1.5 A "for dummies" stability checklist 30

3.1.6 Worked examples (all in terms of parameters) 30

3.1.7 Illustrations: potential and phase portraits (TikZ templates) 31
3.1.8 Optional: small-amplitude frequency near a stable equilibrium 32
3.1.9 Exact solution 33
3.2 Two Particular Cases 35
3.2.1 First particular case: $\dot{x}_0 = 0$. 35
3.2.2 Second Particular Case: $x_0 = 0$. 36
3.3 Summary 37

4 The generalized Duffing oscillator equation 39
4.1 Stability analysis and phase–plane structure of the generalized Duffing–Helmholtz oscillator 39
4.1.1 Energy integral and potential function (the organizing principle) 40
4.1.2 Equilibria: where are the fixed points? 40
4.1.3 Linear stability test (local stability in one line) 40
4.1.4 How many equilibria? (parameter conditions via the cubic discriminant) . 41
4.1.5 Global boundedness vs escape (role of a_3) 41
4.1.6 Phase portraits from the potential: what you will see and why 42
4.1.7 Illustrations (potential shapes and phase portraits) 42
4.1.8 Worked examples (compute equilibria and classify them) 42
4.1.9 Practical algorithm (given parameters and initial data, decide the qualitative dynamics) 44
4.2 Exact Solution–Universal Principle 45
4.2.1 First approach 47
4.2.2 Second approach 49
4.2.3 Third approach 52
4.2.4 Fourth approach 53
4.2.5 Fifth approach 55
4.2.6 Why Duffing–type oscillators are universal models for conservative dynamics 56
4.3 Chaos in the forced–damped Duffing–Helmholtz oscillator: homoclinic orbits and Melnikov analysis 58
4.3.1 From the conservative oscillator to the forced–damped model 58
4.3.2 Phase–plane picture: equilibria, saddles, and separatrices 59
4.3.3 What is a homoclinic orbit and why it matters for chaos? 59
4.3.4 Melnikov method: general formula for the Duffing–Helmholtz oscillator . 60
4.3.5 Classical Duffing case: exact homoclinic orbit and exact Melnikov integrals 60
4.3.6 Why the same mechanism applies to the Duffing–Helmholtz equation . . . 62
4.3.7 Geometry of separatrix splitting (pictures you can actually understand) . 62
4.3.8 Mathematica computation of the Melnikov integrals (exact, when possible) 62
4.3.9 How to locate the saddle and the separatrix in terms of (a_0, a_1, a_2, a_3) . . 64

4.3.10 Parameter maps, explicit examples, and closure with the Melnikov chaos criterion . . . 67

4.4 Global signatures of chaos: bifurcations, Poincaré sections, Lyapunov exponents and a sufficient chaos theorem . . . 71

4.4.1 Bifurcation diagrams in the (γ, ω)–plane . . . 71

4.4.2 Poincaré sections: geometry of chaos . . . 72

4.4.3 Lyapunov exponents and their link to Melnikov theory . . . 73

4.4.4 A sufficient condition for chaos . . . 73

4.5 Summary . . . 74

4.6 Summary of chaotic dynamics in the generalized Duffing–Helmholtz oscillator . . 76

4.6.1 Sufficient conditions for chaos . . . 76

4.6.2 Global view: parameters, dynamics, and analytical tools . . . 77

4.6.3 Conceptual pathway from Duffing to chaos . . . 77

5 Perturbative solution methods 79

5.1 Lindstedt–Poincaré method for conservative oscillators . . . 79

5.2 The undamped and unforced Duffing oscillator . . . 79

5.2.1 The quadratic oscillator . . . 81

5.3 Krylov–Bogoliubov–Mitropolski (KBM) method . . . 82

5.3.1 Dufifng oscillator with forcing and damping . . . 84

5.3.2 Double–Well Duffing oscillator . . . 87

5.3.3 Duffing oscillator with forcing and damping . . . 90

5.3.4 Duffing–Helmholtz oscillator with damping and forcing . . . 92

5.4 The multiple time scales method (MSM) . . . 93

5.4.1 Unforced and undamped Duffing equation . . . 95

5.4.2 The oscillator $\ddot{x} + \omega_0 x + \beta x^2 = 0$ . . . 96

5.4.3 The undamped and unforced cubic-quintic Duffing equation . . . 97

5.4.4 Damped and forced Duffing equation . . . 98

5.4.5 The quadratic oscillatlor $\ddot{x} + 2\varepsilon\dot{x} + \omega_0^2 x + \beta x^2 = f(t)$ . . . 99

5.5 Summary . . . 100

6 Elliptic Averaging Methods 102

6.1 General Idea of Averaging Methods . . . 103

6.2 Trigonometric Averaging Method . . . 104

6.3 Elliptic Averaging Method . . . 106

6.4 Extended Averaging Method . . . 118

6.4.1 Trigonometric form . . . 118

6.4.2 Elliptic form . . . 119

6.5 Cubication-Quintification Method . . . 125

6.6 Summary . . . 133

7 Numerical Methods **135**
7.1 Sixth-Order Runge–Kutta Method 135
7.2 Predictor–Corrector Methods for $\ddot{y} = f(t, y, \dot{y})$ (Explicit Coefficients) 138
7.3 Derivatives 144
7.3.1 General Formulas 147
7.4 Integrals 148
7.4.1 Chebyshev Formulas 148
7.4.2 Singular Integrals 149
7.5 Finite Differences 150
7.6 Fourier series Method 152
7.7 GalerkinMethod 158
7.8 The Collocation Method 164
7.8.1 First Variant 164
7.8.2 Second Variant 166
7.8.3 Third Variant 167
7.9 Summary 168

8 Geodesics and Planetary Motions **170**
8.1 Motivation and overview 170
8.2 From geodesics to a Duffing–Helmholtz Binet equation 171
8.2.1 Massive particles in Reissner–Nordström spacetime 171
8.2.2 Light deflection 171
8.2.3 Unified Duffing–Helmholtz form 171
8.3 Aphelion–perihelion data and the first integral 171
8.3.1 Geometric meaning of the polynomial structure 172
8.4 Illustrations of the orbital families 172
8.5 Exact solution for null geodesics: discriminant splitting 173
8.5.1 Case I: $\Delta = b_1^2 - 4b_0 \geq 0$ (Jacobi elliptic form) 173
8.5.2 Case II: $\Delta = b_1^2 - 4b_0 < 0$ (Weierstrass form) 174
8.6 Particular case: $a_3 = 0$ (quadratic nonlinearity) 174
8.7 Application: Einstein equation for planetary motion (Solar System) 175
8.8 Motion in the Reissner–Nordström metric 176
8.9 Concluding remarks 176

9 Applications of Duffing–Type Oscillators **177**
9.1 Simple pendulum (nonlinear conservative oscillator) 179
9.2 Spring pendulum with nonlinear stiffness 180
9.2.1 Mechanical model and coordinates 181
9.2.2 Lagrangian formulation 181
9.2.3 Reduction to a single Duffing–Helmholtz equation 182
9.2.4 Physical interpretation 182

9.3 The Lagrange top (heavy symmetric top) 183
9.4 Kapitza pendulum (vertically vibrated pivot) 187
9.5 Rolling cylinder inside a fixed circular guide 191
9.6 Pendulum with a Horizontally Moving Support 193
9.6.1 Geometric description 193
9.6.2 Kinetic and potential energy 194
9.6.3 Lagrangian formulation 194
9.6.4 Reduction to an effective nonlinear oscillator 194
9.6.5 Duffing–type approximation 195
9.6.6 Conceptual significance 195
9.7 Relativistic pendulum model 195
9.7.1 Equation of motion 195
9.7.2 Energy integral 196
9.7.3 Quadrature for the period 197
9.7.4 Reduction to a Duffing–type equation 197
9.7.5 Small–angle expansion 198
9.7.6 Averaging the inertial nonlinearity 198
9.7.7 Amplitude–dependent frequency from Duffing theory 199
9.8 The Rayleigh oscillator 200
9.8.1 Derivation and physical motivation 201
9.8.2 Energy balance and limit–cycle formation 201
9.8.3 Phase–space structure 201
9.8.4 Connection with Duffing theory 202
9.8.5 The Rayleigh oscillator: detailed derivation and second–order approximation 202
9.8.6 Rayleigh equation as a Liénard self–excited system 203
9.8.7 Energy balance and amplitude selection (not handwaving) 203
9.8.8 First–order KBM/averaging equations 204
9.8.9 Second–order approximation via Lindstedt–Poincaré (full details) 204
9.8.10 How this plugs into Duffing theory (Rayleigh–Duffing bridge) 207
9.8.11 A picture (qualitative phase portrait) 207
9.8.12 Role within the Duffing centered framework 207
9.9 The Van der Pol–Duffing oscillator: detailed derivation and higher–order analysis 208
9.9.1 Governing equation and physical origin 208
9.9.2 Energy balance and qualitative dynamics 209
9.9.3 First–order averaging: amplitude and phase equations 209
9.9.4 Second–order approximation by the Lindstedt–Poincaré method 210
9.9.5 Elliptic averaging method for the Van der Pol–Duffing oscillator 210
9.9.6 Phase portrait and geometric interpretation 214
9.9.7 Position within Duffing theory 214
9.10 Coupled Van der Pol oscillators 215
9.10.1 First–order form and complex amplitudes 215

9.10.2 Amplitude–phase equations . . . 216
9.10.3 Synchronized solutions . . . 216
9.10.4 In–phase synchronized time series . . . 217
9.10.5 Lissajous synchronization plot . . . 217
9.11 KBM approximation for the coupled pendula and its in–phase/out–of–phase reductions . . . 218
9.11.1 Weakly nonlinear model and modal coordinates . . . 218
9.11.2 First–order KBM approximation (non–resonant case) . . . 219
9.11.3 Second–order KBM approximation . . . 220
9.11.4 Particular solutions: in–phase and out–of–phase modes . . . 221
9.11.5 Discussion . . . 222
9.12 Study of dissipative nonplanar solitons in an electronegative complex plasma . . 222
9.12.1 Damped Cubic-Quintic Equation . . . 223
9.12.2 The Cubic-Quintic Duffing equation and its Applications to plasma physics and rigid rotator . . . 230
9.12.3 Damped Nonlinear Schrodinger equation . . . 230
9.12.4 Traveling wave solutions . . . 231
9.12.5 Cnoidal Wave Solutions . . . 231
9.12.6 Breathers . . . 232
9.12.7 Elliptic Breathers . . . 233
9.13 Damped harmonic oscillator with higher-order nonlinearities . . . 236
9.14 Summary . . . 238

10 Elliptic Method for finding Riemann Zeta Zeros 240
10.1 Introduction . . . 240
10.2 Background and classical approximation . . . 242
10.2.1 Elliptic exponential, elliptic Dirichlet series, and convergence . . . 242
10.2.2 Expected numerical benefits . . . 244
10.2.3 Scope and limitations . . . 244
10.3 An elliptic deformation of the Riemann zeta function . . . 245
10.3.1 Motivation . . . 245
10.3.2 From trigonometric to elliptic phase . . . 245
10.3.3 Scope and limitations . . . 245
10.4 Elliptic França–Leclair seeds . . . 246
10.4.1 Definition of the elliptic family . . . 246
10.4.2 Consistency with the classical approximation . . . 246
10.4.3 Smooth dependence on the elliptic parameter . . . 247
10.4.4 Interpretation as a phase–scale correction . . . 247
10.4.5 Role in numerical root finding . . . 247
10.4.6 Summary . . . 248
10.5 Hardy Z–function and residuals . . . 248

10.5.1 Residuals associated with approximate zeros . . . 248
10.5.2 Smoothness and local structure of the residual . . . 249
10.5.3 Interpretation in terms of phase error . . . 249
10.5.4 Role in Newton–type solvers . . . 249
10.5.5 Preparation for reparametrized root finding . . . 250
10.5.6 Summary . . . 250
10.6 Root finding in the elliptic parameter . . . 250
10.6.1 Elliptic reparametrization: the composed root function . . . 250
10.6.2 Why the reparametrization can improve conditioning . . . 251
10.6.3 Local convergence: a standard sufficient condition . . . 251
10.6.4 Practical derivative control and a diagnostic test . . . 252
10.6.5 Why solving in m can reduce iteration counts . . . 252
10.6.6 Two-level strategy: solve in m, then polish in t . . . 253
10.6.7 Direct numerical implementation at extreme height . . . 253
10.7 Algorithm and implementation details . . . 256
10.7.1 Algorithm (elliptic reparametrization with polishing) . . . 256
10.7.2 Mathematica code (copy–paste) . . . 256
10.7.3 Notes on robustness at large height . . . 257
10.8 Computational complexity . . . 258
10.8.1 Cost model and basic assumptions . . . 258
10.8.2 Cost of evaluating the elliptic seed . . . 258
10.8.3 Cost of Hardy Z–function evaluation . . . 259
10.8.4 Complexity of root finding in the elliptic parameter . . . 259
10.8.5 Comparison with direct Newton iteration in t . . . 260
10.8.6 Overall complexity summary . . . 260
10.8.7 Scope of applicability . . . 260
10.8.8 Conclusion of the complexity analysis . . . 261
10.9 Numerical validation at extreme height . . . 261
10.9.1 Reference data and benchmarking protocol . . . 261
10.9.2 Equation being solved . . . 262
10.9.3 Justification of the search interval $[-1/n, 0]$ . . . 262
10.10 Numerical validation at extreme height . . . 263
10.10.1 Experimental setup . . . 263
10.10.2 Choice of search interval . . . 264
10.10.3 Reference data and comparison protocol . . . 264
10.10.4 Results at height $n \sim 10^9$ . . . 265
10.10.5 Interpretation . . . 265
10.10.6 Numerical significance . . . 266
10.10.7 Scope . . . 266
10.11 Discussion . . . 266
10.12 Conclusion . . . 268

11 Elliptic Method for finding Riemann Zeta Zeros **269**
11.1 Introduction . . . 269
11.2 Background and classical approximation . . . 271
11.2.1 Elliptic exponential, elliptic Dirichlet series, and convergence . . . 271
11.2.2 Expected numerical benefits . . . 273
11.2.3 Scope and limitations . . . 273
11.3 An elliptic deformation of the Riemann zeta function . . . 274
11.3.1 Motivation . . . 274
11.3.2 From trigonometric to elliptic phase . . . 274
11.3.3 Scope and limitations . . . 274
11.4 Elliptic França–Leclair seeds . . . 275
11.4.1 Definition of the elliptic family . . . 275
11.4.2 Consistency with the classical approximation . . . 275
11.4.3 Smooth dependence on the elliptic parameter . . . 276
11.4.4 Interpretation as a phase–scale correction . . . 276
11.4.5 Role in numerical root finding . . . 276
11.4.6 Summary . . . 277
11.5 Hardy Z–function and residuals . . . 277
11.5.1 Residuals associated with approximate zeros . . . 277
11.5.2 Smoothness and local structure of the residual . . . 278
11.5.3 Interpretation in terms of phase error . . . 278
11.5.4 Role in Newton–type solvers . . . 278
11.5.5 Preparation for reparametrized root finding . . . 279
11.5.6 Summary . . . 279
11.6 Root finding in the elliptic parameter . . . 279
11.6.1 Elliptic reparametrization: the composed root function . . . 279
11.6.2 Why the reparametrization can improve conditioning . . . 280
11.6.3 Local convergence: a standard sufficient condition . . . 280
11.6.4 Practical derivative control and a diagnostic test . . . 281
11.6.5 Why solving in m can reduce iteration counts . . . 281
11.6.6 Two-level strategy: solve in m, then polish in t . . . 282
11.6.7 Direct numerical implementation at extreme height . . . 282
11.7 Algorithm and implementation details . . . 285
11.7.1 Algorithm (elliptic reparametrization with polishing) . . . 285
11.7.2 Mathematica code (copy–paste) . . . 285
11.7.3 Notes on robustness at large height . . . 286
11.8 Computational complexity . . . 287
11.8.1 Cost model and basic assumptions . . . 287
11.8.2 Cost of evaluating the elliptic seed . . . 287
11.8.3 Cost of Hardy Z–function evaluation . . . 288
11.8.4 Complexity of root finding in the elliptic parameter . . . 288

11.8.5 Comparison with direct Newton iteration in t . . . 289
11.8.6 Overall complexity summary . . . 289
11.8.7 Scope of applicability . . . 289
11.8.8 Conclusion of the complexity analysis . . . 290
11.9 Numerical validation at extreme height . . . 290
11.9.1 Reference data and benchmarking protocol . . . 290
11.9.2 Equation being solved . . . 291
11.9.3 Justification of the search interval $[-1/n, 0]$ . . . 291
11.10 Numerical validation at extreme height . . . 292
11.10.1 Experimental setup . . . 292
11.10.2 Choice of search interval . . . 293
11.10.3 Reference data and comparison protocol . . . 293
11.10.4 Results at height $n \sim 10^9$ . . . 294
11.10.5 Interpretation . . . 294
11.10.6 Numerical significance . . . 295
11.10.7 Scope . . . 295
11.11 Discussion . . . 295
11.12 Conclusion . . . 297

12 Conclusions and Recommendations **298**
12.1 General Conclusions . . . 298
12.2 Recommendations and Perspectives for Future Work . . . 300
Recommendations and Perspectives for Future Work . . . 300

Glossary **302**

Bibliography **307**

About the Author **316**

Index **317**

List of Symbols

a, b, c Constant coefficients.

A Oscillation amplitude.

$A(t)$ Slowly varying amplitude.

B Auxiliary amplitude or integration constant.

C Constant of integration.

E Energy (mechanical or relativistic).

F External forcing amplitude.

$f(t)$ External forcing function.

g Gravitational acceleration.

H Hamiltonian.

I Action integral.

k Linear stiffness coefficient.

k_2, k_3, k_5 Quadratic, cubic, and quintic stiffness coefficients.

L Lagrangian.

m Mass parameter.

n Integer index or mode number.

p, q, r Polynomial or perturbation parameters.

t Time variable.

T Period of oscillation.

$V(x)$ Potential energy.

$x(t)$ Generalized displacement.

$y(t), z(t)$ Auxiliary state variables.

$\dot{x}$ Velocity.

$\ddot{x}$ Acceleration.

α Fractional order or nonlinearity parameter.

β Cubic nonlinearity coefficient.

γ Damping coefficient.

δ Detuning or small perturbation parameter.

ϵ Bookkeeping perturbation parameter.

η Frequency correction.

κ Effective stiffness.

λ Eigenvalue or Lyapunov exponent.

μ Control or bifurcation parameter.

ν Frequency ratio.

ω Natural angular frequency.

Ω Forcing angular frequency.

ϕ Phase.

ψ Auxiliary phase.

ρ Fractional order or density.

σ Growth or decay rate.

τ Slow time variable.

D_t^α Fractional derivative of order α.

${}^C D_t^\alpha$ Caputo fractional derivative.

${}^R D_t^\alpha$ Riemann–Liouville fractional derivative.

$\Gamma(\cdot)$ Gamma function.

$E_\alpha(\cdot)$ Mittag–Leffler function.

$E_{\alpha,\beta}(\cdot)$ Two-parameter Mittag–Leffler function.

$\mathcal{K}(t)$ Memory kernel.

$*$ Convolution operator.

$\mathrm{sn}(u, k)$ Jacobi elliptic sine function.

$\mathrm{cn}(u, k)$ Jacobi elliptic cosine function.

$\mathrm{dn}(u, k)$ Jacobi elliptic delta function.

(u, k) Jacobi amplitude.

$K(k)$ Complete elliptic integral of the first kind.

$E(k)$ Complete elliptic integral of the second kind.

$\Pi(n, k)$ Elliptic integral of the third kind.

k Elliptic modulus.

$m = k^2$ Elliptic parameter.

$\wp(z)$ Weierstrass elliptic function.

g_2, g_3 Weierstrass invariants.

ω_1, ω_2 Fundamental periods.

$\mathcal{A}$ Attractor.

$\mathcal{B}$ Basin of attraction.

Σ Phase–space manifold.

Φ^t Flow map.

λ_L Largest Lyapunov exponent.

$\mathbf{x}$ State vector.

$\mathbf{J}$ Jacobian matrix.

t_0, t_1, t_2 Multiple time scales.

$\theta(t)$ Slowly varying phase.

$\langle\cdot\rangle$ Time-averaging operator.

$\mathcal{O}(\epsilon^n)$ Order of approximation.

$\Delta\omega$ Frequency shift.

Δt Time step.

N Number of discretization points.

x_n Numerical solution at time step n.

$\|\cdot\|$ Norm.

$\varepsilon_{\mathbf{err}}$ Numerical error.

CPU Computational time.

r Radial coordinate.

θ, φ Angular coordinates.

τ Proper time.

L_z Angular momentum.

$V_{\mathbf{eff}}$ Effective potential.

M Central mass.

Q Electric charge.

c Speed of light.

Chapter 1

Introduction

1.1 Nonlinear dynamics as a unifying framework

Nonlinear dynamical systems constitute one of the central pillars of modern applied mathematics, physics, and engineering. Unlike linear models, which assume proportionality between causes and effects, nonlinear systems capture the intrinsic feedback mechanisms responsible for multistability, self–excitation, resonance saturation, bifurcations, and deterministic chaos. Such phenomena arise naturally in a wide variety of physical contexts, including large–amplitude vibrations of elastic structures, nonlinear electrical circuits, micro– and nano–electromechanical systems (MEMS and NEMS), plasma oscillations, nonlinear wave propagation, and relativistic motion in gravitational fields [34, 69, 94, 120, 134].

Over the past century, nonlinear models have evolved from being regarded as pathological deviations from linear theory into indispensable tools for understanding real physical systems. Classical texts on vibrations and perturbation theory already recognized that even weak nonlinearities may produce qualitative changes in system behaviour, such as amplitude–dependent frequencies and resonance hysteresis [34, 94]. With the advent of chaos theory in the second half of the twentieth century, it became evident that nonlinear systems can exhibit extreme sensitivity to initial conditions, fractal phase–space structures, and long–term unpredictability despite being governed by deterministic equations [28, 91, 100, 120].

Among the many models that have shaped the development of nonlinear science, few have attained the paradigmatic status of the Duffing equation. Its combination of mathematical simplicity and extraordinary dynamical richness has made it a canonical reference model across disciplines.

1.2 The Duffing oscillator: origin and fundamental properties

The Duffing equation was originally introduced by Georg Duffing in 1918 in the context of the buckling of compressed elastic beams [39]. In its most common forced and damped form, it reads

$$\ddot{x} + 2\epsilon\dot{x} + \alpha x + \beta x^3 = \gamma\cos(\omega t), \tag{1.1}$$

where α determines the linear stiffness, β controls the cubic nonlinearity, ϵ represents viscous damping, and γ, ω denote the amplitude and frequency of the external forcing. The sign of β distinguishes softening ($\beta > 0$) from hardening ($\beta < 0$) springs.

Despite its compact form, the Duffing equation exhibits a remarkable spectrum of dynamical behaviours: periodic oscillations with amplitude–dependent frequencies, bistability and hysteresis, homoclinic and heteroclinic orbits, period–doubling cascades, quasiperiodicity, and deterministic chaos [28, 72, 91, 100, 107]. Even in the conservative case ($\epsilon = \gamma = 0$), the cubic term βx^3 introduces separatrices in phase space and energy–dependent periods that have no counterpart in linear oscillation theory [20, 65].

A key reason for the enduring relevance of the Duffing equation lies in its role as a *universal normal form* for weakly nonlinear oscillators near pitchfork and saddle–node bifurcations. As a result, it emerges naturally in mechanical systems, nonlinear RLC circuits with saturable inductors, Josephson junctions, optical cavities, plasma waves, and micro–resonators [12, 14, 60, 80]. In many of these applications, the onset of chaos is not merely a mathematical curiosity but a serious engineering concern, as it may compromise structural integrity, signal stability, or control performance [112, 137].

1.3 Geometric interpretation and phase–space structure

When damping and forcing are absent, equation (9.53) is conservative and can be written as a Newtonian system with potential

$$V(x) = \tfrac{1}{2}\alpha x^2 + \tfrac{1}{4}\beta x^4.$$

Depending on the signs of α and β, this potential describes either a single–well or a double–well configuration. In the double–well case ($\alpha < 0$, $\beta > 0$), the phase space is organized by closed periodic orbits, stable equilibrium points, and a separatrix passing through the unstable saddle at the origin [51, 100].

A geometric overview of the fundamental dynamical mechanisms underlying the Duffing oscillator is provided in Figure 1.1.

1.4 Analytical and numerical challenges

Over the past five decades, an extensive arsenal of analytical techniques has been applied to the Duffing equation. Classical perturbation methods, such as Lindstedt–Poincaré and multiple–time–scale expansions, provide accurate results in weakly nonlinear regimes [20, 65, 94]. Krylov–Bogoliubov–Mitropolsky averaging extends these approaches to forced and damped systems, but often suffers from limited validity near hysteresis and chaos thresholds [107].

Exact solutions expressed in terms of Jacobi and Weierstrass elliptic functions offer an alternative route, yielding closed–form expressions for conservative and certain dissipative variants [44, 72, 114]. More recently, trigonometric and elliptic averaging techniques have been

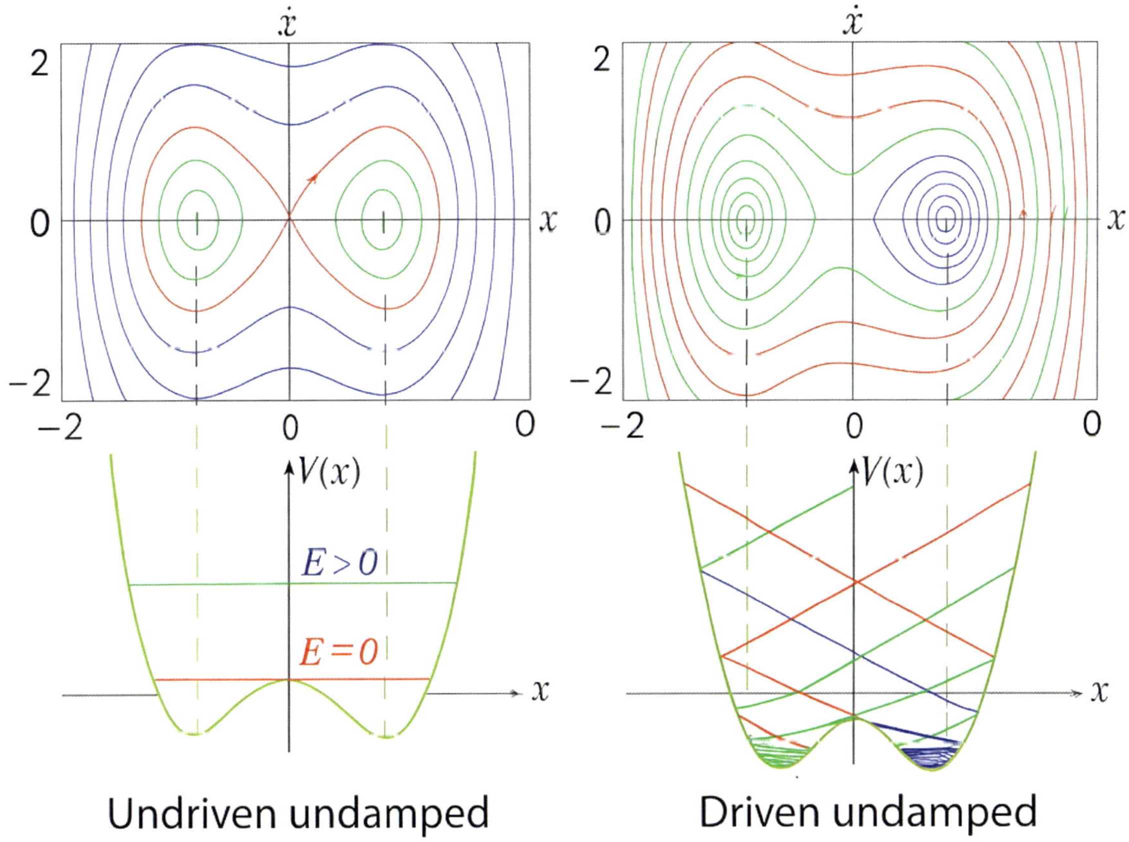

Figure 1.1: Qualitative phase–space and configuration–space structure of the Duffing oscillator. In the undriven and undamped case, trajectories lie on closed invariant curves corresponding to constant total energy, separated by a separatrix passing through the unstable saddle at the origin. In the driven undamped case, energy conservation is destroyed and transport between different energy regions becomes possible.

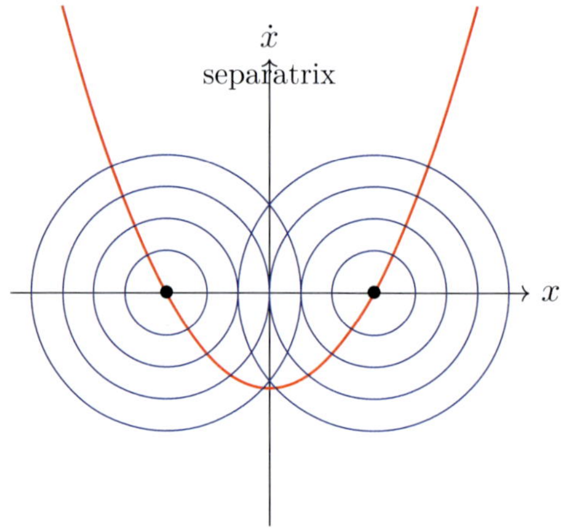

Figure 1.2: Phase portrait of the conservative Duffing oscillator showing two stable centers and the separatrix passing through the saddle at the origin.

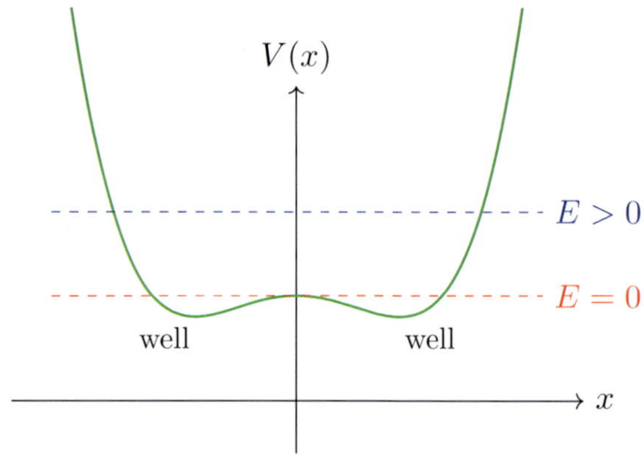

Figure 1.3: Double–well potential of the Duffing oscillator with representative energy levels.

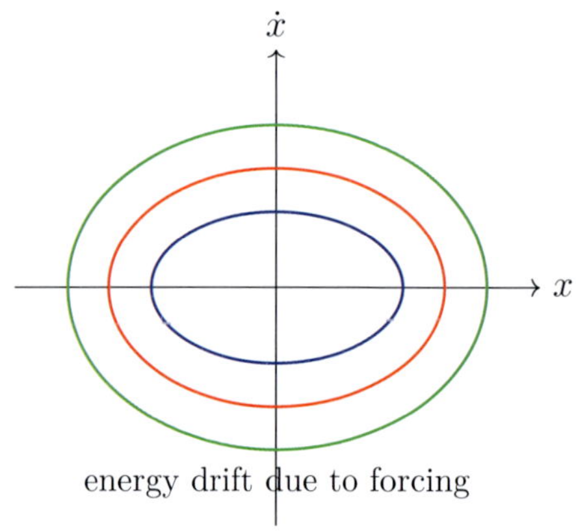

Figure 1.4: Schematic phase–space trajectories of the driven undamped Duffing oscillator illustrating the destruction of closed orbits and energy conservation.

developed to bridge the gap between perturbative and exact methods, achieving high accuracy even in strongly nonlinear regimes [14, 90].

Complementing analytical approaches, high–fidelity numerical methods such as Runge–Kutta schemes, symplectic integrators, spectral methods, and collocation techniques provide reliable reference solutions [12, 22, 57]. Nevertheless, many studies focus on isolated parameter sets, and systematic comparisons between analytical and numerical methods across wide parameter ranges remain scarce.

1.5 Fractional and generalized Duffing–type models

While the classical Duffing equation has proven remarkably successful in describing a wide range of nonlinear oscillatory phenomena, many real physical systems exhibit features that cannot be adequately captured within the framework of integer–order differential equations. In particular, effects such as material memory, hereditary behavior, anomalous dissipation, and long–range temporal correlations are fundamentally nonlocal in time and lie beyond the scope of classical models. These limitations have motivated the development of generalized Duffing–type equations incorporating additional nonlinear terms, higher–order corrections, and fractional–order derivatives.

Generalized Duffing models arise naturally when the restoring force is expanded beyond the cubic approximation. Cubic–quintic Duffing equations, for example, appear in nonlinear beam theory, plasma physics, and optical systems, where higher–order nonlinearities become relevant at moderate or large amplitudes. Similarly, Helmholtz–Duffing oscillators emerge when quadratic and cubic nonlinearities coexist, leading to asymmetric potentials and richer bifurcation structures than those found in the symmetric cubic case. These generalized models retain the essential nonlinear features of the Duffing oscillator while introducing new dynamical regimes, including shifted equilibria, broken symmetries, and modified resonance characteristics.

A further and conceptually deeper extension is obtained by replacing the integer–order time derivatives in the Duffing equation with derivatives of fractional order. Fractional calculus provides a natural mathematical framework for modeling systems with memory and hereditary effects, as the fractional derivative at a given time depends on the entire past history of the state variable. Fractional–order models have been successfully applied to a wide range of physical systems, including viscoelastic materials, dielectric relaxation, anomalous diffusion, plasma dynamics, and complex mechanical systems, where classical integer–order equations fail to capture long–range temporal correlations [137].

In the context of nonlinear oscillations, fractional Duffing and Duffing–Helmholtz equations introduce fundamentally new dynamical behavior. The presence of fractional derivatives modifies the effective damping and inertia of the system, alters phase–space structures, and may suppress, enhance, or qualitatively change the onset of bifurcations and chaos. Even in the absence of explicit damping terms, fractional derivatives can generate effective dissipation due to their nonlocal nature, leading to decay rates and transient dynamics that differ substantially from those predicted by classical models.

A representative fractional Duffing–Helmholtz oscillator can be written in the form

$$D_t^{\rho} x(t) + 2\epsilon D_t^{\rho-1} x(t) + a_0 + a_1 x + a_2 x^2 + a_3 x^3 = \gamma \cos(\omega t), \tag{1.2}$$

where D_t^ρ denotes the Caputo fractional derivative of order $\rho \in (1, 2]$, and the polynomial restoring force generalizes both Duffing and Helmholtz–type nonlinearities. For $\rho = 2$ and appropriate parameter choices, equation (1.2) reduces to the classical integer–order models discussed earlier in this chapter.

The analysis of fractional Duffing–type equations poses significant analytical and numerical challenges. Standard techniques from nonlinear dynamics, such as phase–plane analysis and energy methods, must be reformulated or replaced, since fractional systems are inherently infinite–dimensional and generally lack a classical Hamiltonian structure. Likewise, numerical integration requires specialized algorithms capable of handling memory terms and nonlocal operators with acceptable computational cost and accuracy.

Despite these challenges, fractional Duffing models have attracted increasing attention in recent years due to their ability to reproduce experimentally observed phenomena that are inaccessible to integer–order descriptions. The combination of fractional dynamics with nonlinear restoring forces provides a powerful and flexible modeling framework, capable of capturing both strong nonlinearity and long–memory effects within a unified mathematical structure.

For these reasons, fractional and generalized Duffing–type oscillators occupy a central role in this book. They serve not only as extensions of the classical Duffing equation, but also as testbeds for the analytical and numerical methods developed in subsequent chapters. Their study highlights both the strengths and limitations of existing approximation techniques and motivates the search for new solution strategies capable of addressing the full complexity of nonlinear systems with memory.

1.6 Organization of the book

This monograph develops a unified analytical and computational framework for Duffing–type nonlinear oscillators, ranging from the classical cubic Duffing equation to generalized, cubic–quintic, Duffing–Helmholtz, and fractional formulations. The guiding philosophy throughout is that of *structural continuity*: each extension of the Duffing equation is treated not as an isolated generalization, but as a member of a coherent hierarchy of nonlinear dynamical systems sharing common geometric, analytical, and numerical structures.

A central theme of the book is that low–degree polynomial nonlinearities, elliptic functions, and averaging principles form a unifying backbone across a wide range of problems in applied mathematics, physics, and computational science. The Duffing equation and its variants are used as canonical normal forms through which exact solutions, perturbative approximations, numerical algorithms, and even problems of a nontraditional origin can be systematically analyzed.

The exposition progresses from foundational theory to advanced analytical techniques and, finally, to representative applications that demonstrate both the power and the limitations of Duffing–type models. Analytical developments are consistently benchmarked against high–accuracy numerical computations in order to delineate domains of validity, quantify approximation errors, and identify the mechanisms responsible for qualitative changes in the dynamics.

Chapter 2 establishes the mathematical preliminaries required throughout the book. Special emphasis is placed on Jacobi and Weierstrass elliptic functions, which provide the natural language for exact solutions of conservative Duffing systems. Trigonometric and elliptic approximations, differentiation and integration formulas, and auxiliary analytical tools are developed with a view toward later applications in perturbation theory, averaging methods, and numerical analysis.

Chapter 3 is devoted to the conservative Duffing oscillator. First integrals are derived, phase–space structures are analyzed in detail, and exact periodic solutions are constructed in terms of elliptic functions. Extensions to cubic–quintic nonlinearities are presented to illustrate how higher–order restoring forces modify the geometry of the phase portrait, the existence of separatrices, and the dependence of the oscillation period on energy.

Chapter 4 introduces generalized Duffing equations and the Duffing–Helmholtz oscillator. Several complementary formulations are developed, revealing structural equivalences between apparently distinct nonlinear oscillator models. These equivalence transformations allow results obtained for one class of systems to be systematically transferred to others, highlighting the unifying role of low–degree polynomial restoring forces.

Chapters 5 and 6 are devoted to analytical approximation methods. Classical techniques, including the Lindstedt–Poincaré method, Krylov–Bogoliubov–Mitropolsky averaging, and multiple–time–scale expansions, are developed and applied to conservative, forced, and damped Duffing–type systems. Their accuracy and limitations are assessed through direct comparison with exact or high–precision numerical solutions, with particular attention to regimes near resonance, hysteresis, and bifurcation thresholds.

Chapter 7 presents advanced averaging methods suitable for strongly nonlinear regimes. Trigonometric, elliptic, and extended averaging techniques are introduced, together with cubication and quintification procedures that map complex nonlinearities onto analytically tractable surrogate models. These methods constitute a central methodological contribution of the book and provide a conceptual bridge between exact elliptic solutions and purely numerical approaches.

Chapter 8 is devoted to numerical methods. High–accuracy reference solutions are obtained using Runge–Kutta schemes, predictor–corrector algorithms, spectral and collocation methods, finite–difference discretizations, and Galerkin techniques. These numerical solutions serve both as practical computational tools and as benchmarks for validating the analytical approximations developed in earlier chapters.

Chapter 9 establishes a connection between nonlinear oscillation theory and relativistic dynamics by applying Duffing–Helmholtz–type equations to geodesic motion and planetary dynamics. The geodesic equations in spherically symmetric spacetimes are reduced to nonlinear Binet–type equations, exact solutions are constructed using elliptic functions, and relativistic effects such as perihelion precession and light deflection are analyzed within a unified nonlinear framework.

Chapter 10 is devoted to the fractional Duffing oscillator and fractional Duffing–Helmholtz models. Fractional derivatives in the Caputo sense are introduced, and the effects of memory, hereditary behavior, and nonlocality on nonlinear dynamics are examined in detail. Analytical and numerical solution strategies are developed, and the validity of integer–order surrogate models is critically assessed.

Chapter 11 departs from classical physical applications and applies the analytical and averaging philosophy developed in the book to the numerical computation of the nontrivial zeros of the Riemann zeta function on the critical line. The Hardy Z–function formulation, asymptotic inversion of the Riemann–von Mangoldt counting function, and the França–Leclair approximation are reviewed and extended. An elliptic reparametrization strategy is introduced to improve the conditioning of zero–finding algorithms, providing a novel bridge between nonlinear oscillation theory, elliptic functions, and experimental mathematics. This chapter illustrates how tools originally developed for nonlinear dynamics can be fruitfully transferred to problems in analytic number theory without altering their underlying analytic structure.

Finally, Chapter 12 summarizes the principal results of the book and outlines directions for future research. Open problems, methodological limitations, and emerging applications of nonlinear, fractional, and elliptic oscillator models are discussed, with particular emphasis on cross–disciplinary problems in which strong nonlinearity, memory effects, and large–scale computation coexist.

Taken together, the chapters provide a self–contained and systematic treatment of Duffing–type nonlinear oscillators and their extensions, aimed at researchers and advanced graduate students in applied mathematics, physics, engineering, and computational science who seek both theoretical insight and practical, reproducible analytical tools.

A Unifying Perspective: What This Book Solves for Science

The central scientific contribution of this book is not the study of a single equation, but the identification and systematic exploitation of a *structural universality* shared by a broad class of nonlinear dynamical systems. The Duffing equation and its generalizations serve here as a mathematical backbone through which apparently disparate physical, engineering, and even mathematical problems can be analyzed within a common conceptual and computational framework.

At its core, the book addresses a long–standing fragmentation in nonlinear science. Exact methods, perturbative techniques, numerical simulations, and fractional models are often developed in isolation, each tailored to a narrow class of problems. This work shows that many of these approaches are not competing alternatives but manifestations of a single underlying structure: low–degree polynomial restoring forces combined with conservative, dissipative, or memory–driven dynamics. By placing Duffing–type oscillators at the center, the book provides a coherent language in which these methods can be compared, combined, and systematically extended.

A first unifying insight is that a wide range of nonlinear oscillators— including quadratic, cubic, cubic–quintic, Duffing–Helmholtz, Rayleigh, Van der Pol–Duffing, and fractional oscillators—share the same organizing principles at the level of phase–space geometry. Equilibria, separatrices, energy landscapes, and invariant manifolds arise from the polynomial structure of the restoring force, largely independent of the physical origin of the model. This geometric viewpoint explains why systems from mechanics, electronics, plasma physics, and relativity exhibit strikingly similar dynamical signatures.

A second unifying contribution lies in the systematic role of elliptic functions. Exact solutions in terms of Jacobi and Weierstrass functions are not treated as isolated curiosities, but as fundamental reference solutions that organize approximation theory. Trigonometric approximations, harmonic balance, KBM averaging, multiple–scale expansions, and elliptic averaging are all shown to emerge naturally as controlled deformations of exact elliptic dynamics. This perspective clarifies when and why approximate methods succeed, where they fail, and how their accuracy can be systematically improved.

The book further resolves a persistent gap between conservative and dissipative theories. Classical nonlinear oscillation theory often treats dissipation and forcing as small perturbations of conservative systems, while fractional models introduce memory effects that fundamentally alter the phase–space structure. By embedding dissipative, forced, and fractional Duffing–type oscillators within a single hierarchical framework, the book demonstrates how energy balance, amplitude selection, and resonance mechanisms can be analyzed without abandoning geometric intuition. In particular, fractional derivatives are interpreted not as abstract operators, but as generators of effective damping and inertia that can be compared directly with integer–order models.

A further unifying element is methodological. Throughout the book, analytical results are not presented in isolation but are systematically benchmarked against high–accuracy numerical computations. This dual perspective resolves a common tension in nonlinear science: exact solutions are often viewed as too specialized, while numerical simulations are treated as problem–specific. Here, exact elliptic solutions serve as anchors, perturbative and averaging methods provide scalable analytical tools, and numerical methods supply validation across wide parameter ranges. The result is a reproducible, quantitative workflow rather than a collection of disconnected techniques.

Perhaps most importantly, the book demonstrates that Duffing–type equations extend beyond traditional oscillation problems. Applications to geodesic motion, planetary dynamics, nonlinear wave equations, and even numerical algorithms for locating Riemann zeta zeros reveal that the same mathematical structures recur in contexts far removed from classical mechanics. In this sense, Duffing–type oscillators act as *universal normal forms* not only for weakly nonlinear vibrations, but for a much broader class of nonlinear phenomena governed by polynomial constraints and phase coherence.

In summary, this book solves a conceptual problem rather than a single equation. It provides a unified analytical, geometric, and computational framework in which nonlinear oscillators with polynomial nonlinearities— including their fractional extensions—can be understood, approximated, and applied across disciplines. By doing so, it transforms the Duffing equation from a well–known example into a central organizing principle for nonlinear science.

Chapter 2

Preliminaries

In this chapter we consider the approximation of Jacobi elliptic functions by means trigonometric functions. These elementary approximations constitutes a powerful analytical tool for studying nonlinear oscillatory systems. When the elliptic modulus m is sufficiently small, functions such as $\mathrm{sn}(u|m)$, $\mathrm{cn}(u|m)$, and $\mathrm{dn}(u|m)$ can be expressed in terms of sine and cosine functions with correction terms depending on m. This approximation reduces the mathematical complexity of exact elliptic solutions while preserving the essential periodic structure of the oscillations.

The trigonometric approximation is particularly useful in perturbative methods, such as the method of multiple scales or harmonic balance, where closed-form expressions simplify the derivation of amplitude–frequency relations. It also facilitates the comparison between exact elliptic solutions and approximate solutions of nonlinear equations. An elliptic function is a meromorphic in the complex plane function with double period. That is, F is elliptic if this function is meromorphic in te whole cpmplex plane, non constant and

$$F(z + m_1\omega_1 + m_2\omega_2) = F(z) \text{ for all complex } z.$$

The numbers ω_1 and ω_2 are its periods $(\omega_1 \neq \omega_2)$. The Weierstrass elliptic function and the Jacobi elliptic functions are two kinds of important elliptic functions.

The incomplete elliptic integral of first kind is defined as

$$F(\phi,\ k) = \int_0^{\phi} \frac{d\theta}{\sqrt{1-k^2\sin^2\theta}} = \int_0^{\sin\phi} \frac{dz}{\sqrt{(1-z^2)(1-k^2z^2)}}.$$

The number k is called the parameter and the number

$$m = k^2$$

is the modulus. The elliptic integral of the first kind is given by $K(k) = F(\pi/2,\ k)$, that is,

$$K(k) = \int_0^{\pi/2} \frac{d\theta}{\sqrt{1-k^2\sin^2\theta}} = \int_0^{1} \frac{dz}{\sqrt{(1-z^2)(1-k^2z^2)}}.$$

$$K(m) = \int_0^{\pi/2} \frac{d\theta}{\sqrt{1-m\sin^2\theta}}$$

Equation defines an increasing function, since

$\frac{dF}{d\phi} = \frac{1}{\sqrt{1-k^2 \sin^{\mathfrak{B}} \phi}} > 0$ for any real ϕ nad $0 \le k < 1$.

For a fixed $k \in [O,\ 1)$, the inverse of the function $F(u,\ k)$ is defined as

$$\phi = F^{-1}(u,\ k) = \mathrm{am}(u,\ k)\ ,$$

and it nis called the Jacobi namplitude.

2.1 Jacobian Elliptic Functions

The Jacobi elliptic functions sn, cn and dn are defined as follows :

$$\mathrm{cn}(u,\ k) = \cos(\mathrm{am}(u,\ k)) = \mathrm{cn}(u,\ m)\ .$$
$$\mathrm{sn}(u,\ k) = \sin(\mathrm{am}(u,\ k)) = \mathrm{sn}(u,\ m)\ .$$
$$\mathrm{dn}(u,\ k) = \sqrt{1 - k^2\mathrm{sn}^2(u, k^2)} = \sqrt{1 - m\sin^2(\phi)} = \mathrm{dn}(u,\ m)\ .$$

being

$$m = k^2.$$

The functions sn and cn are periodic with period

$$T = 4K(k) = 4K(m)\ .$$

The function dn has period $T/2 = 2K(m)$. See Figure 1 for a graphical illustration.See Figure 2.1.

The other Jacobi elliptic functions are

$\mathrm{nc}(u,\ m) = \frac{1}{\mathrm{cn}(u,m)}$, $\mathrm{ns}(u,\ m) = \frac{1}{\mathrm{sn}(u,m)}$, $\mathrm{nd}(u,\ m) = \frac{1}{\mathrm{dn}(u,m)}$.
$\mathrm{cd}(u,\ m) = \frac{\mathrm{cn}(u,m)}{\mathrm{dn}(u,m)}$, $\mathrm{sd}(u,\ m) = \frac{\mathrm{sn}(u,m)}{\mathrm{dn}(u,m)}$,$\mathrm{dc}(u,\ m) = \frac{1}{\mathrm{cd}(u,m)} = \frac{\mathrm{dn}(u,m)}{\mathrm{cn}(u,m)}$,
$\mathrm{ds}(u,\ m) = \frac{\mathrm{dn}(u,m)}{\mathrm{sn}(u,m)}$, $\mathrm{sc}(u,\ m) = \frac{\mathrm{sn}(u,m)}{\mathrm{cn}(u,m)}$, $\mathrm{cs}(u,\ m) = \frac{\mathrm{cn}(u,m)}{\mathrm{sn}(u,m)}$.

Derivatives :

$$\frac{d}{du}\mathrm{cn}(u|m) = -\mathrm{dn}(u|m)\mathrm{sn}(u|m),\ \frac{d}{du}\mathrm{sn}(u|m) = \mathrm{cn}(u|m)\mathrm{dn}(u|m),$$
$$\frac{d}{du}\mathrm{dn}(u|m) = -m\mathrm{cn}(u|m)\mathrm{sn}(u|m).$$
$$\frac{d}{du}\mathrm{nc}(u|m) = \mathrm{dc}(u|m)\mathrm{sc}(u|m),$$
$$\frac{d}{du}\mathrm{ns}(u|m) = -\mathrm{cs}(u|m)\mathrm{ds}(u|m),$$
$$\frac{d}{du}\mathrm{nd}(u|m) = m\mathrm{cd}(u|m)\mathrm{sd}(u|m).$$
$$\frac{d}{du}\mathrm{cd}(u|m) = (m-1)\mathrm{nd}(u|m)\mathrm{sd}(u|m),$$
$$\frac{d}{du}\mathrm{dc}(u|m) = (1-m)\mathrm{nc}(u|m)\mathrm{sc}(u|m),.$$
$$\frac{d}{du}\mathrm{sc}(u|m) = \mathrm{dc}(u|m)\mathrm{nc}(u|m),$$
$$\frac{d}{du}\mathrm{cs}(u|m) = -\mathrm{ds}(u|m)\mathrm{ns}(u|m).$$
$$\frac{d}{du}\mathrm{sd}(u|m) = \mathrm{cd}(u|m)\mathrm{nd}(u|m),$$
$$\frac{d}{du}\mathrm{ds}(u|m) = -\mathrm{cs}(u|m)\mathrm{ns}(u|m).$$

The three main Jacobi elliptic functions cn, sn and dn obey the following nonlinear differential equations :

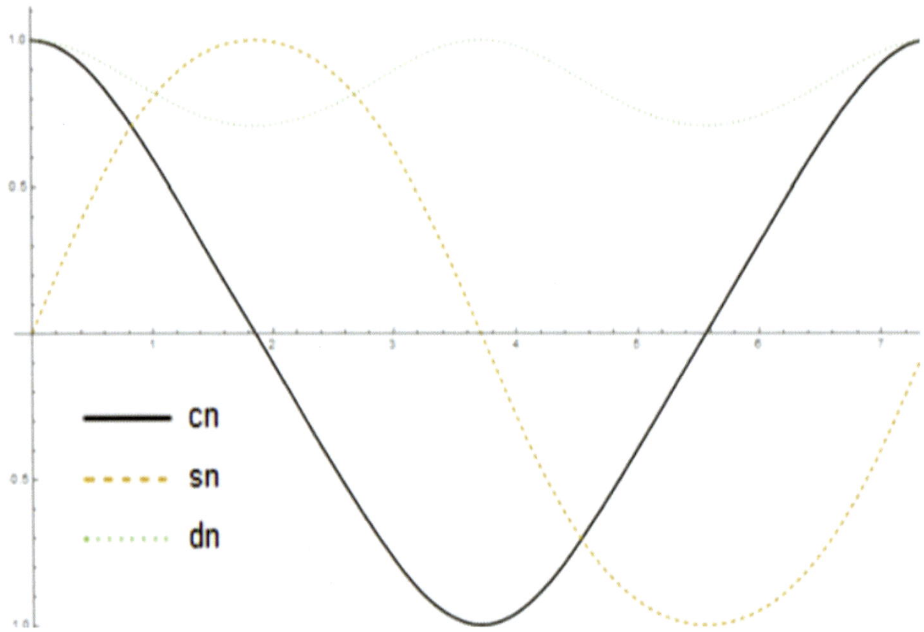

Figure 2.1: The Jacobi elliptic functions cn , sn and dn

$$\frac{d^2}{du^2}\mathrm{cn}(u,\ m) + (1-2m)\mathrm{cn}(u,\ m) + 2m\mathrm{cn}^3(u,\ m)\ .$$
$$\frac{d^2}{du^2}\mathrm{sn}(u,\ m) + (1+m)\mathrm{sn}(u,\ m) - 2m\mathrm{sn}^3(u,\ m)\ .$$

As $m \to 0$ the Jacobi elliptic functions degenerate in the trigonometric functions cos and sin :

$$\mathrm{cn}(u,\ 0) = \cos u \text{ and } \mathrm{sn}(u,\ 0) = \sin u.$$

On the other hand, the Duffing equation

$$x''(t) + \alpha x(t) + \beta x^3(t) = 0$$

may be solved in terms of the Jacobi elliptic functions as shown in Table 1.

We saw that the solution to the Duffing equation (9.53) may be expressed in terms of the jacobian function cn. We may obtain approximate elementary solutions by approximating the functions cn and sn by means of trigonometric functions. There are several ways to do this. First, observe that

$$\mathrm{cn}(z,m) = \frac{2\pi}{\sqrt{m}K(m)} \sum_{n=0}^{\infty} \frac{q(m)^{n+\frac{1}{2}}}{1+q(m)^{2n+1}} \cos\left(\frac{2n+1}{2K(m)}\pi z\right), \tag{2.1}$$

and

$$\mathrm{sn}(z,m) = \frac{2\pi}{\sqrt{m}K(m)} \sum_{n=0}^{\infty} \frac{q(m)^{n+1/2}}{1-q(m)^{1+2n}} \sin\left(\frac{2n+1}{2K(m)}\pi z\right) \tag{2.2}$$

where the elliptic nome $q = q(m)$ is defined as

$$q(m) = \exp\left(-\pi K(1-m)/K(m)\right). \tag{2.3}$$

The elliptic nome q function (2.3) was introduced by C.G. J. Jacobi in 1828 for real arguments and by K. Weierstrass in 1883 for complex arguments.Thus,

$$\operatorname{cn}(z,m) \approx \frac{2\pi}{\sqrt{m}K(m)} \sum_{n=0}^{N} \frac{q(m)^{n+\frac{1}{2}}}{1+q(m)^{2n+1}} \cos\left(\frac{2n+1}{2K(m)}\pi z\right).$$

and

$$\operatorname{sn}(z,m) \approx \frac{2\pi}{\sqrt{m}K(m)} \sum_{n=0}^{N} \frac{q(m)^{n+1/2}}{1-q(m)^{1+2n}} \sin\left(\frac{2n+1}{2K(m)}\pi z\right).$$

The elliptic nome q may be approximated on $-0.95 \leq m \leq 0.95$ with max error 0.00033 as follows:

$$\begin{aligned} &q(m) = \exp\left(-\pi K(1-m)/K(m)\right) \approx 0.754619m(m+1.13413)\\ &(m^2-2.02311m+1.05886) \times (m^2-1.4673m+0.701797)\\ &(m^2-0.568272m+0.353466) \times (m^2+0.492655m+0.35302)\\ &(m^2+1.41983m+0.710968)\,(m^2+2.046m+1.1136)\,. \end{aligned}$$

The inverse elliptic nome q is approximated on $0 \leq m \leq 1$ with max error 0.000011 as follows :

$$\begin{aligned} &q^{-1}(m) = -1499.m(1.m-1.23108)\,(m^2-2.32734m+1.40291)\\ &(m^2-1.94902m+1.10822) \times (m^2-1.3997m+0.737379)\\ &(m^2-0.785884m+0.410802)\,(m^2-0.279268m+0.184691)\\ &(m^2+0.156286m+0.0996614) \end{aligned}$$

However, the expressions (2.1)-(2.2) are difficult to evaluatc. Using Chebyshev approximation techniques we may obtain several trigonometric approximations.

2.2 Approximations for the function cn

$$\operatorname{cn}(t,\ m) \approx \cos\left(\frac{\pi}{2K(m)}t\right), -0.1 \leq m \leq 0.1 \quad \text{with max error} \quad \mathcal{E} = 0.0102.$$

$$\begin{aligned} &\operatorname{cn}(t,\ m) \approx \left(\tfrac{7}{29\left(\frac{9m}{34}+1\right)} + \tfrac{22}{29}\right) \cos\left(\left(1-\tfrac{16m}{63}\right)t\right) + \tfrac{1}{16} m \cos\left(3\left(1-\tfrac{16m}{63}\right)t\right)\\ &+m\left(\tfrac{7}{29\left(m+\frac{151}{40}\right)} - \tfrac{1}{16}\right)\cos\left(5\left(1-\tfrac{10m}{63}\right)t\right) \end{aligned}$$

for $-0.5 \leq m \leq 0.5$ with max error $\mathcal{E} = 0.08$.

$\operatorname{cn}\,(t,m) \approx A\cos\left(\frac{\pi}{2K(m)}t\right) + (1-A)\cos\left(\frac{3\pi}{2K(m)}t\right)$ for $0 \leq m \leq 0.5$,
where $A = \frac{1-0.665388m}{1-0.603851m}$ with max error $\mathcal{E} = 0.0025$.

$\operatorname{cn}\,(t,m) \approx A\cos\left(\frac{\pi}{2K(m)}t\right) + B\cos\left(\frac{3\pi}{2K(m)}t\right) +$
$(1-A-B)\cos\left(\frac{5\pi}{2K(m)}t\right)$ for $0 \leq m \leq 0.75$,
where $A = \frac{1-0.749909m}{1-0.690944m}$ and $B = \frac{0.0585492m}{1-0.641696m}$
with max error $\mathcal{E} = 0.001344$.

$$\text{cn}\ (t,m) \approx A\cos\left(\frac{\pi}{2K(m)}t\right) + B\cos\left(\frac{3\pi}{2K(m)}t\right) + C\cos\left(\frac{5\pi}{2K(m)}t\right) + (1-A-B-C)\cos\left(\frac{7\pi}{2K(m)}t\right) \text{ for } 0 \le m \le 0.95,$$

where $A = \frac{1-0.819m}{1-0.7624m}$, $B = \frac{0.016m}{1-0.32m}$ and $C = \frac{0.05m}{1-0.75m}$
with max error $E = 0.0155$.

$$\text{cn}(t,m) \approx \frac{\sqrt{1+\lambda}\cos(\frac{\pi}{2K(m)}t)}{\sqrt{1+\lambda\cos^2(\frac{\pi}{2K(m)}t)}}, -1 \le m \le 0.5,$$

$$\text{where} \quad \lambda = -\frac{m/2}{1-3/10m} \quad \text{with max error} \quad \mathcal{E} = 0.0014.$$

$$\text{cn}(t,m) \approx \frac{\sqrt{1+\lambda+\mu}\cos(\frac{\pi}{2K(m)}t)}{\sqrt{1+\lambda\cos^2(\frac{\pi}{2K(m)}t)+\mu\cos^4(\frac{\pi}{2K(m)}t)}}, -1 \le m \le 0.95,$$

$\lambda = \frac{-0.29706m-0.0337822}{1-0.787028m}$, $\mu = \frac{0.0506755m-0.00262007}{1-0.940446m}$ with max error $\mathcal{E} = 0.018$.

$\text{cn}(t,\ m) \approx |\cos(\omega t)|^\rho \text{sign}(\cos(\omega t))$, $-1 \le m \le 0.9$ with max error $\mathcal{E} = 0.052$,
where $\omega = \frac{\pi}{2K(m)}$ and $\rho = \frac{1-0.581726m-0.0769795m^2}{1-0.813146m-0.0468663m^2}$.

2.2.1 Approximation for the function sn

$$\text{sn}(t,\ m) \approx \sin\left(\frac{\pi}{2K(m)}t\right), -0.1 \le m \le 0.1$$

$$\text{with max error} \quad \mathcal{E} = 0.0102.$$

$\text{sn}(t,\ m) \approx A\sin\left(\frac{\pi}{2K(m)}t\right) + B\sin\left(\frac{3\pi}{2K(m)}t\right), 0 \le m \le 0.9$
with max error $\mathcal{E} = 0.04$,where
$A = \frac{0.562142m+0.983271}{0.3906m+1}$ and $B = \frac{0.158371m-0.0180858}{0.0445769m+1}$.

$$\text{sn}(t,\ m) \approx \frac{\sqrt{1+\frac{31m}{61}+\frac{12m^2}{29}}\sin\left(\frac{\pi}{2K(m)}t\right)}{\sqrt{1+\left(\frac{31m}{61}+\frac{12m^2}{29}\right)\sin^2\left(\frac{\pi}{2K(m)}t\right)}} \quad \text{for} -1 \le m \le 0.9 \quad \text{with max error } E = 0.05$$

$\text{sn}(t,\ m) \approx |\sin(\omega t)|^\rho \text{sign}(\sin(\omega t)), -1 \le m \le 0.9$ with max error $E = 0.08$,
where $\omega = \frac{\pi}{2K(m)}$ and $\rho = \frac{1+0.047204m-0.474324m^2}{1+0.284777m-0.31321m^2}$.

2.3 Approximation for the function $\frac{d}{dt}\mathbf{cn}\ (t) = -\mathbf{sn}(t)\mathbf{dn}(t)$.

$\frac{d}{dt}\mathrm{cn}(t) \approx a\sin\left(\frac{\pi}{2K(m)}t\right) + b\sin\left(\frac{3\pi}{2K(m)}t\right),\ -0.5 \le m \le 0.5,$
where $a = \frac{0.63305m-0.999865}{1-0.31398m}$ and $b = \frac{-0.19444m-0.00748375}{0.0485259m+1}$,
with max error $\mathcal{E} = 0.014$.

For small m, say $|m| \le 0.1$,

$$\frac{d}{dt}\mathrm{cn}\ (t) \approx -\sqrt{1-m}\sin\left(\frac{\pi}{2K(m)}t\right).$$

2.4 Approximation for the function dn

$\mathrm{dn}(t,\ m) \approx 1 + (\sqrt{1-m} - 1)\sin^2\left(\frac{\pi}{2K(m)}t\right)$
for $-0.25 \le m \le 0.25$, with max error $\mathcal{E} = 0.00241$.

$\mathrm{dn}(t,\ m) \approx 1 + A\sin^2\left(\frac{\pi}{2K(m)}t\right) + B\sin^4\left(\frac{\pi}{2K(m)}t\right)$
for $-0.5 \le m \le 0.5$ with max error $\mathcal{E} = 0.011$, where
$A = \frac{-0.541107m-0.0258998}{1-0.12315m}$ and $B = \frac{0.0287749m+0.0176944}{1-0.0214768m}$.

2.5 Applications

2.5.1 Differentiation with variable modulus

Let

$$\frac{d}{dt}m(t) = \dot{m} \text{ and } \frac{d^2}{dt^2}m(t) = \ddot{m}.$$

Assuming that $m = m(t)$ is small, say $|m(t)| \le 0.25$, we will have

$$K(m(t)) \approx \frac{\pi}{2} + 0.39227m(t),\ \frac{d}{dt}K(m(t)) \approx \frac{0.033445m+0.392711}{1-1.0391m}\dot{m},$$

and

$$\mathrm{cn}(t,\ m(t)) \approx \cos\left((1-\tfrac{8}{31}m(t))t\right),\ \mathrm{sn}(t,\ m(t)) \approx \sin\left((1-\tfrac{6}{23}m(t))t\right),$$

so that

$$\frac{d}{dt}\mathrm{cn}(t,\ m(t)) \approx -\left(1-\tfrac{8}{31}\left(m(t)+\dot{m}t\right)\right)\sin\left(\left(1-\tfrac{8}{31}m(t)\right)t\right),$$

and

$$\frac{d}{dt}\mathrm{sn}(t,\ m(t)) \approx \left(1-\tfrac{6}{23}\left(m(t)+\dot{m}t\right)\right)\cos\left(\left(1-\tfrac{6}{23}m(t)\right)t\right).$$

Also,
$\frac{d^2}{dt^2}\mathrm{cn}(t,\ m(t)) \approx -\frac{1}{961}\cos\left(\left(1-\frac{8m(t)}{31}\right)t\right)(-31+8m(t)+8t\dot{m})^2 +$
$\frac{8}{31}\left(2\dot{m}+t\ddot{m}\right)\sin\left(\left(1-\frac{8m(t)}{31}\right)t\right)$.
$\frac{d^2}{dt^2}\mathrm{sn}(t,\ m(t)) \approx \frac{1}{23}(-6)\cos\left(\left(1-\frac{6m(t)}{23}\right)t\right)(2\dot{m}+t\ddot{m}) -$
$\frac{1}{529}\left(-23+6m(t)+6t\dot{m}\right)^2\sin\left(\left(1-\frac{6m(t)}{23}\right)t\right)$
This is easier than evaluating, for example,

$$\frac{d}{dt}\mathrm{cn}(t,\ m(t)) = \frac{\mathrm{dn}(t|m)\mathrm{sn}(t|m)(\dot{m}(m\mathrm{cd}(t|m)\mathrm{sn}(t|m)-\mathcal{E}(t|m)-mt+t)-2(m-1)m)}{2(m-1)m}.$$

2.5.2 Integration

Let

$$\omega = \frac{\pi}{2K(m)}.$$

We have

$$\int \mathrm{cn}(t,\ m)dt \approx \int A\cos(\omega t) + (1-A)\cos(3\omega t)dt = \frac{-3(A+B-1)\sin(5t\omega)+15A\sin(t\omega)+5B\sin(3t\omega)}{15\omega} + C,$$

which is easier than

$$\int \mathrm{cn}(t,\ m)dt = -\frac{1}{\sqrt{m}}\tan^{-1}\left(\frac{\mathrm{dn}(t|m)}{\sqrt{m}\mathrm{sn}(t|m)}\right) + C.$$

Assuming that m is small,

$$\int \frac{dt}{1+\lambda\mathrm{cn}(t,m)} \approx \int \frac{dt}{1+\lambda\cos(\omega t)} = \frac{2}{\sqrt{\lambda^2-1}\omega}\tanh^{-1}\left(\frac{(\lambda-1)}{\sqrt{\lambda^2-1}}\tan\left(\frac{\omega}{2}t\right)\right),$$

while the exact integral is quite difficult to evaluate in closed form.

2.5.3 Differential equations

The differential equation

$$\ddot{x} + 2\varepsilon\dot{x} + \omega_0^2 x = \mathrm{cn}(t,\ m), x(0) = x_0 \text{ and } x'(0) = \dot{x}_0 \tag{2.4}$$

We replace the i.v.p. (2.4) with the following one

$$\ddot{x} + 2\varepsilon\dot{x} + \omega_0^2 x = \gamma(A\cos wt + B\cos 3wt + (1-A-B)\cos 5wt),\ x(0) = x_0 \text{ and } x'(0) = \dot{x}_0$$

The solution is obtained as

$$\begin{aligned} x(t) &= \exp(-\varepsilon t)\left(x_0\cos\left(\sqrt{\omega_0^2-\varepsilon^2 t}\right) + \frac{\varepsilon x_0+\dot{x}_0}{\sqrt{\omega_0^2-\varepsilon^2}}\sin\left(\sqrt{\omega_0^2-\varepsilon^2 t}\right)\right) \\ &+ \frac{1}{\sqrt{\omega_0^2-\varepsilon^2}}\int_0^t \exp(-\varepsilon(t-\eta))\sin\left(\sqrt{\omega_0^2-\varepsilon^2}(t-\eta)\right)f(\eta)d\eta, \end{aligned} \tag{2.5}$$

where

$$f(\eta) = \gamma(A\cos w\eta + B\cos 3w\eta + (1-A-B)\cos 5w\eta),\ w = \frac{\pi}{2K(m)}.$$

The integral in the r.h.s. of (2.5) may be easily evaluated. A similar approach is used for solving the i.v.p.

$$\ddot{x} + 2\varepsilon\dot{x} + \omega_0^2 x = \gamma\mathrm{sn}(t,m),\ x(0) = x_0 \text{ and } x'(0) = \dot{x}_0$$

2.5.4 Laplace transforms

$$\mathcal{L}_t[\text{cn}(t,m)](s) \approx \frac{As}{s^2+w^2} + \frac{(1-A)s}{s^2+9w^2}$$
$$\mathcal{L}_t[\text{cn}(t,m)](s) \approx s\left(\frac{A}{s^2+w^2} + \frac{B}{s^2+9w^2} + \frac{1-A-B}{s^2+25w^2}\right)$$
$$\mathcal{L}_t[\text{sn}(t,m)](s) \approx \frac{Aw}{s^2+w^2} + \frac{3Bw}{s^2+9w^2}.$$
$$w = \frac{\pi}{2K(m)}.$$

2.5.5 Generalized elliptic functions

These functions are defined as

$$\text{CN}(t) = \text{cn}(\xi(t)), m(t)), \text{SN}(t) = \text{sn}(\xi(t)), m(t)), \text{DN}(t) = \text{dn}(\xi(t)), m(t)) ,$$

where

$$\xi(t) = \int_0^t [1 + \frac{m'(t)}{2} \int_0^t \frac{d\tau}{1-m(\tau)}]dt, \;\; m(t) \ll 1.$$

We have that

$$\text{CN}'(t) \approx -\text{sn}(t)\text{dn}(t) ,$$

and

$$\text{CN}''(t) \approx -(1-m(t))\text{cn}(t) - 2m(t)\text{CN}^3(t) .$$

These functions may be approximated as follows :
$\text{CN}(t) \approx \cos\left(\frac{\pi}{2K(m(t))}\xi(t)\right)$,$\text{SN}(t) \approx \sin\left(\frac{\pi}{2K(m(t))}\xi(t)\right)$,$\text{DN}(t) \approx 1+(\sqrt{1-m(t)}-1)\sin^2\left(\frac{\pi}{2K(m(t))}\xi(t)\right)$

.

$$\text{CN}'(t) \approx -\text{sn}(t)\text{dn}(t) \approx -\sqrt{1-m(t)}\sin\left(\frac{\pi}{2K(m(t))}\xi(t)\right) .$$

For example, if

$$m(t) = \frac{a\exp(-\delta t)}{2(1+a\exp(-\delta t))},$$

then

$$\xi(t) = \frac{a-(2a+3)\log(a+1)}{a+1} + 3\log(a+e^{\delta t}) - \frac{a(\log(a+e^{\delta t})-3\log(e^{\delta t})+1)}{a+e^{\delta t}} + 5\delta t.$$

2.6 Summary

In this chapter we introduced the analytical tools that will be used throughout the rest of the thesis, with particular emphasis on Jacobi elliptic functions and their approximations. The main points can be summarized as follows:

1. **Jacobi elliptic functions as a natural language for nonlinear oscillations.** The Jacobi elliptic functions $\mathrm{sn}(u|m)$, $\mathrm{cn}(u|m)$ and $\mathrm{dn}(u|m)$ were defined and their basic properties, periods, derivative formulas and interrelations were established. This provides a precise functional framework for describing periodic motions in nonlinear oscillations, especially those governed by double–well or cubic–quintic potentials.

2. **Trigonometric and series representations that reduce analytical complexity.** Several representations of Jacobi elliptic functions were derived, including trigonometric approximations, nome–based Fourier series and transformation formulas between different elliptic functions. These show that, for sufficiently small modulus m, elliptic functions can be replaced by combinations of elementary trigonometric functions while preserving the essential periodic structure of the solutions.

3. **Accurate approximations for** cn, dn **and their derivatives.** Dedicated approximations were obtained for $\mathrm{cn}(t,m)$, $\mathrm{dn}(t,m)$ and for the derivative

$$\frac{\mathrm{d}}{\mathrm{d}t}\,\mathrm{cn}(t,m) = -\,\mathrm{sn}(t,m)\,\mathrm{dn}(t,m).$$

   The proposed formulas include explicit ranges of the modulus m and quantified maximum errors, which are typically very small. This allows elliptic-type responses to be handled with a simplicity comparable to harmonic oscillations, but with significantly higher accuracy.

4. **Tools for problems with variable modulus and generalized elliptic functions.** The above approximations were extended to the case of a slowly varying modulus $m(t)$, leading to practical formulas for $\mathrm{sn}(t,m(t))$, $\mathrm{cn}(t,m(t))$ and their derivatives, as well as for generalized elliptic functions defined through a reparametrization $\xi(t)$. This shows how the elliptic description can be adapted to systems with slowly changing parameters or amplitudes.

5. **Illustrative applications to differential equations and integral transforms.** The usefulness of these approximations was illustrated by applications to initial–value problems for second–order differential equations with elliptic forcing terms, and to the computation of Laplace transforms of $\mathrm{sn}(t,m)$ and $\mathrm{cn}(t,m)$. In each case, the approximations yield explicit closed–form expressions that would be difficult to obtain starting from the exact elliptic functions alone.

6. **Foundations for the subsequent analysis of Duffing-type oscillators.** Overall, Chapter 2 establishes the analytical toolkit required in the remainder of the thesis. The definitions, identities and approximations for elliptic functions developed here will be systematically employed in later chapters to construct exact and approximate solutions of Duffing-type oscillators, to derive amplitude–frequency relations, and to analyze the qualitative behaviour of nonlinear oscillations.

Chapter 3

The conservative Duffing oscillator equation

The Duffing oscillator is one of the most fundamental models in nonlinear dynamics, originally introduced to describe the oscillations of a nonlinear spring system. In its conservative form, the Duffing equation captures the behavior of oscillators with nonlinear restoring forces in the absence of damping or external forcing. This conservative variant is of particular importance because it reveals intrinsic properties of nonlinear oscillations such as amplitude–frequency dependence, energy conservation, and the existence of periodic and quasi-periodic solutions.

A classical form of the conservative Duffing oscillator is given by

$$\ddot{x} + \alpha x + \beta x^3 = 0, \tag{3.1}$$

where $\alpha, \beta \in \mathbb{R}$ are parameters that determine the linear stiffness and the nonlinear cubic contribution, respectively. This system admits exact solutions in terms of Jacobi elliptic functions and provides an archetypal example of a nonlinear conservative oscillator.

Beyond the cubic nonlinearity, higher-order restoring forces often appear in practical systems. The cubic–quintic Duffing oscillator is defined as

$$\ddot{x} + \alpha x + \beta x^3 + \gamma x^5 = 0, \tag{3.2}$$

where the additional quintic term γx^5 allows the model to capture more complex nonlinearities, including stiffening and softening effects beyond the cubic approximation. Such extensions are particularly relevant in structural mechanics, nonlinear optics, and the modeling of phase transitions.

The generalized Duffing equation encompasses a broader class of polynomial nonlinearities, often expressed as

$$\ddot{x} + \sum_{k=1}^{n} \alpha_k x^{2k-1} = 0, \tag{3.3}$$

where nonlinear restoring forces of arbitrary odd order can be incorporated. This formulation

highlights the Duffing model as a unifying framework for the study of nonlinear oscillations. It not only provides insights into the qualitative behavior of nonlinear systems but also establishes connections with soliton theory, nonlinear lattices, and fractional-order extensions.
This equation is written as

$$\ddot{x} + \alpha x + \beta x^3 = 0,\ x(0) = x_0 \text{ and } x'(0) = \dot{x}_0$$

Assume the solution in the ansatz form

$$x = c_0 \mathrm{cn}(\sqrt{\omega} t + c_1, m),$$

where cn stands for the Jacobian elliptic function. The function $y = \mathrm{cn}(t, m)$ is a solution to the nonlinear ode

$$\ddot{y} + (1 - 2m)y + 2my^3 = 0.$$

We have:

$$\begin{aligned}
&\ddot{x} + \alpha x + \beta x^3 = \\
&c_0 \mathrm{cn} \left(\alpha + 2m\omega - \omega + \left(\beta c_0^2 - 2m\omega\right) \mathrm{cn}^2\right), \\
&\text{where} \\
&\mathrm{cn} = \mathrm{cn}(\sqrt{\omega} t + c_1, m)
\end{aligned}$$

Equating to zero the coefficients of cn and solving the resulting algebraic system gives

$$\omega = \alpha + \beta c_0^2,\ m = \frac{\beta c_0^2}{2\left(\alpha + \beta c_0^2\right)}.$$

The exact solution is

$$x = x(t) = c_0 \mathrm{cn}\left(\sqrt{\alpha + \beta c_0^2} t + c_1, \frac{\beta c_0^2}{2\left(\alpha + \beta c_0^2\right)}\right).$$

The constants c_0 and c_1 are obtained form the initial conditions as follows. Define

$$c_1 = \mathrm{cn}^{-1}\left(\frac{x_0}{c_0}\right).$$

Then $x(0) = x_0$. From the condition $x'(0) = \dot{x}_0$ we obtain

$$c_0^2 \omega \left(1 - \frac{x_0^2}{c_0^2}\right)\left(\frac{m x_0^2}{c_0^2} - m + 1\right) - \dot{x}_0^2 = 0$$

from where

$$c_0 = \pm \sqrt{\frac{-\alpha \pm \sqrt{\left(\alpha + \beta x_0^2\right)^2 + 2\beta \dot{x}_0^2}}{\beta}},\ \beta \neq 0. \tag{3.4}$$

Assuming that at least one of the numbers in (3.4) is real, we will obtain a real solution. In the case when $\left(\alpha + \beta x_0^2\right)^2 + 2\beta \dot{x}_0^2 < 0$ we won't get a real solution. The number

$$\Delta = \left(\alpha + \beta x_0^2\right)^2 + 2\beta \dot{x}_0^2 \tag{3.5}$$

is called the discriminant of the Duffing equation (3.5). Assuming that $\Delta > 0$ and using the addition formula

$$\mathrm{cn}(u+v) = \frac{\mathrm{cn}(u)\mathrm{cn}(v) - \mathrm{dn}(u)\mathrm{sn}(u)\mathrm{dn}(v)\mathrm{sn}(v)}{1 - m\,\mathrm{sn}^2(u)\mathrm{sn}^2(v)}$$

we may write the solution in the form

$$x(t) = \frac{x_0 \mathrm{cn}\left(\sqrt{\omega}t|\, m\right) + d_1 \mathrm{dn}\left(\sqrt{\omega}t|\, m\right)\mathrm{sn}\left(\sqrt{\omega}t|\, m\right) +}{1 - d_2 \mathrm{sn}^2\left(\sqrt{\omega}t|\, m\right)},$$

where

$$d_1 = \tfrac{\dot{x}_0}{\sqrt{\omega}},\ d_2 = -\tfrac{\alpha + \beta x_0^2 - \omega}{2\omega}.$$

$$\omega = \tfrac{\alpha}{1-2m},\ m = \tfrac{1}{2}\left(1 - \tfrac{\alpha}{\sqrt{(\alpha+\beta x_0^2)^2 + 2\beta \dot{x}_0^2}}\right).$$

Assuming that the solution is periodic, its period equals

$$T_{\text{period}} = \frac{4K(m)}{\sqrt{\omega}} \text{ if } m < 1.$$

or

$$T_{\text{period}} = \frac{2K(1/m)}{\sqrt{m\omega}} \text{ if } m > 1.$$

Here,

$$K(m) = \int_0^{\pi/2} \frac{d\theta}{\sqrt{1 - m\sin^2\theta}} = \int_0^1 \frac{dt}{\sqrt{(1-t^2)(1-mt^2)}}.$$

We have the following approximations:

$K(m) \approx \frac{\pi}{2} + \frac{13}{33}m$ for $-0.1 \le m \le 0.1$ with max error$E = 0.0023$.

$K(m) \approx \frac{\pi}{2} + \frac{73}{183}m + \frac{7}{31}m^2$ for $-0.2 \le m \le 0.2$ with max error$E = 0.000524$.

$K(m) \approx \frac{\pi}{2} + \frac{51}{130}m + \frac{5}{23}m^2 + \frac{9}{58}m^3 + \frac{12}{77}m^4 + \frac{11}{95}m^5$ for $-0.5 \le m \le 0.5$ with max error$E =$ 0.00015.

$K(m) \approx \frac{79 - 46\pi}{4(13m-23)} + \frac{55}{64}$ for $-1 \le m \le 0.5$ with max error$E = 0.0033$.

$K(m) \approx \frac{\frac{\pi}{2} - \frac{18}{25}m - \frac{1}{19}m^2}{1 - \frac{17}{24}m}$ for $-1 \le m \le 0.5$ with max error$E = 0.0009$.

$\frac{1}{K(m)} \approx \frac{7}{11} - \frac{4}{25}m$ for $-0.1 \le m \le 0.1$ with max error$E = 0.0003$.

$\frac{1}{K(m)} \approx \frac{7}{11} - \frac{4}{25}m - \frac{1}{18}m^2 - \frac{1}{36}m^3$ for $-0.5 \le m \le 0.5$ with max error$E = 0.001$.

$\frac{1}{K(m)} \approx \frac{-\frac{m^2}{22} - \frac{5m}{26} + \frac{7}{11}}{1 - \frac{m}{22}}$ for $-0.5 \le m \le 0.5$ with max error$E = 0.002$.

For the values $-1 \leq m \leq 0.75$ we have the following approximation:

$$K(m) \approx \frac{-\frac{57\pi\varepsilon}{2048} - \frac{5\pi m}{32} + \frac{\pi}{2}}{-\frac{57\varepsilon}{1024} - \frac{9m}{16} + 1},$$

$$\text{Error } \mathcal{E}(m) = 0.41793m^7 + 0.835652m^6 + 0.413744m^5 + 0.116163m^4$$

$$+0.562721m^3 + 1.36871m^2 + 0.0150267m + 0.000931047 \leq 0.000422.$$

Example 1. Let

$$\ddot{x} - 6x + x^3 = 0, \quad x(0) = -1, \quad x'(0) = 1.$$

The exact solution is

$$x(t) = \frac{0.438691\text{dn}(2.27951t|1.07735)\text{sn}(2.27951t|1.07735) - 1.\text{cn}(2.27951t|1.07735)}{1. - 0.981125\text{sn}(2.27951t|1.07735)^2}$$

with period $T = 2.31189$. See Figure 3.1 for a comparison with the Runge–Kutta numerical solution. The solution may also be expressed as

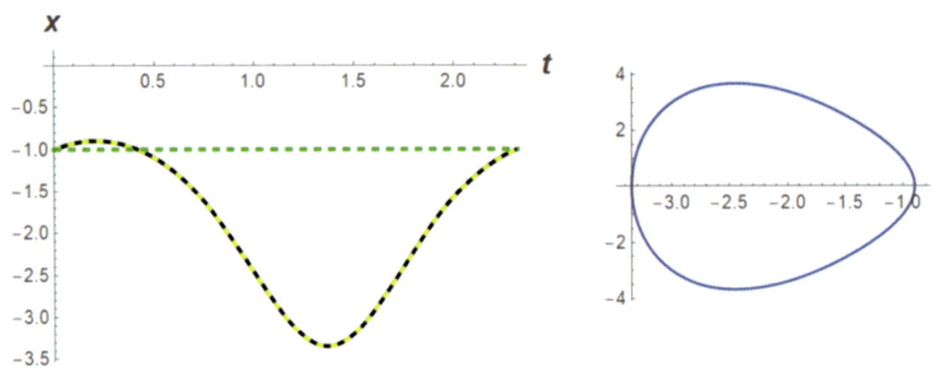

Figure 3.

Figure 3.1: Runge–Kutta numerical solution

$$x(t) = -3.34607\text{cn}(2.27951t + 2.1577|1.07735).$$

A particular case is

$$\ddot{x} + \alpha x + \beta x^3 = 0, \; x(0) = x_0 \text{ and } x'(0) = 0$$

with exact solution

$$x(t) = x_0 \text{cn}\left(\sqrt{\alpha + \beta x_0^2}t, \frac{\beta x_0^2}{2(\alpha + \beta x_0^2)}\right)$$

and period

$$T_{\text{period}} = \frac{4K(m)}{\sqrt{\alpha + \beta x_0^2}}, \ m = \frac{\beta x_0^2}{\alpha + \beta x_0^2} \text{ and } \alpha + \beta x_0^2 > 0.$$

Yet another particular case :

$$\ddot{x} + \alpha x + \beta x^3 = 0, \, x(0) = 0 \text{ and } x'(0) = \dot{x}_0.$$

The exact solution is given by

$$x(t) = \frac{\dot{x}_0}{\sqrt{\omega}} \operatorname{sd}\left(\sqrt{\omega} t \middle| m\right), \ m = \frac{1}{2} - \frac{\alpha}{2\sqrt{\alpha^2 + 2\beta \dot{x}_0^2}}, \ \omega = \sqrt{\alpha^2 + 2\beta \dot{x}_0^2}, \, \alpha^2 + 2\beta \dot{x}_0^2 > 0.$$

Assume that $\Delta = 0$. In this case,

$$\dot{x}_0 = \pm \frac{\alpha + \beta x_0^2}{\sqrt{2}\sqrt{-\beta}}.$$

The solution is expressed in the form

$$x(t) = \sqrt{-\frac{\alpha}{\beta}} \tanh\left(\sqrt{\frac{\alpha}{2}} t + \tanh^{-1}\left(\sqrt{\frac{-\beta}{\alpha}} x_0\right)\right).$$

Example 2.

$$\ddot{x} + x - 2x^3 = 0, x(0) = 1 \text{ and } x'(0) = -\frac{1}{2}.$$

The exact solution reads (see Figure 3.2)

$$x(t) = \frac{1}{\sqrt{2}} \tanh\left(\frac{t}{\sqrt{2}} + \tanh^{-1}\left(\sqrt{2}\right)\right)$$

Now, let $\Delta < 0$. We assume the solution in the ansatz form

$$x(t) = A - \frac{2A}{1 + y(t)}, \ A > 0,$$

where the function $y = y(t)$ is a solution to some Duffing equation

$$\ddot{y} + py + qy^3 = 0, \, y(0) = y_0 \text{ and } y'(0) = \dot{y}_0 \text{ with } \left(p + qy_0^2\right)^2 + 2q\,\dot{y}_0^2 > 0. \tag{3.6}$$

We are given to find the suitable constants A, p and q. Observe that

$$y_0 = \tfrac{A + x_0}{A - x_0}, \ \dot{y}_0 = \tfrac{2A\dot{x}_0}{(A - x_0)^2}.$$

$$y'(t)^2 = 2C - py^2(t) - \tfrac{1}{2} qy^4(t).$$

$$C = \tfrac{py_0^2}{2} + \tfrac{qy_0^4}{4} + \tfrac{\dot{y}_0^2}{2}.$$

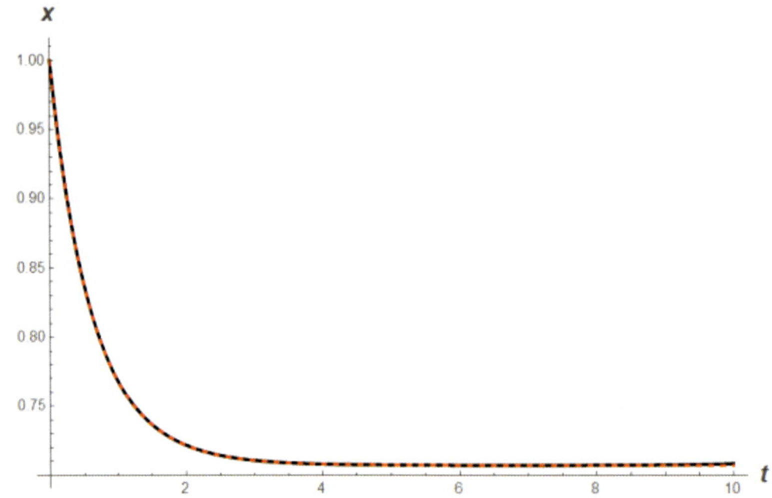

Figure 3.2: The conservative Duffing oscillator equation

We have:

$$\begin{array}{l} \ddot{x}+\alpha x+\beta x^3= \\ \frac{A}{y^3(t)}\left[-\alpha-A^2\beta-8C+\left(-\alpha+3A^2\beta-2p\right)y(t)\right. \\ \left.+\left(\alpha-3A^2\beta+2p\right)y^2(t)+\left(\alpha+A^2\beta-2q\right)y^3(t)\right]. \end{array}$$

Equating to zero the coefficients of $y^j(t)$ gives an algebraic system whose solution is

$$\begin{array}{l} p=\frac{1}{2}\left(3A^2\beta-\alpha\right) \\ q=\frac{1}{2}\left(\alpha+A^2\beta\right) \\ A=\sqrt[4]{\frac{2\alpha x_0^2+\beta x_0^4+2\dot{x}_0^2}{-\beta}}. \end{array}$$

Observe that

$$\frac{2\alpha x_0^2+\beta x_0^4+2\dot{x}_0^2}{-\beta}=\frac{\alpha^2-\Delta}{\beta^2}>0.$$

On the other hand, the discriminant $\delta=\left(p+qy_0^2\right)^2+2q\,\dot{y}_0^2$ of the Duffing equation (3.6) is positive.

Example 3. Let

$$\ddot{x}+x-x^3=0\wedge x_0=\frac{1}{4}\wedge\dot{x}_0=1.$$

The solution equals

$$x(t)=\frac{\sqrt[4]{543}}{4}-\frac{\sqrt[4]{543}/2}{1+y(t)}$$

where

$$y(t)=\frac{1.52257\text{cn}(1.63551t|1.00182)+1.612\text{dn}(1.63551t|1.00182)\text{sn}(1.63551t|1.00182)}{1.-1.1007\text{sn}^2(1.63551t|1.00182)}.$$

The solution is periodic with period $T = 5.55202$. From Figure 3.3 we can see that it is unbounded.

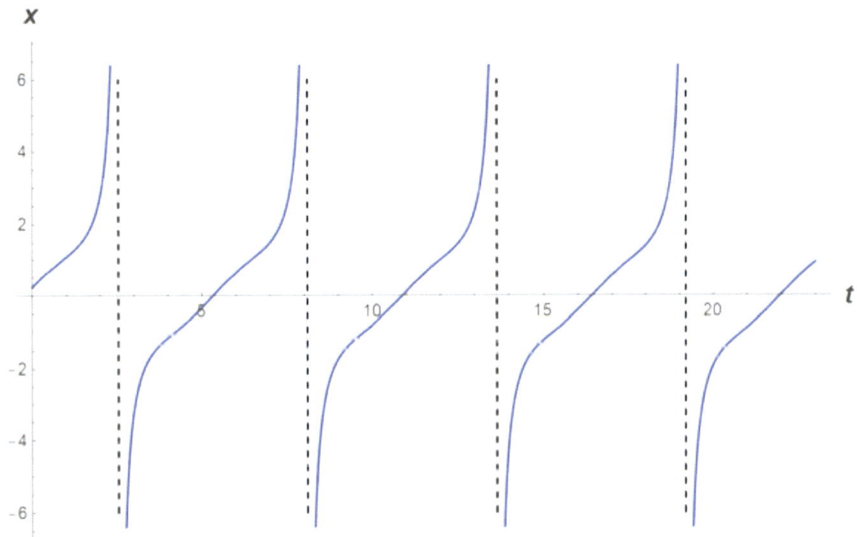

Figure 3.3: Unbounded solution

We also may approximate the solution to the Duffing equation

$$\ddot{u} + \alpha u + \beta u^3 = 0,\ u(0) = u_0 \text{ and } u'(0) = \dot{u}_0$$

as follows:

$$u(t) = u_0 \cos(\omega t) + \nu \sin(\omega t),$$

where

$$\omega = \frac{\sqrt{4\alpha + \sqrt{(4\alpha + 3\beta u_0^2)^2 + 48\beta u_0^2} + 3\beta u_0^2}}{2\sqrt{2}}$$

and

$$\nu = \frac{2\sqrt{2}\dot{u}_0}{\sqrt{4\alpha + \sqrt{(4\alpha + 3\beta u_0^2)^2 + 48\beta \dot{u}_0^2} + 3\beta u_0^2}}.$$

A more precise formula is

$$u(t) = u_0 \left(A \left(v \sin\left(t\sqrt{w}\right) + \cos\left(t\sqrt{w}\right)\right) + (1 - A)\left(v \sin\left(3t\sqrt{w}\right) + \cos\left(3t\sqrt{w}\right)\right)\right),$$

where

$$A = [512\alpha^4 + 6\beta^2\left(381\alpha^2 u_0^4 + 242\alpha \dot{u}_0^2 u_0^2 + 45\dot{u}_0^4\right) + 48\alpha^2\beta\left(37\alpha u_0^2 + 17\dot{u}_0^2\right) + \beta^3 u_0^4\left(1291\alpha u_0^2 + 639\dot{u}_0^2\right) + 269\beta^4 u_0^8][512\alpha^4 + 2\beta^2\left(1165\alpha^2 u_0^4 + 690\alpha \dot{u}_0^2 u_0^2 + 117\dot{u}_0^4\right) + 256\alpha^2\beta\left(7\alpha u_0^2 + 3\dot{u}_0^2\right) + \beta^3 u_0^4\left(1331\alpha u_0^2 + 615\dot{u}_0^2\right) + 281\beta^4 u_0^8],$$

and

$$w = 1/8(4\alpha + 3(A(2A-3)+2)\beta u_0^2 + \frac{1}{(3-2A)^2}\sqrt{\begin{array}{c}16\alpha^2(3-2A)^4 + 24(A(2A-3)+2)(3-2A)^2\beta(\alpha(3-2A)^2u_0^2 + 2\dot{u}_0^2) + \\ 9(A(2A-3)+2)^2(3-2A)^4\beta^2u_0^4\end{array}}.$$

Example 4. Let

$$5v(t)^3 + 0.3v(t) + \dot{u} = 0 \wedge v(0) = 3 \wedge v'(0) = 0.1$$

The solutions are (see Figures 3.4 and 3.5):

$$u_1(t) = 0.017137\sin(5.83533t) + 3\cos(5.83533t),$$

and

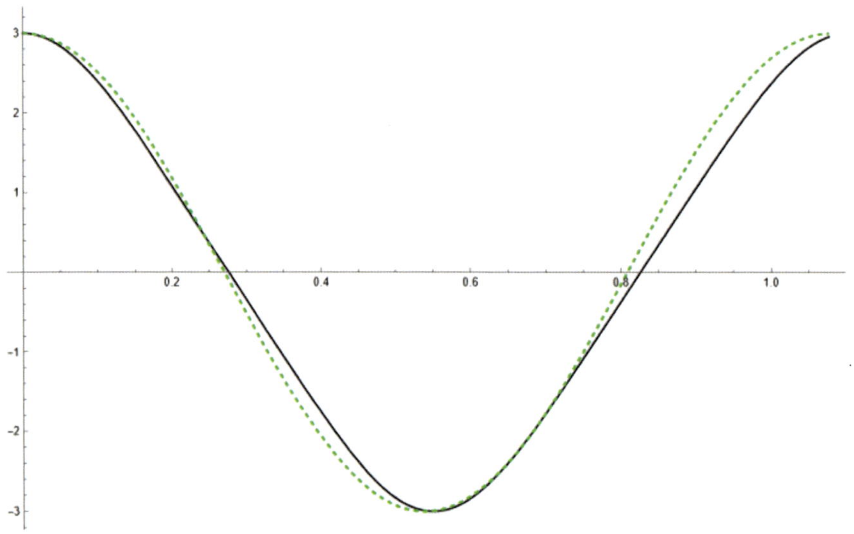

Figure 3.4: Solutions

$$u_2(t) = 3(0.957695(0.00537081\sin(5.72223t) + \cos(5.72223t))$$
$$+0.0423046(0.00537081\sin(17.1667t) + \cos(17.1667t))).$$

3.1 The cubic-quintic Duffing oscillator.

This is the oscillator

$$\ddot{x} + px + qx^3 + rx^5 = 0,\ x(0) = x_0 \text{ and } x'(0) = \dot{x}_0 \tag{3.7}$$

This model extends the conservative Duffing oscillator by adding a quintic restoring term. It is a conservative (autonomous) system and admits a Hamiltonian formulation.

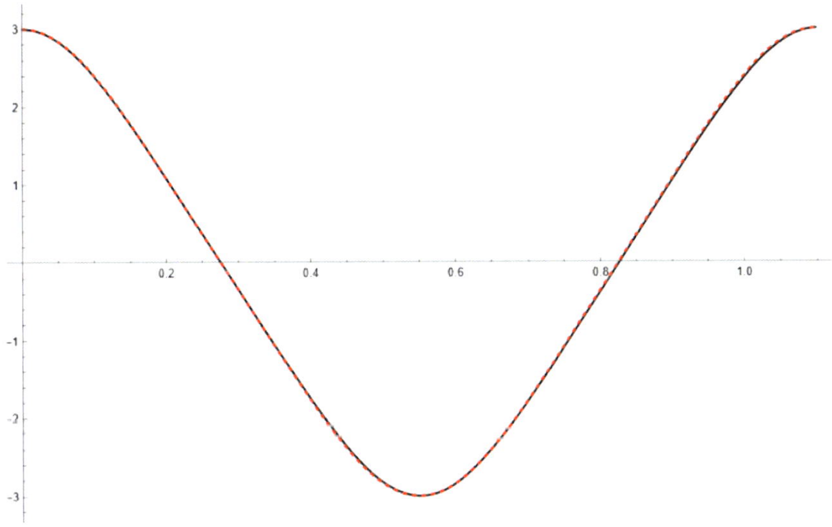

Figure 3.5: Solutions

3.1.1 Mechanical interpretation and energy integral

Equation (3.7) can be written as a Newtonian equation

$$\ddot{x} = -V'(x),$$

with potential

$$V(x) = \frac{p}{2}x^2 + \frac{q}{4}x^4 + \frac{r}{6}x^6. \tag{3.8}$$

Therefore, the energy (first integral) is

$$E = \frac{1}{2}\dot{x}^2 + V(x) = \frac{1}{2}\dot{x}^2 + \frac{p}{2}x^2 + \frac{q}{4}x^4 + \frac{r}{6}x^6, \qquad \frac{dE}{dt} = 0. \tag{3.9}$$

This is the fundamental reason why the dynamics is organized by energy level sets in the phase plane $(x, \dot{x})$.

Easy to understand translation. Think of $x(t)$ as the position of a particle sliding without friction on the curve $V(x)$. Energy E is constant. The particle oscillates if it is trapped in a potential well. It escapes to infinity if the potential does not confine it and the energy is high enough.

3.1.2 Equilibria and their stability in terms of p, q, r

Step 1: find equilibria

Equilibria are constant solutions $x(t) \equiv x_*$, so $\ddot{x} = 0$ and

$$px_* + qx_*^3 + rx_*^5 = 0 \quad \Longleftrightarrow \quad x_*\left(p + qx_*^2 + rx_*^4\right) = 0. \tag{3.10}$$

Hence one equilibrium is always

$$x_* = 0.$$

Nonzero equilibria satisfy

$$p + qx_*^2 + rx_*^4 = 0. \tag{3.11}$$

Let amplitude variable $y = x_*^2 \geq 0$. Then (3.11) becomes

$$ry^2 + qy + p = 0. \tag{3.12}$$

This is just a quadratic equation in y.

Step 2: existence of nonzero equilibria

The discriminant of (3.12) is

$$\Delta_y = q^2 - 4rp. \tag{3.13}$$

If $\Delta_y < 0$ then there are *no real* y, hence no nonzero equilibria. If $\Delta_y \geq 0$ then

$$y_\pm = \frac{-q \pm \sqrt{q^2 - 4rp}}{2r} \qquad (r \neq 0). \tag{3.14}$$

A nonzero equilibrium exists *iff* at least one of $y_\pm$ is nonnegative. Whenever $y_\pm > 0$, we obtain equilibria

$$x_* = \pm\sqrt{y_\pm}.$$

Step 3: stability classification (the one-line test)

Linearize (3.7) about $x = x_*$. Put $x = x_* + \xi$ with $|\xi| \ll 1$. Then the linearized equation is

$$\ddot{\xi} + V''(x_*)\,\xi = 0, \qquad V''(x) = p + 3qx^2 + 5rx^4. \tag{3.15}$$

Thus:

- If $V''(x_*) > 0$, the equilibrium is a center (linearly stable).
- If $V''(x_*) < 0$, the equilibrium is a saddle (unstable).
- If $V''(x_*) = 0$, the equilibrium is degenerate and we must use higher-order terms.

This is the cleanest "in terms of parameters" stability test.

Important simplification at a nonzero equilibrium. If $x_* \neq 0$ satisfies $p + qx_*^2 + rx_*^4 = 0$, then

$$\begin{aligned} V''(x_*) &= p + 3qx_*^2 + 5rx_*^4 \\ &= \underbrace{(p + qx_*^2 + rx_*^4)}_{=0} + 2qx_*^2 + 4rx_*^4 \\ &= 2x_*^2(q + 2rx_*^2). \end{aligned} \tag{3.16}$$

Since $x_*^2 > 0$, the sign of $V''(x_*)$ is exactly the sign of

$$q + 2rx_*^2.$$

So a nonzero equilibrium is stable iff $q + 2rx_*^2 > 0$.

Stability of the origin

At $x_* = 0$,

$$V''(0) = p.$$

Hence:

- $p > 0$: $x = 0$ is a stable center.
- $p < 0$: $x = 0$ is an unstable saddle.
- $p = 0$: degenerate (higher-order terms decide).

3.1.3 Global boundedness vs blow-up: when is the motion confined?

Because the system is conservative, solutions are confined by energy (3.9). The key question is whether the potential $V(x)$ grows to $+\infty$ as $|x| \to \infty$.

- If $r > 0$, then $V(x) \sim \frac{r}{6}x^6 \to +\infty$ as $|x| \to \infty$. So *all* trajectories with finite energy are bounded.
- If $r < 0$, then $V(x) \to -\infty$ as $|x| \to \infty$. Then sufficiently large energy trajectories can escape to infinity. This produces unbounded motion.
- If $r = 0$, we recover the classical cubic Duffing potential $V(x) = \frac{p}{2}x^2 + \frac{q}{4}x^4$, which is confining iff $q > 0$.

This is the clean "big-picture" classification:

$$\boxed{\text{The quintic term controls global confinement.}}$$

3.1.4 Phase portraits and bifurcations: parameter cases

We now list the main qualitative regimes in terms of p, q, r. To keep the discussion physically meaningful, we first assume $r > 0$ (globally confining).

Case A: $r > 0$, $p > 0$

Then $x = 0$ is a stable center. Depending on q, there may exist additional equilibria:

- If $q^2 - 4rp < 0$ then there are no nonzero equilibria. The potential has a single well and all motions are periodic oscillations about $x = 0$.

- If $q^2 - 4rp \geq 0$ and at least one $y_\pm > 0$, then nonzero equilibria exist. Because $p > 0$, they typically correspond to additional wells only if $q < 0$ is sufficiently negative. This case leads to multiple wells and multistability.

Case B: $r > 0$, $p < 0$

Then $x = 0$ is an unstable saddle. In this regime, the potential often becomes a double-well (or triple-well) shape, and nonzero equilibria typically exist and are stable centers. This is the conservative source of separatrices and energy thresholds.

Case C: $r > 0$, $p = 0$ **(degenerate origin)**

Then the leading restoring term near the origin is $qx^3 + rx^5$.

- If $q > 0$, the origin is still stable (but with nonlinear stiffness).
- If $q < 0$, the origin is unstable.
- If $q = 0$, the leading term is rx^5 and the origin is stable for $r > 0$.

Here the linear test fails because $V''(0) = 0$.

3.1.5 A "for dummies" stability checklist

1. **Find equilibria.** Solve $x_*(p + qx_*^2 + rx_*^4) = 0$.
2. **Check stability.** Compute $V''(x_*) = p + 3qx_*^2 + 5rx_*^4$:
   - $V'' > 0$ stable center,
   - $V'' < 0$ unstable saddle,
   - $V'' = 0$ degenerate (need nonlinear terms).
3. **Check global confinement.** If $r > 0$ then bounded for all finite energies. If $r < 0$ then possible escape.
4. **Use energy to classify motion.** Compare E with potential barrier heights $V(x_{\text{saddle}})$.

3.1.6 Worked examples (all in terms of parameters)

Example 1: single-well, only one stable equilibrium

Take

$$p = 1, \qquad q = 1, \qquad r = 1.$$

Then $\Delta_y = q^2 - 4rp = 1 - 4 = -3 < 0$ so the only equilibrium is $x_* = 0$. Also $V''(0) = p = 1 > 0$, so $x = 0$ is a stable center. All solutions are bounded and oscillate periodically about 0.

Example 2: double-well with an unstable origin

Take

$$p = -1, \qquad q = 0, \qquad r = 1.$$

Equilibria solve $x(-1 + x^4) = 0$, so $x_* = 0$ and $x_* = \pm 1$. Stability:

$$V''(0) = p = -1 < 0 \quad \Rightarrow \quad x = 0 \text{ is a saddle (unstable)},$$

$$V''(\pm 1) = p + 5r = -1 + 5 = 4 > 0 \quad \Rightarrow \quad x = \pm 1 \text{ are stable centers.}$$

Thus there are two stable wells and a separatrix through the origin.

Example 3: triple-well type behaviour (possible)

Take

$$r > 0, \quad p > 0, \quad q < 0 \text{ large in magnitude.}$$

Then the quartic term tends to create side wells while the sextic term stabilizes growth. Here one can obtain three minima (one at 0 and two symmetric side minima), which implies three stable centers separated by saddles. The exact regime is obtained by solving $ry^2 + qy + p = 0$ and checking which roots $y_\pm > 0$ exist and which satisfy $q + 2ry_\pm > 0$ (stability condition).

3.1.7 Illustrations: potential and phase portraits (TikZ templates)

Below are TikZ templates you can compile directly (no external data). They are schematic (qualitative) plots: they illustrate the typical shapes.

(A) Single-well potential ($r > 0$, $p > 0$, $\Delta_y < 0$)

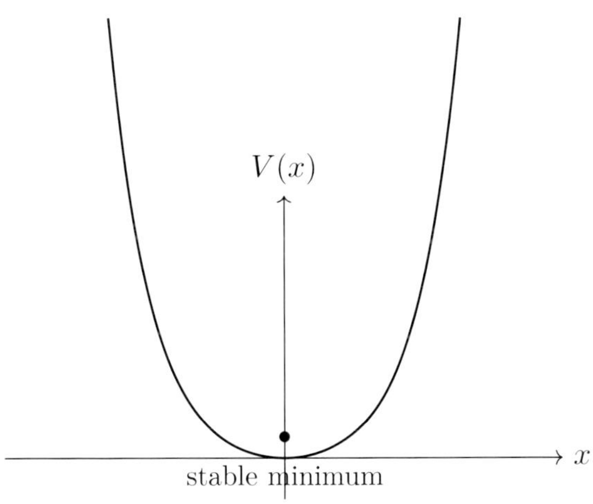

Figure 3.6: Potential shape: single well (typical for $r > 0$, $p > 0$).

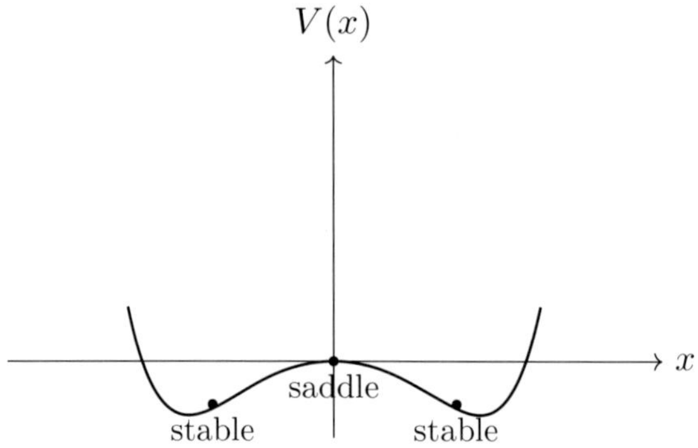

Figure 3.7: Double-well potential: unstable origin, two stable wells.

(B) Double-well potential ($r > 0$, $p < 0$)

(C) Phase portrait sketch (double-well)

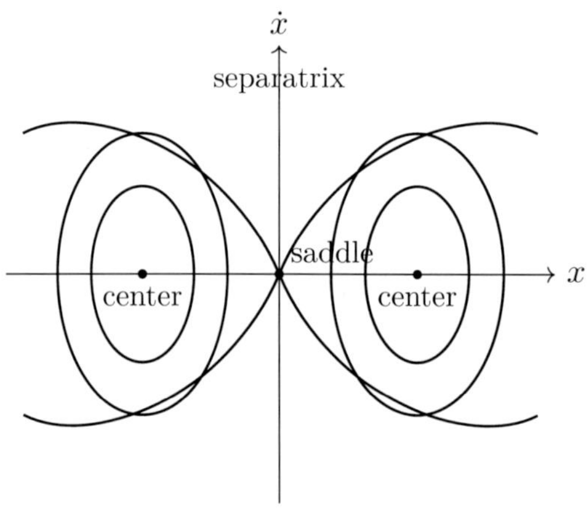

Figure 3.8: Phase portrait sketch for a conservative double-well cubic–quintic Duffing oscillator.

3.1.8 Optional: small-amplitude frequency near a stable equilibrium

If x_* is a stable equilibrium ($V''(x_*) > 0$), then for small oscillations about x_* we have approximately

$$x(t) \approx x_* + A\cos(\Omega t), \qquad \Omega = \sqrt{V''(x_*)} = \sqrt{p + 3qx_*^2 + 5rx_*^4}.$$

This gives the linearized small-amplitude frequency in terms of parameters and the equilibrium location.

Special case: oscillations about 0. If $p > 0$, then $x_* = 0$ is stable and

$$\Omega_0 = \sqrt{p}.$$

Nonlinearity then shifts the frequency with amplitude; that is treated later with perturbation or averaging methods.

3.1.9 Exact solution

The solution is assumed in the ansatz form

$$x(t) = \frac{y(t)}{\sqrt{1+\lambda y^2(t)+\mu y^4(t)}},$$

where the function $y = y(t)$ is the solution to some Duffing oscillator

$$\ddot{y} + \alpha y + \beta y^3 = 0,\ y(0) = y_0 \text{ and } y'(0) = \dot{y}_0$$

Observe that

$$y'(t)^2 = 2C - \alpha y^2(t) - \frac{1}{2}\beta y^4(t)$$

and

$$\frac{px^2}{2} + \frac{qx^4}{4} + \frac{rx^6}{6} + \frac{\dot{x}^2}{2} - D = 0,$$

being

$$D = \frac{px_0^2}{2} + \frac{qx_0^4}{4} + \frac{rx_0^6}{6} + \frac{\dot{x}_0^2}{2}.$$

We have

$$\frac{px^2}{2} + \frac{qx^4}{4} + \frac{rx^6}{6} + \frac{\dot{x}^2}{2} - D =$$
$$-\frac{1}{12\left(\mu y^4+\lambda y^2+1\right)^3}[12D - 12C + (6\alpha + 36D\lambda - 6p)y^2 +$$
$$\left(3\beta + 12C\mu + 36D\lambda^2 + 36D\mu - 12\lambda p - 3q\right)y^4 +$$
$$\left(-12\alpha\mu + 12D\lambda^3 + 72D\lambda\mu - 6\lambda^2 p - 12\mu p - 3\lambda q - 2r\right)y^6 +$$
$$\left(-6\beta\mu - 6C\mu^2 + 36D\lambda^2\mu + 36D\mu^2 - 12\lambda\mu p - 3\mu q\right)y^8 +$$
$$\left(6\alpha\mu^2 + 36D\lambda\mu^2 - 6\mu^2 p\right)y^{10} + \left(3\beta\mu^2 + 12D\mu^3\right)y^{12}]$$

We equate to zero the coefficients of y^j in the last expression and we will obtain a system of algebraic equations. Solving it we get

$$C = D.$$

$$\alpha = p - 6D\lambda,\ \beta = -12D\lambda^2 - 20D\mu + 4\lambda p + q.$$

$$\mu = -\frac{6\lambda^2 p + 3\lambda q + 2r - 12D\lambda^3}{24(p - 6D\lambda)}.$$

The number λ is obtained from the cubic

$$192D^2\lambda^3 - 96Dp\lambda^2 + 12\left(p^2 - qD\right)\lambda + 4Dr + 3pq = 0.$$

From the initial conditions we obtain

$$\dot{y}_0 = -\frac{\dot{x}_0 \left(1 + \lambda y_0^2 + \mu y_0^4\right)^{3/2}}{\mu y_0^4 - 1}.$$

and

$$y_0 = \pm\sqrt{\frac{1 - \lambda x_0^2 + \pm\sqrt{x_0^4 \left(\lambda^2 - 4\mu\right) - 2\lambda x_0^2 + 1}}{2\mu x_0^2}}, \; \mu \neq 0.$$

Letting $\mu = 0$ may give other solutions. The numbers λ and y_0 are chosen so that

$$\left(\alpha + \beta y_0^2\right)^2 + 2\beta \dot{y}_0^2 \geq 0.$$

Example 5. Let

$$\ddot{x} - 0.777954x - 1.78885x^3 + 0.737354x^5 = 0 \wedge \mathrm{x}(0) = 0.611716 \wedge \mathrm{x}'(0) = 0.658052$$

The solution equals

$$x(t) = \frac{y(t)}{\sqrt{1 - 1.4476y^2(t)}},$$

where

$$\ddot{y} + 2.3441y^3 - 0.649566y = 0 \wedge y(0) = 0.492665 \wedge y'(0) = 0.343769.$$

$$y(t) = -0.612601\mathrm{cn}(0.531802t + 2.45388| - 0.0262994).$$

$$x(t) = \frac{0.772334\mathrm{cn}(1.00677 - 0.865268t|0.933803)}{\sqrt{1 - 0.863492\mathrm{cn}^2(1.00677 - 0.865268t|0.933803)}}.$$

See Figure 3.9 for the graph of $y(t)$.In Figure 3.10 we plotted the solution $x(t)$.

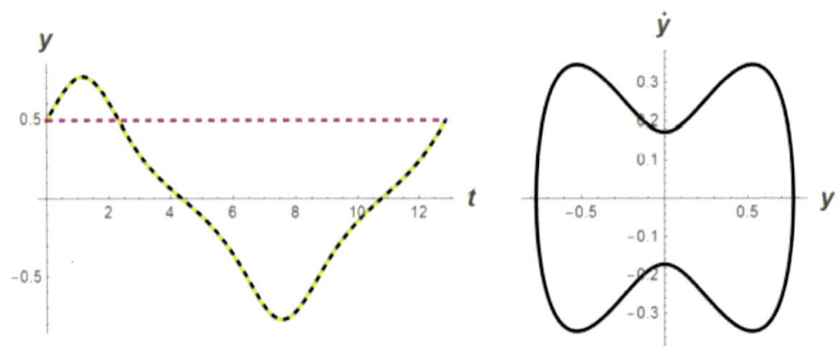

Figure 3.9

3.2 Two Particular Cases

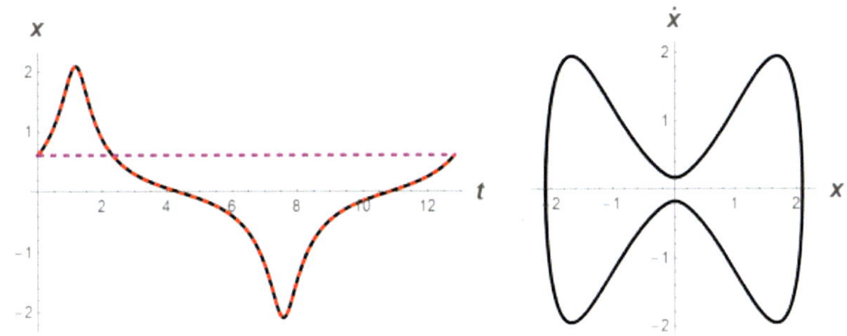

Figure 3.10

Let

$$x''(t) + px(t) + qx(t)^3 + rx(t)^5 = 0 \text{ given that } x(0) = x_0 \text{ and } x'(0) = \dot{x}_0.$$

We will assume that

$$(p + qx_0^2 + rx_0^4)r \neq 0.$$

3.2.1 First particular case: $\dot{x}_0 = 0$.

Let

$$x''(t) + px(t) + qx(t)^3 + rx(t)^5 = 0 \text{ given that } x(0) = x_0 \text{ and } x'(0) = 0.$$

Define

$$\Delta_1 = 3q^2 - 16pr - 4qrx_0^2 - 4r^2x_0^4.$$

If $\Delta_1 > 0$ then the exact solution is given by

$$x(t) = \frac{x_0\sqrt{1+\lambda}\,\text{cn}\left(\sqrt{\omega}t\middle|\, m\right)}{\sqrt{1+\lambda\,\text{cn}^2\left(\sqrt{\omega}t\middle|\, m\right)}}.$$

where

$$\omega = \frac{6D\lambda+4D-\lambda px_0^2-px_0^2}{(\lambda+1)x_0^2}.$$

$$m = \frac{6D\lambda+2D-\lambda px_0^2-px_0^2}{6D\lambda+4D-\lambda px_0^2-px_0^2}.$$

$$\lambda = \frac{x_0^2}{12\left(p+qx_0^2+rx_0^4\right)}\left(-3q - 6rx_0^2 \pm \sqrt{3\Delta_1}\right)$$

Assume now that $\Delta_1 < 0$. The exact solution reads

$$x(t) = \frac{x_0\sqrt{1+\lambda+\mu}\,\text{cn}\left(\sqrt{\omega}t\middle|\, m\right)}{\sqrt{1+\lambda\,\text{cn}^2\left(\sqrt{\omega}t\middle|\, m\right)+\mu\,\text{cn}^4\left(\sqrt{\omega}t\middle|\, m\right)}},$$

where

$$\omega = -\frac{12\lambda p+9\lambda qx_0^2+6qx_0^2+6\lambda rx_0^4+4rx_0^4}{12\lambda}.$$

$$m = \frac{12\lambda p+9\lambda qx_0^2+3qx_0^2+6\lambda rx_0^4+2rx_0^4}{12\lambda p+9\lambda qx_0^2+6qx_0^2+6\lambda rx_0^4+4rx_0^4}.$$

$$\mu = -\frac{12\lambda p+9\lambda qx_0^2+3qx_0^2+6\lambda rx_0^4+2rx_0^4}{x_0^2\left(3q+2rx_0^2\right)}.$$

$$\lambda = \frac{2\left(3q+2rx_0^2\right)\left(12p+9qx_0^2+6rx_0^4\pm 2\sqrt{6}\sqrt{\Delta_2}\right)}{3x_0^2\left(4r\left(4p+rx_0^4\right)-3q^2+4qrx_0^2\right)}.$$

$$\Delta_2 = \left(p+qx_0^2+rx_0^4\right)\left(6p+3qx_0^2+2rx_0^4\right)$$

We claim that $\Delta_2 > 0$. Indeed,

$$\Delta_2 = \frac{\left(3\left(q+2rx_0^2\right)^2-\Delta_1\right)\left(\left(3q+2rx_0^2\right)^2-3\Delta_1\right)}{128r^2}.$$

From the last identity it is clear that $\Delta_2 > 0$.
Assume now that $\Delta_1 = 0$.In this case we have the following solutions :

$$y(t) = \frac{y_0\sqrt{3q+2ry_0^2}\cos\left(\frac{1}{4}\sqrt{\frac{(q+2ry_0^2)(3q+2ry_0^2)}{r}}t\right)}{\sqrt{3q+4ry_0^2-2ry_0^2\cos\left(\frac{1}{4}\sqrt{\frac{(q+2ry_0^2)(3q+2ry_0^2)}{r}}t\right)}} \text{ for } \frac{(q+2ry_0^2)(3q+2ry_0^2)}{r} > 0.$$

$$y(t) = \frac{y_0\sqrt{3q+2ry_0^2}\cosh\left(\frac{1}{4}\sqrt{-\frac{(q+2ry_0^2)(3q+2ry_0^2)}{r}}t\right)}{\sqrt{3q+4ry_0^2-2ry_0^2\cosh\left(\frac{1}{4}\sqrt{-\frac{(q+2ry_0^2)(3q+2ry_0^2)}{r}}t\right)}} \text{ for } \frac{(q+2ry_0^2)(3q+2ry_0^2)}{r} < 0. \quad (3.17)$$

The solutions (3.17) are called solitons. They arise in Soliton Theory.

3.2.2 Second Particular Case: $x_0 = 0$.

Let

$$x''(t) + px(t) + qx(t)^3 + rx(t)^5 = 0 \text{ given that } x(0) = 0 \text{ and } x'(0) = \dot{x}_0 \neq 0.$$

The solution is given by

$$x(t) = \frac{y(t)}{\sqrt{1+\lambda y^2(t)+\frac{1}{8\dot{x}_0^2}\left(4\lambda p+q-6\lambda^2\dot{x}_0^2\right)y^4(t)}},$$

where the function $y = y(t)$ is the solution to the following Duffing equation :

$$\ddot{y} + \left(p-3\lambda\dot{x}_0^2\right)y - \frac{1}{4}\left(4\lambda p+q-6\lambda^2\dot{x}_0^2\right)y^3 = 0,\ y(0) = 0 \text{ and } y'(0) = \dot{x}_0.$$

The number λ obeys the cubic

$$48\dot{x}_0^4\lambda^3 - 48p\dot{x}_0^2\lambda^2 + \left(12p^2 - 6q\dot{x}_0^2\right)\lambda + 3pq + 2r\dot{x}_0^2 = 0.$$

3.3 Summary

Chapter 3 is devoted to the study of the conservative Duffing oscillator and some of its extensions. The main results can be summarized as follows:

1. **Formulation of the conservative Duffing oscillator.** The chapter introduces the classical conservative Duffing equation

$$\ddot{x} + \alpha x + \beta x^3 = 0,$$

together with its cubic–quintic generalization and the more general polynomial Duffing model. Emphasis is placed on the physical interpretation of the parameters, on the associated potential function and on the Hamiltonian (energy first integral), which makes explicit the conservative character of the system.

2. **Exact elliptic-function solution of the cubic Duffing oscillator.** By assuming an ansatz in terms of the Jacobi elliptic function $\mathrm{cn}(\cdot, m)$, the conservative cubic Duffing oscillator is solved in closed form. Algebraic relations are derived between the physical parameters (α, β) and the elliptic parameters (amplitude, frequency and modulus m). From these relations an explicit expression for the period in terms of the complete elliptic integral $K(m)$ is obtained, making the amplitude–frequency dependence completely transparent.

3. **Qualitative analysis via energy and phase portraits.** Using the energy integral, the chapter describes the geometry of the trajectories in the phase plane $(x, \dot{x})$ for different signs of α and β. The existence of centres, saddles, separatrices and double–well potentials is discussed, and the relation between energy levels and phase portraits is made explicit. This analysis clarifies how periodic and unbounded motions arise in the conservative Duffing oscillator.

4. **Transformation to the cubic–quintic Duffing oscillator.** A nonlinear transformation of the form

$$x(t) = \frac{y(t)}{\sqrt{1 + \lambda y^2(t) + \mu y^4(t)}}$$

is introduced, where $y(t)$ solves a cubic Duffing equation. By choosing the parameters λ and μ appropriately, this transformation maps the cubic Duffing oscillator onto the cubic–quintic Duffing oscillator

$$\ddot{x} + px + qx^3 + rx^5 = 0.$$

The chapter derives the algebraic relations linking (p, q, r) with $(\alpha, \beta, \lambda, \mu)$ and shows that, in this way, solutions of the cubic–quintic model can be constructed from those of the simpler cubic equation.

5. **Closed-form solutions for particular parameter choices.** Two particular cases of the cubic–quintic Duffing oscillator are analysed in detail. For suitable parameter values, the transformation above leads to explicit expressions of the solutions in terms of elementary and/or elliptic functions. Numerical examples illustrate the procedure and confirm the agreement between the transformed solutions and direct numerical integration of the original equation.

6. **Foundations for generalized Duffing equations.** Overall, Chapter 3 shows how the conservative Duffing oscillator serves as a prototype for more complicated nonlinear conservative systems. The exact solutions, phase-plane analysis and transformation techniques developed here provide the basis for the generalized Duffing equations treated in Chapter 4 and for the perturbative and averaging methods applied in later chapters.

Chapter 4

The generalized Duffing oscillator equation

The source of the nonlinearity in a mechanical system that results in its dynamic behaviour being modelled by the Duffing equation is the stiffness. In the case considered here, the stiffness is generally a function of position. This means that the force applied to the spring Fs, and the resulting displacement y have a nonlinear relationship. If the stiffness is not symmetric about y then the force–deflection character istic requires terms in the series of y, in which the exponents of y are integers of even order. Thus, the expression for the spring force has the form

$$F_s = k_1 u \pm k_2 u^2 \pm k_3 u^3.$$

This situation is described by an equation of the form

$$\ddot{u} + a_0 + a_1 u + a_2 u^2 + a_3 u^3 = 0, u(0) = u_0 \text{ and } u'(0) = \dot{u}_0, \tag{4.1}$$

which is called a generalized Duffing oscillator equation. This equation is also called Duffing-Helmholtz equation or mixed parity Duffing equation. This oscillator may be solved in several ways.

4.1 Stability analysis and phase–plane structure of the generalized Duffing–Helmholtz oscillator

We study the generalized Duffing oscillator equation Duffing–Helmholtz equation (or mixed–parity Duffing equation)

$$\ddot{u} + a_0 + a_1 u + a_2 u^2 + a_3 u^3 = 0, \qquad u(0) = u_0, \qquad u'(0) = \dot{u}_0, \tag{4.2}$$

which coincides with (4.1). This is a conservative second–order nonlinear oscillator with a polynomial restoring force of mixed parity (constant + linear + quadratic + cubic terms).

4.1.1 Energy integral and potential function (the organizing principle)

Equation (4.2) can be written as Newton's law

$$\ddot{u} = -V'(u),$$

with potential function

$$V(u) = a_0 u + \frac{a_1}{2}u^2 + \frac{a_2}{3}u^3 + \frac{a_3}{4}u^4. \tag{4.3}$$

Therefore the energy first integral is

$$E = \frac{1}{2}\dot{u}^2 + V(u) = \frac{1}{2}\dot{u}^2 + a_0 u + \frac{a_1}{2}u^2 + \frac{a_2}{3}u^3 + \frac{a_3}{4}u^4, \qquad \frac{dE}{dt} = 0. \tag{4.4}$$

Hence the motion in the phase plane $(u, \dot{u})$ occurs on the level sets $E = \text{const}$.

For easy understanding. You can think of $u(t)$ as a particle sliding (without friction) on the curve $V(u)$. Energy is conserved. If $V(u)$ has a well (a minimum), the particle oscillates in that well. If the potential opens downward and has no barrier, the particle can escape and the solution becomes unbounded.

4.1.2 Equilibria: where are the fixed points?

Equilibria are constant solutions $u(t) \equiv u_*$. Substituting in (4.2) gives

$$a_0 + a_1 u_* + a_2 u_*^2 + a_3 u_*^3 = 0. \tag{4.5}$$

Thus equilibria are exactly the real roots of the cubic polynomial

$$P(u) := a_3 u^3 + a_2 u^2 + a_1 u + a_0. \tag{4.6}$$

Depending on the parameters (a_0, a_1, a_2, a_3), the cubic can have one equilibrium (one real root) or three equilibria (three real roots). The number of equilibria determines the possible shapes of the phase portrait: one well, double well, tilted wells, or escape.

4.1.3 Linear stability test (local stability in one line)

Write the system as a first–order planar system

$$\dot{u} = v, \qquad \dot{v} = -(a_0 + a_1 u + a_2 u^2 + a_3 u^3). \tag{4.7}$$

A fixed point is $(u_*, 0)$ where u_* solves (4.5). The Jacobian matrix at $(u_*, 0)$ is

$$J(u_*, 0) = \begin{pmatrix} 0 & 1 \\ -(a_1 + 2a_2 u_* + 3a_3 u_*^2) & 0 \end{pmatrix}.$$

Its eigenvalues satisfy

$$\lambda^2 + (a_1 + 2a_2u_* + 3a_3u_*^2) = 0.$$

Define the effective stiffness

$$\kappa(u_*) := a_1 + 2a_2u_* + 3a_3u_*^2 = V''(u_*). \tag{4.8}$$

Then:

- If $\kappa(u_*) > 0$, then $\lambda = \pm i\sqrt{\kappa(u_*)}$ and $(u_*, 0)$ is a center (linearly stable).
- If $\kappa(u_*) < 0$, then $\lambda = \pm\sqrt{-\kappa(u_*)}$ and $(u_*, 0)$ is a saddle (unstable).
- If $\kappa(u_*) = 0$, the equilibrium is degenerate and higher-order terms decide; in phase portraits this corresponds to a separatrix collision or a flat inflection of $V(u)$.

Key identity. Because the system is conservative, **stability is equivalent to the type of critical point of** $V(u)$:

$$V'(u_*) = 0 \iff u_* \text{ is an equilibrium}, \qquad V''(u_*) > 0 \iff \text{stable center (minimum of } V\text{)}.$$

Similarly, $V''(u_*) < 0$ means a local maximum of $V(u)$ and gives an unstable saddle.

4.1.4 How many equilibria? (parameter conditions via the cubic discriminant)

The cubic (4.6) may have one or three real roots. A standard way to decide this is through the discriminant of a cubic. For $P(u) = a_3u^3 + a_2u^2 + a_1u + a_0$ with $a_3 \neq 0$, define

$$\Delta_3 = 18a_3a_2a_1a_0 - 4a_2^3a_0 + a_2^2a_1^2 - 4a_3a_1^3 - 27a_3^2a_0^2. \tag{4.9}$$

Then:

- $\Delta_3 > 0$: three distinct real equilibria (three real roots).
- $\Delta_3 = 0$: multiple root (at least two equilibria collide).
- $\Delta_3 < 0$: one real equilibrium (one real root + two complex).

This criterion is purely algebraic in (a_0, a_1, a_2, a_3).

4.1.5 Global boundedness vs escape (role of a_3)

The quartic term dominates for large $|u|$:

$$V(u) \sim \frac{a_3}{4}u^4 \quad \text{as } |u| \to \infty.$$

Hence:

- If $a_3 > 0$, then $V(u) \to +\infty$ as $|u| \to \infty$ and all finite–energy motions are bounded. Phase portraits consist of closed curves around centers, separatrices through saddles, and outer closed curves depending on the number of wells.
- If $a_3 < 0$, then $V(u) \to -\infty$ as $|u| \to \infty$ and sufficiently large energy trajectories can escape to infinity, producing unbounded solutions.
- If $a_3 = 0$, the leading term becomes cubic/quadratic; boundedness depends on a_2 and a_1 and typically allows escape in one direction due to the odd term (tilt).

This is the global picture: the sign of a_3 decides whether the oscillator is confining at large amplitudes.

4.1.6 Phase portraits from the potential: what you will see and why

Every phase portrait feature can be read from $V(u)$:

- Closed orbits correspond to energy levels trapped inside a well (around a minimum).
- Separatrices occur at the energy level equal to a local maximum of $V(u)$; they pass through the corresponding saddle equilibrium.
- Unbounded trajectories occur if the energy level is above the last barrier and the potential does not confine motion on at least one side.

Very important formula. If u_s is a saddle (a local maximum of V), then the separatrix energy is

$$E_{\text{sep}} = V(u_s). \tag{4.10}$$

Solutions with $E < E_{\text{sep}}$ are trapped in the well; solutions with $E > E_{\text{sep}}$ can cross the saddle and move to the other region (or escape if no outer wall exists).

4.1.7 Illustrations (potential shapes and phase portraits)

(A) One stable equilibrium: single tilted well (typical when $\Delta_3 < 0$ and $a_3 > 0$)

(B) Two wells (asymmetric double well): two stable centers and one saddle

(C) Phase portrait sketch: two centers + one saddle + separatrix

(D) Escape case: $a_3 < 0$ (potential opens downward)

4.1.8 Worked examples (compute equilibria and classify them)

Example 1 (one stable equilibrium; tilted single well)

Choose

$$a_3 = 1, \qquad a_2 = 1, \qquad a_1 = 1, \qquad a_0 = -1.$$

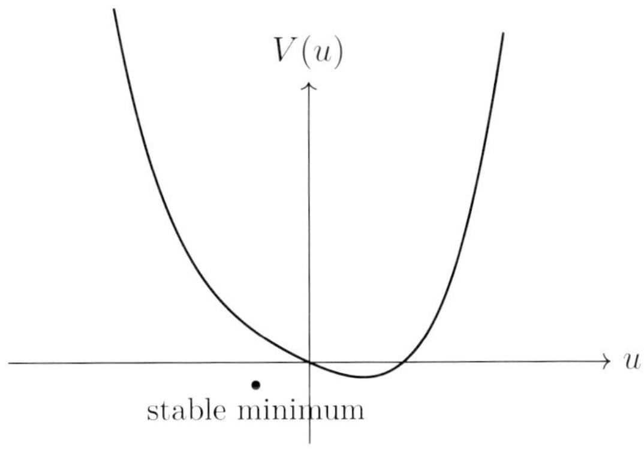

Figure 4.1: Potential with one minimum: one center in the phase plane.

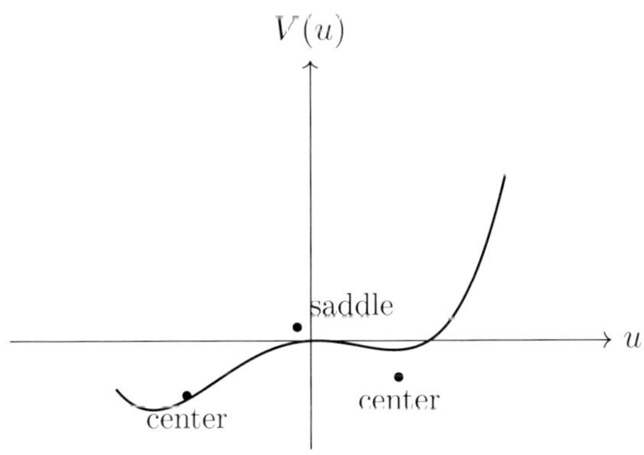

Figure 4.2: Asymmetric double-well potential (mixed parity) producing two centers and one saddle.

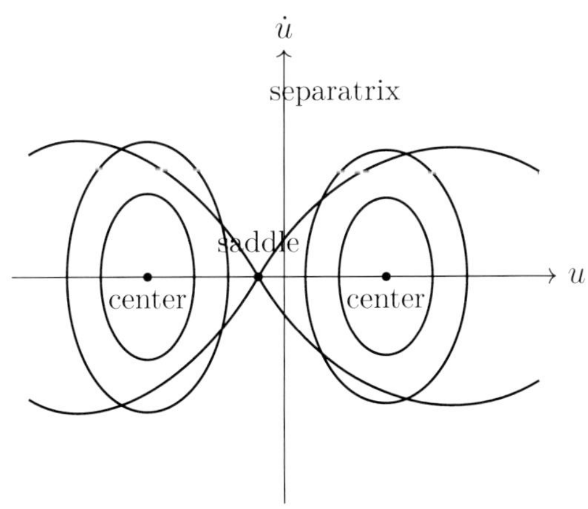

Figure 4.3: Phase portrait sketch for the mixed–parity Duffing–Helmholtz oscillator.

Equilibria solve $u^3 + u^2 + u - 1 = 0$. This cubic has one real root (numerically $u_* \approx 0.5437$), hence a single equilibrium. Stability is decided by

$$\kappa(u_*) = a_1 + 2a_2u_* + 3a_3u_*^2 = 1 + 2u_* + 3u_*^2 > 0,$$

so $(u_*, 0)$ is a stable center. Hence all small motions around u_* are periodic.

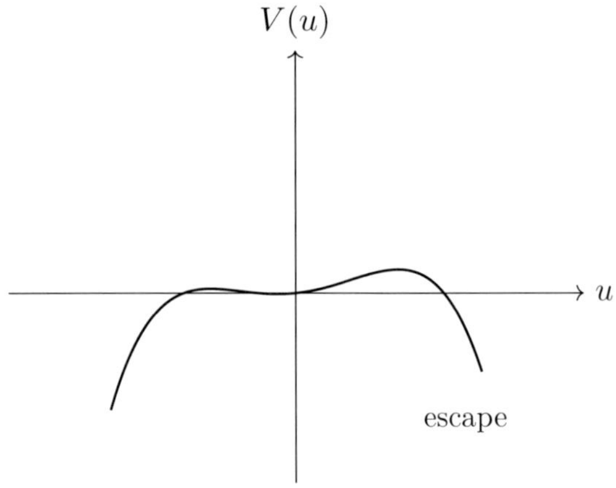

Figure 4.4: If $a_3 < 0$, then $V(u) \to -\infty$ and unbounded solutions are possible.

Example 2 (asymmetric double well; two centers and one saddle)

Choose

$$a_3 = 1, \qquad a_2 = -1, \qquad a_1 = -1, \qquad a_0 = 0.$$

Equilibria solve $u(u^2 - u - 1) = 0$, so

$$u_* \in \left\{0, \frac{1-\sqrt{5}}{2}, \frac{1+\sqrt{5}}{2}\right\}.$$

Classify using $\kappa(u_*) = a_1 + 2a_2u_* + 3a_3u_*^2$:

$$\kappa(0) = a_1 = -1 < 0 \quad \Rightarrow \quad (0,0) \text{ is a saddle,}$$

and

$$\kappa\left(\frac{1 \pm \sqrt{5}}{2}\right) = -1 - 2\left(\frac{1 \pm \sqrt{5}}{2}\right) + 3\left(\frac{1 \pm \sqrt{5}}{2}\right)^2 > 0,$$

so the other two equilibria are stable centers. Thus the phase portrait has two families of closed orbits separated by a separatrix through the saddle.

Example 3 (escape; potential opens downward)

Choose

$$a_3 = -1, \qquad a_2 = 0, \qquad a_1 = 1, \qquad a_0 = 0,$$

so $V(u) = \frac{1}{2}u^2 - \frac{1}{4}u^4 \to -\infty$ as $|u| \to \infty$. Then sufficiently large energy initial conditions yield unbounded trajectories.

4.1.9 Practical algorithm (given parameters and initial data, decide the qualitative dynamics)

Given (a_0, a_1, a_2, a_3) and $(u_0, \dot{u}_0)$:

1. Compute equilibria by solving $P(u) = a_3u^3 + a_2u^2 + a_1u + a_0 = 0$.

2. For each equilibrium u_*, compute $\kappa(u_*) = a_1 + 2a_2u_* + 3a_3u_*^2$:

   - $\kappa > 0 \Rightarrow$ stable center,
   - $\kappa < 0 \Rightarrow$ unstable saddle.

3. Compute the conserved energy

$$E = \frac{1}{2}\dot{u}_0^2 + a_0u_0 + \frac{a_1}{2}u_0^2 + \frac{a_2}{3}u_0^3 + \frac{a_3}{4}u_0^4.$$

4. Identify saddle equilibria u_s (where $\kappa(u_s) < 0$) and compute barrier energies $E_{\text{sep}} = V(u_s)$.

5. Compare E with barrier energies: if E is below the barrier around a center, motion is trapped and periodic; if E exceeds the relevant barrier, the solution crosses the separatrix and may jump wells or escape (depending on a_3 and the outer shape of V).

This completes a full stability and phase–portrait analysis in terms of the parameters (a_0, a_1, a_2, a_3) and initial data $(u_0, \dot{u}_0)$.

4.2 Exact Solution–Universal Principle

Theorem 4.1 (Universal quadratic reduction of conservative . *mixed–parity polynomial oscillators] Consider the conservative generalized Duffing–Helmholtz oscillator with a polynomial restoring force containing even and odd powers,*

$$\ddot{u} + P(u) = 0, \qquad P(u) := \sum_{j=0}^{N} a_j u^j, \qquad a_j \in \mathbb{R},\ N \geq 1. \tag{4.11}$$

Assume there exists a real number A such that

(i) *A is an equilibrium of (4.11), i.e.*

$$P(A) = 0, \tag{4.12}$$

(ii) *the shifted restoring force is even about A, i.e.*

$$P(A+z) = P(A-z) \quad \textit{for all } z \textit{ in a neighborhood of } 0, \tag{4.13}$$

equivalently, in the Taylor expansion $P(A+z) = \sum_{\ell \geq 0} c_\ell z^\ell$ all odd coefficients vanish: $c_{2k+1} = 0$ for every k.

Then there exist constants $B \in \mathbb{R} \setminus \{0\}$ and a function $x(t)$ such that the solution of (4.11) can be written in the quadratically reduced form

$$u(t) = A + B\,x^2(t), \tag{4.14}$$

where $x(t)$ satisfies a Duffing–type odd polynomial oscillator

$$\ddot{x} + \alpha_1 x + \alpha_3 x^3 + \alpha_5 x^5 + \cdots + \alpha_{2m-1} x^{2m-1} = 0, \qquad m \le \left\lfloor \frac{N+1}{2} \right\rfloor, \tag{4.15}$$

with real coefficients α_{2k-1} depending algebraically on $(a_0, \ldots, a_N)$ and on (A, B).

Meaning (for easy understanding). *Condition* (4.13) *says:* after shifting to the equilibrium A, the restoring force becomes symmetric. *Once the model is symmetric, writing $u - A$ as a* square *($u - A = Bx^2$) automatically produces an* odd *Duffing–type equation in x. That is why odd-parity Duffing models (cubic, cubic–quintic, etc.) are the natural normal forms for analysis.*

Proof. Step 1 (shift to an equilibrium). Let $u(t) = A + z(t)$. Using (4.12) we obtain the shifted equation

$$\ddot{z} + P(A + z) = 0. \tag{4.16}$$

By (4.13), the polynomial $P(A + z)$ is an even function of z, hence it contains only even powers:

$$P(A + z) = c_0 + c_2 z^2 + c_4 z^4 + \cdots + c_{2M} z^{2M}, \qquad c_0 = P(A) = 0, \tag{4.17}$$

for some $M \le \lfloor N/2 \rfloor$.

Step 2 (quadratic substitution). Fix any $B \ne 0$ and set

$$z(t) = B\, x^2(t). \tag{4.18}$$

Then

$$\dot{z} = 2B\, x\dot{x}, \qquad \ddot{z} = 2B(\dot{x}^2 + x\ddot{x}).$$

Substituting into (4.16) and using (4.17) yields

$$2B(\dot{x}^2 + x\ddot{x}) + \sum_{k=1}^{M} c_{2k}\, B^{2k} x^{4k} = 0. \tag{4.19}$$

Step 3 (energy removes the $\dot{x}^2$ term and produces an odd restoring force). Equation (4.16) is conservative with potential

$$V(z) = \int_0^z P(A + s)\, ds = \sum_{k=1}^{M} \frac{c_{2k}}{2k+1}\, z^{2k+1},$$

so the energy first integral is

$$E = \frac{1}{2}\dot{z}^2 + V(z) = \text{const.} \tag{4.20}$$

Using $z = Bx^2$ and $\dot{z} = 2Bx\dot{x}$ in (4.20) we obtain

$$2B^2 x^2 \dot{x}^2 + \sum_{k=1}^{M} \frac{c_{2k}}{2k+1}\, B^{2k+1} x^{4k+2} = E. \tag{4.21}$$

Differentiate (4.21) with respect to t and divide by $2Bx$ (for $x \neq 0$) to eliminate the $\dot{x}^2$ term in (4.19). The result is precisely an equation of the form (4.15) with odd restoring force (only odd powers of x). Finally, since $u = A + z = A + Bx^2$, the representation (4.14) follows. □

Remark 4.1 (Why odd Duffing models are the right normal form). The mixed–parity oscillator (4.11) is geometrically less transparent because the restoring force is not symmetric. When a symmetry point A exists in the sense of (4.13), the dynamics near and across wells can be studied through the odd Duffing hierarchy (4.15), where energy methods, phase portraits, separatrix geometry, and (after adding small damping and forcing) Melnikov analysis become systematic. This explains why the book repeatedly reduces generalized Duffing–Helmholtz problems to odd-parity Duffing (cubic, cubic–quintic, ...): odd models are the *canonical* conservative building blocks for both exact elliptic solutions and chaos diagnostics.

4.2.1 First approach

We assume that

$$u(t) = A + (u_0 - A)x^2(t),$$

where the function $x = x(t)$ is the solution to some cubic-quintic Duffing oscillator

$$\ddot{x} + px + qx^3 + rx^5 = 0,\ x(0) = 1 \text{ and } x'(0) = \dot{x}_0 := -\frac{\dot{u}_0}{2\,(A - u_0)}$$

Observe that

$$\dot{x}^2 = 2D - px^2 - \tfrac{qx^4}{2} - \tfrac{rx^6}{3},$$
where
$$D = \tfrac{1}{24}\left(\tfrac{3\dot{u}_0^2}{(A-u_0)^2} + 12p + 6q + 4r\right).$$

Using the above relations we obtain

$$\begin{aligned}
&\ddot{u} + a_0 + a_1 u + a_2 u^2 + a_3 u^3 = \\
&a_3 A^3 + a_2 A^2 + a_1 A + a_0 - 4AD + 4Du_0 \\
&+ (A - u_0)\left(-3a_3 A^2 - 2a_2 A - a_1 + 4p\right) x^2 \\
&+ (A - u_0)\left(3a_3 A^2 - 3a_3 A u_0 + a_2 A - a_2 u_0 + 3q\right) x^4 \\
&+ \tfrac{1}{3}(A - u_0)\left(-3a_3 A^2 + 6a_3 A u_0 - 3a_3 u_0^2 + 8r\right) x^6.
\end{aligned}$$

Equating to zero the coefficents of x^j and solving the resulting algebraic system one finds that

$$p = \frac{1}{4}\left(3a_3 A^2 + 2a_2 A + a_1\right),\ q = -\frac{1}{3}\left(3a_3 A + a_2\right)(A - u_0) \text{ and } r = \frac{3}{8} a_3 (A - u_0)^2.$$

The number A is found from the quartic

$$-\left[3a_3u_0^4+4a_2u_0^3+6a_1u_0^2+12a_0u_0+6\dot{u}_0^2\right]+12a_0A+6a_1A^2+4a_2A^3+3a_3A^4=0.$$

Example 6. Let

$$\ddot{u}-0.430638-1.19926u+0.998757u^2+0.0478466u^3=0, u(0)=0.116892, u'(0)=0.848492.$$

Then

$$u(t)=A+(0.116892\ -A)x^2(t)\text{ for }A=2.32891,$$

and the function $x=x(t)$ is the solution to the following cubic-quintic Duffing equation :

$$\ddot{x}+1.05783x-0.982908x^3+0.0877929x^5=0\wedge x(0)=1\wedge x'(0)=-0.191791.$$

The solution to has the form

$$x(t)=\frac{y(t)}{\sqrt{1+\lambda y^2(t)+\mu y^4(t)}},$$

where

$$\lambda=0.801572,\ \mu=-0.00578368,$$

and the function $y=y(t)$ is the solution to the following Duffing equation :

$$\ddot{y}+0.00731548y^3-0.46297y=0\wedge \mathrm{y}(0)=2.11184\wedge \mathrm{y}'(0)=-1.62002$$

The explicit expression for $y(t)$ is

$$y(t)=11.3104\mathrm{cn}(0.687646t+2.30156|0.989545)\text{ with period }T_y=21.3694$$

See Figure 4.5.

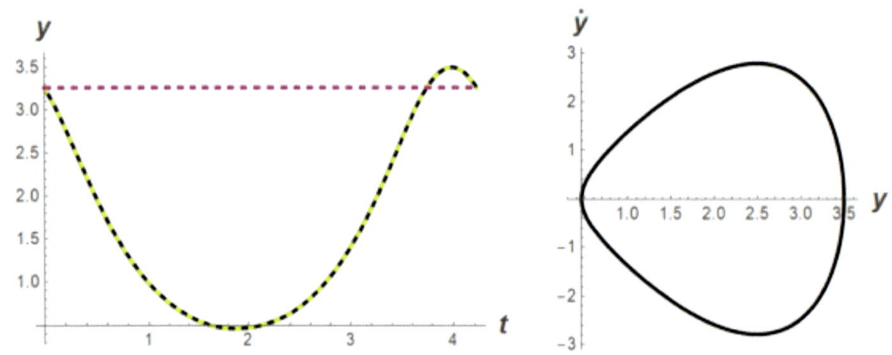

Figure 4.5

The explict expression for $x = x(t)$ is

$$x(t) = 2.32891 + \frac{1}{-0.362371 + 0.334479\mathrm{cn}(0.687646t + 2.30156|0.989545)^2 - \frac{0.00353394}{\mathrm{cn}(0.687646t+2.30156|0.989545)^2}},$$

See Figure 4.6.

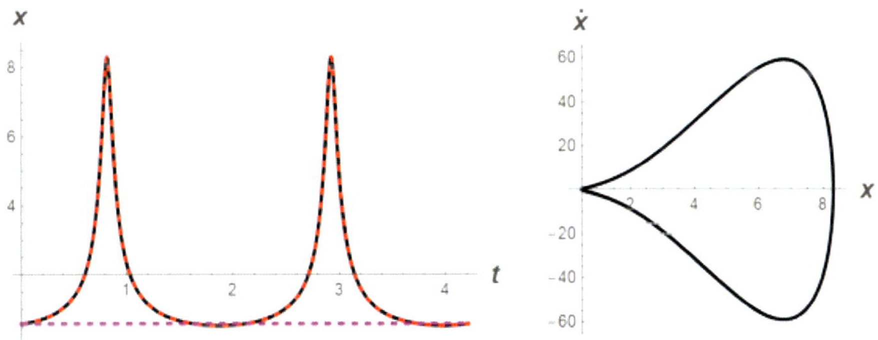

Figure 4.6

4.2.2 Second approach

The solution is assumed in the ansatz form

$$u = A + \frac{B}{1 + C\wp(t + D; g_2, g_3)},$$

where $\wp = \wp(t) = \wp(t + D; g_2, g_3)$ stands for the elliptic Weierstrass function with elliptic invariants g_2 and g_3. This function obeys the ode

$$\wp'(t)^2 = 4\wp^3(t) - g_2\wp(t) - g_3. \tag{4.22}$$

We have :

$2(Cw + 1)^3 \left(\ddot{u} + a_0 + a_1 u + a_2 u^2 + a_3 u^3\right) =$

$2a_3A^3 + 6a_3A^2B + 2a_2A^2 + 6a_3AB^2 + 4a_2AB + 2a_1A + 2a_3B^3 + 2a_2B^2 + 2a_1B + 2a_0 - 4BC^2g_3 + BCg_2-$

$C\left(-6a_3A^3 - 12a_3A^2B - 6a_2A^2 - 6a_3AB^2 - 8a_2AB - 6a_1A - 2a_2B^2 - 4a_1B - 6a_0 + 3BCg_2\right)\wp+$

$2C\left(3a_3A^3C + 3a_3A^2BC + 3a_2A^2C + 2a_2ABC + 3a_1AC + a_1BC + 3a_0C - 6B\right)\wp^2+$

$2C^2\left(a_3A^3C + a_2A^2C + a_1AC + a_0C + 2B\right)\wp^3.$

Equating to zero the coefficients of $\wp^j$ in the last expression we obtain an algebraic system.

Solving it gives

$$B = -\frac{6(A(A(a_3A+a_2)+a_1)+a_0)}{A(3a_3A+2a_2)+a_1}.$$

$$C = \frac{12}{3a_3A^2+2a_2A+a_1}.$$

$$g_2 = \frac{1}{12}\left(-3a_3^2A^4 - 2a_3A\left(A\left(2a_2A+3a_1\right)+6a_0\right)+a_1^2-4a_0a_2\right).$$

$$g_3 = \frac{1}{216}\begin{pmatrix} 18a_3a_1^2A^2 + \left(a_2A^2+3a_0\right)\left(9a_0a_3 - a_2A\left(3a_3A+4a_2\right)\right) + \\ a_1\left(9a_3^2A^4 + 12a_3A\left(a_2A^2+3a_0\right) - 6a_2\left(a_2A^2+a_0\right)\right) + a_1^3 \end{pmatrix}.$$

The constants A and D are obtained from the initilal conditions as follows. Define

$$D = \wp^{-1}\left(\frac{-A-B+u_0}{C\left(A-u_0\right)}; g_2, g_3\right).$$

Clearly, $u(0) = u_0$. Now, from the condition $u'(0) = \dot{u}_0$ we obtain the following quartic for the value of A :

$$3a_3A^4 + 4a_2A^3 + 6a_1A^2 + 12a_0A - (3a_3u_0^4 + 4a_2u_0^3 + 6a_1u_0^2 + 12a_0u_0 + 6\dot{u}_0^2) = 0$$

Example 7. Let

$$\ddot{u} - 0.430638 - 1.19926u + 0.998757u^2 + 0.0478466u^3 = 0, u(0) = 0.116892, u'(0) = 0.848492.$$

Then

$$u(t) = A + (0.116892\ - A)x^2(t) \text{ for } A = 2.32891,$$

and the function $x = x(t)$ is the solution to the following cubic-quintic Duffing equation :

$$\ddot{x} + 1.05783x - 0.982908x^3 + 0.0877929x^5 = 0 \wedge x(0) = 1 \wedge x'(0) = -0.191791.$$

The solution has the form

$$x(t) = \frac{y(t)}{\sqrt{1+\lambda y^2(t)+\mu y^4(t)}},$$

where

$$\lambda = 0.801572,\ \mu = -0.00578368,$$

and the function $y = y(t)$ is the solution to the following Duffing equation :

$$\ddot{y} + 0.00731548y^3 - 0.46297y = 0 \wedge \mathrm{y}(0) = 2.11184 \wedge \mathrm{y}'(0) = -1.62002$$

The explicit expression for $y(t)$ is

$$y(t) = 11.3104\text{cn}(0.687646t + 2.30156|0.989545) \text{ with period } T_y = 21.3694$$

See Figure 4.7. The explict expression for $x = x(t)$ is

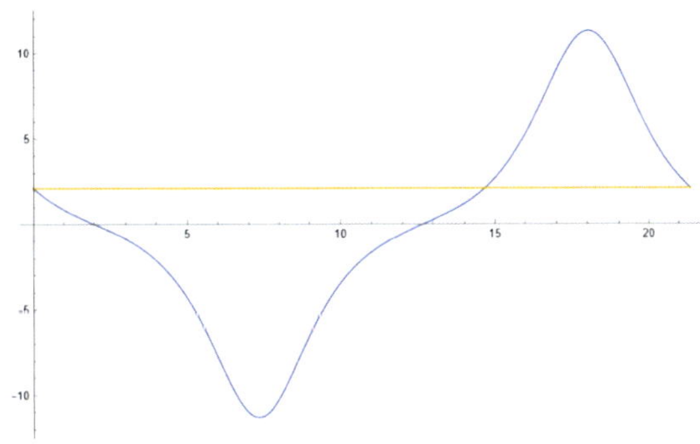

Figure 4.7

$$x(t) = 2.32891 + \frac{1}{-0.362371 + 0.334479\text{cn}(0.687646t + 2.30156|0.989545)^2 - \frac{0.00353394}{\text{cn}(0.687646t+2.30156|0.989545)^2}},$$

See Figure 4.8.

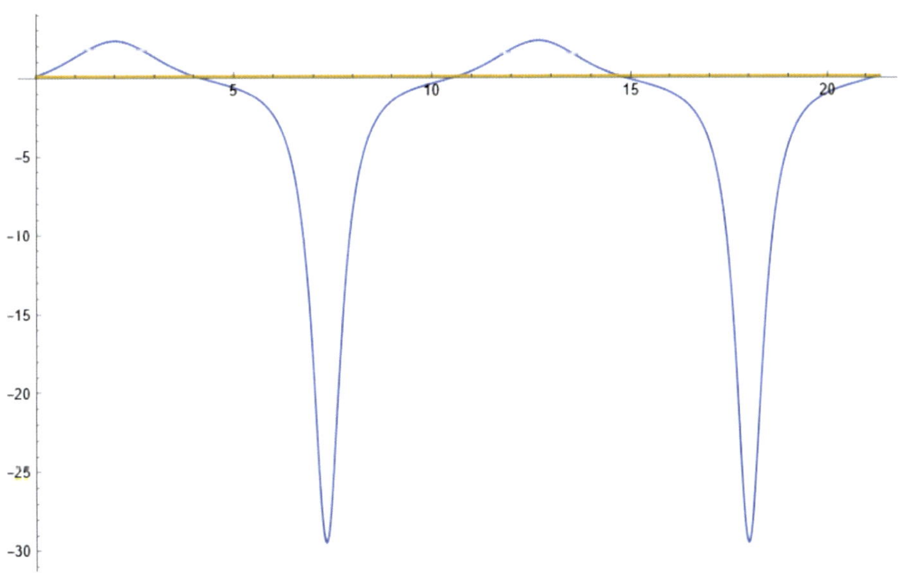

Figure 4.8

If $a_3 = 0$ then we assume that $a_2 \neq 0$ and then we will have the following cubic for A :

$$2a_2A^3 + 3a_1A^2 + 6a_0A - 2a_2u_0^3 - 3a_1u_0^2 - 6a_0u_0 - 3\dot{u}_0^2$$

In this last case we will solve the following oscillator, called Helmholtz or quadratic oscillator equation:

$$\ddot{u} + a_0 + a_1 u + a_2 u^2 = 0, u(0) = u_0 \text{ and } u'(0) = \dot{u}_0$$

This differential equation is used to model the free vibration of an eardrum and also in the ship rolling vibration [114]

Example 8. Let us consider the Duffing-Helmholtz oscillator in previous example :

$$\ddot{u} - 0.430638 - 1.19926u + 0.998757u^2 + 0.0478466u^3 = 0, u(0) = 0.116892, u'(0) = 0.848492.$$

The solution reads

$$x(t) = 2.32891 - \frac{3.96735}{1 + 2.83601\wp(-t - (3.34696 - 2.29032i); 0.248771, -0.0408264)}$$

In this case, $A = 2.32891$. We have a periodic solution with period $T = 10.6847$. See Figure 4.9.

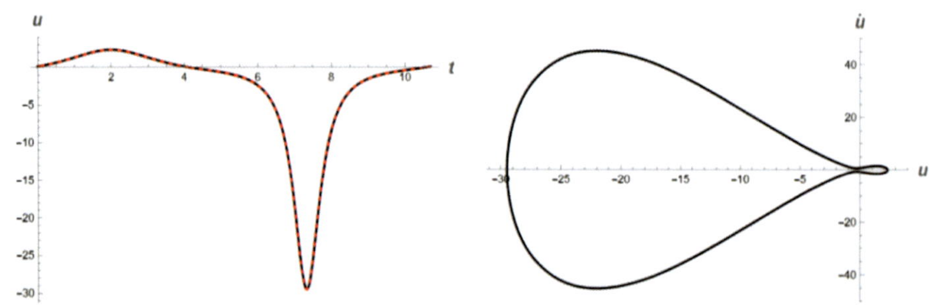

Figure 4.9

If $a_3 = 0$ then we assume that $a_2 \neq 0$ and then we will have the following cubic for A :

$$2a_2 A^3 + 3a_1 A^2 + 6a_0 A - 2a_2 u_0^3 - 3a_1 u_0^2 - 6a_0 u_0 - 3\dot{u}_0^2$$

In this last case we will solve the following oscillator, called Helmholtz or quadratic oscillator equation:

$$\ddot{u} + a_0 + a_1 u + a_2 u^2 = 0, u(0) = u_0 \text{ and } u'(0) = \dot{u}_0$$

This differential equation is used to model the free vibration of an eardrum and also in the ship rolling vibration [114]

4.2.3 Third approach

The solution to the oscillator (4.1) may also be expressed in terms of the solution to some Duffing equation. Indeed, let us prove that this holds for the solutionn to the ode (4.22) for any given initial condition $\wp(0) = \wp_0$. Assume that

$$\wp(t) = R + \frac{S}{1 + x(t)},$$

where the function $x = x(t)$ is the solution to some Duffing equation

$$\ddot{x} + \alpha x + \beta x^3 = 0,\ x(0) = x_0 \text{ and } x'(0) = \dot{x}_0,$$

where

$$x_0 = \frac{\wp_0 - R - S}{R - \wp_0},\ \dot{x}_0 = \frac{S\sqrt{4\wp_0^3 - g_2\wp_0 - g_3}}{(R - \wp_0)^2}$$

Then

$$x'(t)^2 = 2C - \alpha y^2(t) - \frac{1}{2}\beta y^4(t).$$

We have:

$\wp'(t)^2 - [4\wp^3(t) \quad g_2\wp(t) - g_3] = \frac{1}{2(1+x^4)} \times$
$[(2CS^2 + 2g_2R + 2g_2S + 2g_3 - 8R^3 - 24R^2S - 24RS^2 - 8S^3)x^0 +$
$(8g_2R + 6g_2S + 8g_3 - 32R^3 - 72R^2S - 48RS^2 - 8S^3)\, x^1 +$
$(12g_2R + 6g_2S + 12g_3 - 48R^3 - 72R^2S - 24RS^2 - 2\alpha S^2)\, x^2 +$
$(8g_2R + 2g_2S + 8g_3 - 32R^3 - 24R^2S)\, x^3 +$
$(2g_2R + 2g_3 - 8R^3 - \beta S^2)\, x^4].$

Equating to zero the coefficients of x^j in the last expression we obtain an algebraic system. A solution to it reads

$$\alpha = -\tfrac{3}{S^2}\left(-2g_2R - g_2S - 2g_3 + 8R^3 + 12R^2S + 4RS^2\right).$$

$$\beta = -\tfrac{2}{S^2}\left(-g_2R - g_3 + 4R^3\right).$$

$$S = \tfrac{4}{12R^2 - g_2}\left(g_2R + g_3 - 4R^3\right).$$

$$C = \tfrac{272g_2R^4 - 256g_3R^3 - 172g_2^2R^2 - 320g_2g_3R + 3g_2^3 - 256g_3^2 - 64R^6}{16(12R^2 - g_2)(-g_2R - g_3 + 4R^3)}.$$

The constant R is obtained by solving the following ix[sextic equation (for R or P)]sextic :

$$64R^6 - 80g_2R^4 - 320g_3R^3 - 20g_2^2R^2 - 16g_2g_3R + \left(g_2^3 - 32g_3^2\right) = 0.$$

The discriminant of this ix[Sextic equation (for R or P)]sextic equals

$$\delta = 288230376151711744\left(27g_3^2 - g_2^3\right)^5.$$

4.2.4 Fourth approach

We assume the solution to the i.v.p.

$$\ddot{u} + a_0 + a_1u + a_2u^2 + a_3u^3 = 0, u(0) = u_0 \text{ and } u'(0) = \dot{u}_0,$$

in the ansatz form

$$u(t) = P + \frac{Q}{1 + x(t)},$$

whre the function $x = x(t)$ is the solution to some Duffing equation

$$\ddot{x} + \alpha x + \beta x^3 = 0,\ x(0) = x_0 \text{ and } x'(0) = \dot{x}_0,$$

being

$$x_0 = \frac{u_0 - P - Q}{P - u_0},\ \dot{x}_0 = -\frac{Q}{(P - u_0)^2}\dot{u}_0.$$

Observe that

$$\tfrac{a_3}{4}u^4 + \tfrac{a_2}{3}u^3 + \tfrac{a_1}{2}u^2 + a_0 u + \tfrac{\dot{u}^2}{2} = C.$$

$$\tfrac{1}{2}\dot{x}^2 + \tfrac{1}{2}\alpha x^2 + \tfrac{1}{4}\beta x^4 = D.$$

$$C = \tfrac{1}{4}a_3 u_0^4 + \tfrac{1}{3}a_2 u_0^3 + \tfrac{1}{2}a_1 u_0^2 + a_0 u_0 + \tfrac{\dot{u_0}^2}{2},\ D = \tfrac{\alpha x_0^2}{2} + \tfrac{\beta x_0^4}{4} + \tfrac{\dot{x_0}^2}{2}.$$

Using the relation
$\dot{x}^2 = 2D - \frac{\beta}{2}x^4 - \alpha x^2$ we obtain
$\frac{a_3}{4}u^4 + \frac{a_2}{3}u^3 + \frac{a_1}{2}u^2 + a_0 u + \frac{1}{2}\dot{u}^2 - C = -\frac{1}{12(x+1)^4}($
$[3a_3P^4 + 12a_3P^3Q + 4a_2P^3 + 18a_3P^2Q^2 + 12a_2P^2Q + 6a_1P^2 + 12a_3PQ^3 + 12a_2PQ^2 +$
$12a_1PQ + 12a_0P + 3a_3Q^4 + 4a_2Q^3 + 6a_1Q^2 + 12a_0Q - 12C + 12DQ^2]x^0$
$-4[-3a_3P^4 - 9a_3P^3Q - 4a_2P^3 - 9a_3P^2Q^2 - 9a_2P^2Q - 6a_1P^2 - 3a_3PQ^3 - 6a_2PQ^2 - 9a_1PQ -$
$12a_0P - a_2Q^3 - 3a_1Q^2 - 9a_0Q + 12C]x^1 -$
$6[-3a_3P^4 - 6a_3P^3Q - 4a_2P^3 - 3a_3P^2Q^2 - 6a_2P^2Q - 6a_1P^2 - 2a_2PQ^2 - 6a_1PQ - 12a_0P - a_1Q^2 -$
$6a_0Q + 12C + \alpha Q^2]x^2 -$
$4[-3a_3P^4 - 3a_3P^3Q - 4a_2P^3 - 3a_2P^2Q - 6a_1P^2 - 3a_1PQ - 12a_0P - 3a_0Q + 12C]x^3 +$
$[3a_3P^4 + 4a_2P^3 + 6a_1P^2 + 12a_0P - 12C - 3\beta Q^2]x^4)$.

Equating to zero the coefficients of x^j in the above expression and solving the resulting algebraic system we obtain the following solutions :

$$\alpha = \frac{36a_3CP^2 + 24a_2CP + 12a_1C + a_2^2P^4 - 3a_1a_3P^4 + 2a_1a_2P^3 - 18a_0a_3P^3 + 3a_1^2P^2 - 6a_0a_2P^2 + 6a_0a_1P + 9a_0^2}{-3a_3P^4 - 4a_2P^3 - 6a_1P^2 - 12a_0P + 12C}.$$

$$\beta - \frac{3\left(a_3P^3 + a_2P^2 + a_1P + a_0\right)^2}{3a_3P^4 + 4a_2P^3 + 6a_1P^2 + 12a_0P - 12C}.$$

$$Q = -\frac{3a_3P^4 + 4a_2P^3 + 6a_1P^2 + 12a_0P - 12C}{3(a_3P^3 + a_2P^2 + a_1P + a_0)}.$$

The number P obeys the following ix[Sextic equation (for R or P)]sextic :

$$18\left(8a_2C^2+6a_1a_0C+3a_0^3\right)$$

$$+18\left(24a_3C^2+6a_1^2C-4a_0a_2C+3a_0^2a_1\right)P$$

$$+90\left(-6a_3a_0C+2a_1a_2C+a_2a_0^2\right)P^2$$

$$+30\left(4a_2^2C+9a_3a_0^2\right)P^3$$

$$-15\left(-12a_3a_2C+2a_0a_2^2-9a_0a_1a_3\right)P^4$$

$$+3\left(36a_3^2C+9a_3a_1^2-2a_2^2a_1-6a_0a_2a_3\right)P^5$$

$$+\left(-2a_2^3+9a_1a_3a_2-27a_0a_3^2\right)P^6=0.$$

4.2.5 Fifth approach

We assume the solution to the i.v.p.

$$\ddot{u}+a_0+a_1u+a_2u^2+a_3u^3=0, u(0)=u_0 \text{ and } u'(0)=\dot{u}_0,$$

in the ansatz form

$$u(t)=P+\frac{Q}{1+Rx^2(t)},\quad R=\frac{u_0-P-Q}{P-u_0},$$

where the function $x=x(t)$ is the solution to some Duffing equation

$$\ddot{x}+\alpha x+\beta x^3=0,\ x(0)=1 \text{ and } x'(0)=\dot{x}_0.$$

$$\dot{x}_0=\tfrac{Q\dot{u}_0}{2(P-u_0)(P+Q-u_0)}.$$

Using the relation $\dot{x}^2=2D-\frac{\beta}{2}x^4-\alpha x^2$ we obtain

$\frac{a_3}{4}u^4+\frac{a_2}{3}u^3+\frac{u_1}{2}u^2+a_0u+\frac{1}{2}u^2-C=-\frac{1}{12(Rx^2+1)^4}\times$
$[(3a_3P^4+12a_3P^3Q+4a_2P^3+18a_3P^2Q^2+12a_2P^2Q+6a_1P^2+12a_3PQ^3+12a_2PQ^2+12a_1PQ+$
$12a_0P+3a_3Q^4+4a_2Q^3+6a_1Q^2+12a_0Q-12C)x^0-$
$4R(-3a_3P^4-9a_3P^3Q-4a_2P^3-9a_3P^2Q^2-9a_2P^2Q-6a_1P^2-3a_3PQ^3-6a_2PQ^2-9a_1PQ-$
$12a_0P-$
$a_2Q^3-3a_1Q^2-9a_0Q+12C-12DQ^2R)x^2-$
$6R^2(-3a_3P^4-6a_3P^3Q-4a_2P^3-3a_3P^2Q^2-6a_2P^2Q-6a_1P^2-2a_2PQ^2-6a_1PQ-12a_0P-$
$a_1Q^2-6a_0Q+12C+4\alpha Q^2)x^4-$
$4R^2(-3a_3P^4R-3a_3P^3QR-4a_2P^3R-3a_2P^2QR-6a_1P^2R-3a_1PQR-12a_0PR$
$-3a_0QR+12CR+3\beta Q^2)x^6-$

$R^4\left(-3a_3P^4-4a_2P^3-6a_1P^2-12a_0P+12C\right)x^8]$.

Equating to zero the coefficients of x^j in the above expression and solving the resulting algebraic system we obtain the following solutions :

$$\alpha=\frac{6a_3P^3+3a_3P^2Q+6a_2P^2+2a_2PQ+6a_1P+a_1Q+6a_0}{4Q}.$$

$$\beta=\frac{a_3P^3R+a_2P^2R+a_1PR+a_0R}{Q}.$$

The numbers P and Q obey the following algebraic equations :

$12a_3P^3+12a_2P^2+12a_1P+12a_0+\left(18a_3P^2+12a_2P+6a_1\right)Q+\left(12a_3P+4a_2\right)Q^2+3a_3Q^3=0.$

$6a_3P^7+\left(9a_3Q-24a_3u_0+6a_2\right)P^5+\left(6a_3Q^2-36a_3Qu_0+6a_2Q+36a_3u_0^2-24a_2u_0+6a_1\right)P^6+$
$(3a_3Q^3-24a_3Q^2u_0+2a_2Q^2+54a_3Qu_0^2-24a_2Qu_0+3a_1Q-24a_3u_0^3+36a_2u_0^2-24a_1u_0+6a_0)P^4+$
$(-6a_3Q^3u_0+2a_2Q^3+36a_3Q^2u_0^2-8a_2Q^2u_0-36a_3Qu_0^3+36a_2Qu_0^2-12a_1Qu_0+6a_3u_0^4-24a_2u_0^3$
$+36a_1u_0^2-24a_0u_0)P^3+$
$(9a_3Q^3u_0^2+3a_1Q^3-24a_3Q^2u_0^3+12a_2Q^2u_0^2+9a_3Qu_0^4-24a_2Qu_0^3+18a_1Qu_0^2+6a_2u_0^4-24a_1u_0^3+$
$36a_0u_0^2)P^2+$
$\left(-6a_3Q^3u_0^3+6a_0Q^3+6a_3Q^2u_0^4-8a_2Q^2u_0^3+6a_2Qu_0^4-12a_1Qu_0^3+6a_1u_0^4-24a_0u_0^3\right)P+$
$\left(-2a_2Q^3u_0^3-3a_1Q^3u_0^2-6a_0Q^3u_0+2a_2Q^2u_0^4+3a_1Qu_0^4+6a_0u_0^4-3Q^3\dot{u}_0^2\right)=0$.

$12a_3P^3+Q\left(18a_3P^2+12a_2P+6a_1\right)+12a_2P^2+Q^2\left(12a_3P+4a_2\right)+12a_1P+3a_3Q^3+12a_0=0.$

$6a_3P^7+\left(9a_3Q-24a_3u_0+6a_2\right)P^6+\left(6a_3Q^2-36a_3Qu_0+6a_2Q+36a_3u_0^2-24a_2u_0+6a_1\right)P^5+$
$(3a_3Q^3-24a_3Q^2u_0+2a_2Q^2+54a_3Qu_0^2-24a_2Qu_0+3a_1Q-24a_3u_0^3+36a_2u_0^2-24a_1u_0+6a_0)P^4$
$(-6a_3Q^3u_0+2a_2Q^3+36a_3Q^2u_0^2-8a_2Q^2u_0-36a_3Qu_0^3+36a_2Qu_0^2-12a_1Qu_0+6a_3u_0^4$
$-24a_2u_0^3+36a_1u_0^2-24a_0u_0)P^3+$
$(9a_3Q^3u_0^2+3a_1Q^3-24a_3Q^2u_0^3+12a_2Q^2u_0^2+9a_3Qu_0^4-24a_2Qu_0^3+18a_1Qu_0^2+6a_2u_0^4-24a_1u_0^3+$
$36a_0u_0^2)P^2+$
$\left(-6a_3Q^3u_0^3+6a_0Q^3+6a_3Q^2u_0^4-8a_2Q^2u_0^3+6a_2Qu_0^4-12a_1Qu_0^3+6a_1u_0^4-24a_0u_0^3\right)P+$
$\left(-2a_2Q^3u_0^3-3a_1Q^3u_0^2-6a_0Q^3u_0+2a_2Q^2u_0^4+3a_1Qu_0^4+6a_0u_0^4-3Q^3\dot{u}_0^2\right)=0.$

4.2.6 Why Duffing–type oscillators are universal models for conservative dynamics

Consider a general one–dimensional conservative oscillator

$$\ddot{u}+F(u)=0, \tag{4.23}$$

where $F\in C^1(\mathbb{R})$ is an arbitrary restoring force. Equation (4.23) admits the energy first integral

$$E=\frac{1}{2}\dot{u}^2+V(u),\qquad V'(u)=F(u),$$

which shows that the motion takes place on level sets of the potential function $V(u)$.

General solvability and its limitations. In the generic case, the solution of (4.23) is given implicitly by

$$t - t_0 = \int_{u_0}^{u} \frac{ds}{\sqrt{2(E - V(s))}},$$

which leads to hyperelliptic integrals whenever $V(u)$ is a polynomial of degree greater than four. Therefore, exact closed–form solutions are exceptional rather than generic.

Key observation (bounded motion). If the motion is bounded, then there exist constants $A < B$ such that

$$A \leq u(t) \leq B \quad \text{for all } t.$$

In this case, the oscillator only "samples" the restoring force $F(u)$ on the compact interval $[A, B]$. Consequently, only the restriction $F|_{[A,B]}$ is dynamically relevant.

Polynomial approximation and canonical reduction. Since F is continuous on $[A, B]$, it admits a uniform approximation by Chebyshev polynomials. Truncating this approximation at the lowest order capable of reproducing asymmetry, multiple equilibria, and separatrix structures yields

$$F(u) \approx a_0 + a_1 u + a_2 u^2 + a_3 u^3,$$

which leads to the generalized Duffing–Helmholtz oscillator

$$\ddot{u} + a_0 + a_1 u + a_2 u^2 + a_3 u^3 = 0. \tag{4.24}$$

This equation constitutes the *lowest–degree universal surrogate* that preserves the essential phase–space geometry of bounded conservative motion: equilibria, potential wells, separatrices, and escape thresholds.

Why odd Duffing equations emerge as normal forms. As proved earlier in this chapter, solutions of the conservative Duffing–Helmholtz oscillator can be reduced, via quadratic transformations, to solutions of odd Duffing–type equations (cubic or cubic–quintic). This explains why odd–parity Duffing models play a privileged role: they act as normal forms for a broad class of conservative polynomial oscillators.

Connection with algebraic equations. A remarkable consequence of this viewpoint is that certain quintic algebraic equations, which are not solvable by radicals in general, can be solved *indirectly* by means of second–order Duffing–type equations. Indeed, equilibria of a conservative oscillator satisfy

$$F(u) = 0,$$

which, for polynomial F, yields an algebraic equation. When the dynamics is reduced to a cubic or cubic–quintic Duffing equation, the corresponding equilibria and turning points can be expressed explicitly in terms of elliptic functions, thereby providing a constructive representation of selected real roots of higher–degree algebraic equations.

Conceptual importance. The role of Duffing and Duffing–Helmholtz oscillators is therefore not merely that of special solvable examples. They constitute *canonical analytical models* that:

- approximate arbitrary conservative oscillators on invariant sets,
- preserve the qualitative phase–space structure,
- admit explicit elliptic solutions,
- provide a natural framework for perturbation, averaging, and chaos analysis via Melnikov theory.

This perspective justifies the central position of Duffing–type models in the analytical study of nonlinear conservative dynamics.

4.3 Chaos in the forced–damped Duffing–Helmholtz oscillator: homoclinic orbits and Melnikov analysis

In this section we explain, in a fully constructive way, how chaos arises in the Duffing–Helmholtz oscillator once damping and external forcing are introduced. The analysis generalizes the classical chaotic mechanism of the forced Duffing equation to the mixed–parity Duffing equation (4.1), by showing that the decisive ingredient is the existence of a homoclinic orbit (a separatrix loop) attached to a saddle equilibrium. Under weak time–periodic forcing and weak damping, this separatrix typically splits transversely, creating a Smale horseshoe and producing chaos in the Poincaré map.

4.3.1 From the conservative oscillator to the forced–damped model

Start from the conservative mixed–parity oscillator (4.1),

$$\ddot{u} + a_0 + a_1 u + a_2 u^2 + a_3 u^3 = 0,$$

and add linear damping and a time–periodic forcing:

$$\ddot{u} + 2\epsilon\dot{u} + a_0 + a_1 u + a_2 u^2 + a_3 u^3 = \gamma\cos(\omega t), \qquad 0 < \epsilon \ll 1, \quad 0 < \gamma \ll 1. \tag{4.25}$$

Equation (4.25) is a non-autonomous nonlinear oscillator. When $\epsilon = \gamma = 0$ it reduces to a Hamiltonian system with conserved energy.

Why this generalizes the classical Duffing chaos. The classical forced Duffing equation is recovered by taking $a_0 = a_2 = 0$ and $(a_1, a_3) = (\alpha, \beta)$:

$$\ddot{u} + 2\epsilon\dot{u} + \alpha u + \beta u^3 = \gamma\cos(\omega t).$$

The generalized model (4.25) adds the constant and quadratic terms (a_0 and a_2), which break the odd symmetry $u \mapsto -u$ and produce asymmetric potentials. The chaotic mechanism, however, remains the same: *damping plus time–periodic forcing splits a conservative separatrix.*

4.3.2 Phase–plane picture: equilibria, saddles, and separatrices

Write (4.25) as the planar system

$$\dot{u} = v, \qquad \dot{v} = -2\epsilon v - (a_0 + a_1 u + a_2 u^2 + a_3 u^3) + \gamma \cos(\omega t). \tag{4.26}$$

For $\epsilon = \gamma = 0$, equilibria are $(u_*, 0)$ where u_* solves

$$a_0 + a_1 u_* + a_2 u_*^2 + a_3 u_*^3 = 0.$$

An equilibrium is a saddle if

$$a_1 + 2a_2 u_* + 3a_3 u_*^2 < 0,$$

and a center if the same quantity is > 0. A separatrix loop (homoclinic orbit) can occur only if at least one saddle exists.

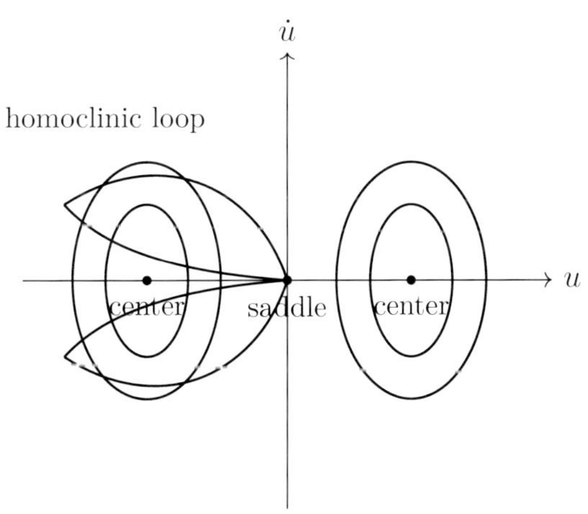

Figure 4.10: Schematic phase portrait of the conservative system ($\epsilon = \gamma = 0$) showing a homoclinic orbit attached to a saddle.

4.3.3 What is a homoclinic orbit and why it matters for chaos?

A homoclinic orbit is a nontrivial solution $(u_h(t), v_h(t))$ of the *conservative* system ($\epsilon = \gamma = 0$) such that

$$(u_h(t), v_h(t)) \to (u_s, 0) \quad \text{as } t \to \pm\infty,$$

where $(u_s, 0)$ is a saddle equilibrium. In the phase plane this means the trajectory leaves the saddle along its unstable manifold and returns to the same saddle along its stable manifold, forming a separatrix loop.

For dummies (the punchline). In the conservative system, the stable and unstable manifolds coincide exactly (they are the same curve). When damping and forcing are added, these manifolds usually *no longer coincide*. If they intersect *transversely*, the dynamics contains a Smale horseshoe and the system exhibits chaos (infinitely many periodic orbits, sensitive dependence, etc.). The Melnikov method gives an explicit first–order condition for this transverse intersection.

4.3.4 Melnikov method: general formula for the Duffing–Helmholtz oscillator

Set $\epsilon = \mu\varepsilon$ and $\gamma = \mu\Gamma$ with $0 < \mu \ll 1$ to emphasize small perturbations:

$$\ddot{u} + a_0 + a_1 u + a_2 u^2 + a_3 u^3 = \mu\Big(\Gamma\cos(\omega t) - 2\varepsilon\dot{u}\Big).$$

Write the planar system as

$$\dot{\mathbf{z}} = \mathbf{f}(\mathbf{z}) + \mu\,\mathbf{g}(\mathbf{z}, t), \qquad \mathbf{z} = (u, v),$$

with

$$\mathbf{f}(u, v) = (v,\ -(a_0 + a_1 u + a_2 u^2 + a_3 u^3)), \qquad \mathbf{g}(u, v, t) = (0,\ \Gamma\cos(\omega t) - 2\varepsilon v).$$

Let $\mathbf{z}_h(t) = (u_h(t), v_h(t))$ be a homoclinic orbit of the unperturbed system. Then the Melnikov function is

$$M(t_0) = \int_{-\infty}^{\infty} \mathbf{f}(\mathbf{z}_h(t)) \wedge \mathbf{g}(\mathbf{z}_h(t), t + t_0)\,dt = \int_{-\infty}^{\infty} v_h(t)\Big[\Gamma\cos(\omega(t + t_0)) - 2\varepsilon v_h(t)\Big]\,dt. \tag{4.27}$$

A simple zero of $M(t_0)$ (i.e. $M(t_0) = 0$ and $M'(t_0) \neq 0$) implies that the perturbed stable and unstable manifolds of the saddle intersect transversely, which yields chaos in the Poincaré map.

Immediate simplification. Since $v_h(t)$ is odd and $\cos(\omega t)$ is even, one has

$$\int_{-\infty}^{\infty} v_h(t)\cos(\omega t)\,dt = 0,$$

so the forcing contribution becomes purely sinusoidal in t_0:

$$\int_{-\infty}^{\infty} v_h(t)\cos(\omega(t + t_0))\,dt = -\sin(\omega t_0)\int_{-\infty}^{\infty} v_h(t)\sin(\omega t)\,dt.$$

Therefore

$$M(t_0) = -2\varepsilon\int_{-\infty}^{\infty} v_h^2(t)\,dt - \Gamma\,\sin(\omega t_0)\int_{-\infty}^{\infty} v_h(t)\sin(\omega t)\,dt. \tag{4.28}$$

Hence $M(t_0)$ has simple zeros iff the forcing amplitude is large enough to overcome the damping term.

4.3.5 Classical Duffing case: exact homoclinic orbit and exact Melnikov integrals

To make everything explicit (and to show the mechanism with *exact integrals*), consider the standard double–well Duffing potential case:

$$\ddot{x} - ax + bx^3 = 0, \qquad a > 0, \quad b > 0, \tag{4.29}$$

whose potential is $V(x) = -\frac{1}{2}ax^2 + \frac{1}{4}bx^4$ and whose origin is a saddle. Equation (4.29) has the explicit homoclinic orbit

$$x_h(t) = \pm\sqrt{\frac{2a}{b}}\ \mathrm{sech}(\sqrt{a}\,t), \qquad \dot{x}_h(t) = \mp\sqrt{\frac{2a}{b}}\ \sqrt{a}\ \mathrm{sech}(\sqrt{a}\,t)\tanh(\sqrt{a}\,t). \tag{4.30}$$

Now add weak damping and forcing:

$$\ddot{x} - ax + bx^3 = \mu\Big(\Gamma\cos(\omega t) - 2\varepsilon\dot{x}\Big). \tag{4.31}$$

The Melnikov function is

$$M(t_0) = \int_{-\infty}^{\infty} \dot{x}_h(t)\Big[\Gamma\cos(\omega(t+t_0)) - 2\varepsilon\dot{x}_h(t)\Big]\,dt.$$

Exact damping integral

Using (4.30) one finds the exact identity

$$\int_{-\infty}^{\infty} \dot{x}_h^2(t)\,dt = \frac{4}{3}\frac{a^{3/2}}{b}. \tag{4.32}$$

Exact forcing integral

By integration by parts,

$$\int_{-\infty}^{\infty} \dot{x}_h(t)\sin(\omega t)\,dt = -\omega\int_{-\infty}^{\infty} x_h(t)\cos(\omega t)\,dt,$$

and the Fourier transform identity

$$\int_{-\infty}^{\infty} \mathrm{sech}(\lambda t)\cos(\omega t)\,dt = \frac{\pi}{\lambda}\ \mathrm{sech}\Big(\frac{\pi\omega}{2\lambda}\Big)$$

yields

$$\int_{-\infty}^{\infty} \dot{x}_h(t)\sin(\omega t)\,dt = -\pi\omega\sqrt{\frac{2}{b}}\ \mathrm{sech}\Big(\frac{\pi\omega}{2\sqrt{a}}\Big). \tag{4.33}$$

Closed form Melnikov function and chaos threshold

Substituting (4.32) and (4.33) into (4.28) gives

$$M(t_0) = -\frac{8}{3}\,\varepsilon\,\frac{a^{3/2}}{b} + \Gamma\,\pi\omega\sqrt{\frac{2}{b}}\ \mathrm{sech}\Big(\frac{\pi\omega}{2\sqrt{a}}\Big)\ \sin(\omega t_0). \tag{4.34}$$

Therefore $M(t_0)$ has simple zeros (hence transverse homoclinic intersection and chaos) provided

$$|\Gamma| > \Gamma_{\mathrm{crit}} := \frac{4\sqrt{2}}{3\pi}\,\frac{\varepsilon\,a^{3/2}}{\omega\sqrt{b}}\ \cosh\Big(\frac{\pi\omega}{2\sqrt{a}}\Big). \tag{4.35}$$

This is an explicit chaos criterion in terms of parameters: damping competes against forcing, and the frequency dependence enters through the cosh term.

Interpretation (again, for dummies). If forcing is too weak, damping destroys the separatrix splitting and no transverse intersections occur. If forcing is strong enough, the stable and unstable manifolds intersect in a complicated tangle, producing chaos.

4.3.6 Why the same mechanism applies to the Duffing–Helmholtz equation

For the mixed–parity oscillator, the unperturbed dynamics is still conservative with potential

$$V(u) = a_0 u + \frac{a_1}{2}u^2 + \frac{a_2}{3}u^3 + \frac{a_3}{4}u^4.$$

If the parameters are such that $V(u)$ has a saddle u_s and a separatrix (homoclinic loop) exists at energy $E_{sep} = V(u_s)$, then the forced–damped system (4.25) falls exactly into the Melnikov setting. The Melnikov function is still given by (4.27)–(4.28), but now the orbit $u_h(t)$ is typically not a simple sech function.

Local normal form near the saddle (why Duffing appears). Let $u = u_s + x$, where u_s solves $a_0 + a_1 u_s + a_2 u_s^2 + a_3 u_s^3 = 0$. Then expanding the restoring force gives

$$\ddot{x} + \kappa x + \eta x^2 + \zeta x^3 = 0,$$

with

$$\kappa = a_1 + 2a_2 u_s + 3a_3 u_s^2, \qquad \eta = a_2 + 3a_3 u_s, \qquad \zeta = a_3.$$

A saddle means $\kappa < 0$, so at leading order this is exactly a Duffing–type instability ($\ddot{x} - |\kappa| x + \cdots = 0$). Thus the separatrix splitting mechanism that produces chaos in the classical Duffing oscillator also produces chaos here; the only difference is that the separatrix and Melnikov integrals depend on (a_0, a_1, a_2, a_3) through the shifted coefficients above.

4.3.7 Geometry of separatrix splitting (pictures you can actually understand)

(A) Conservative case: stable and unstable manifolds coincide

(B) Forced–damped case: manifolds split and intersect transversely

4.3.8 Mathematica computation of the Melnikov integrals (exact, when possible)

For the classical Duffing homoclinic orbit, Mathematica confirms the exact formulas (4.32) and (4.33). The following code computes them symbolically.

```
(* Mathematica: exact Melnikov integrals for the homoclinic orbit *)
ClearAll[a,b,,t,xh,dxh,I1,I2];


(* parameters: a>0, b>0 *)
```

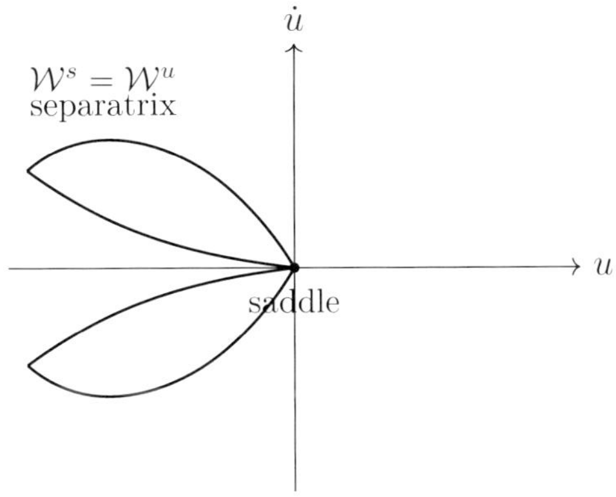

Figure 4.11: Conservative case: the stable manifold and unstable manifold coincide along the separatrix.

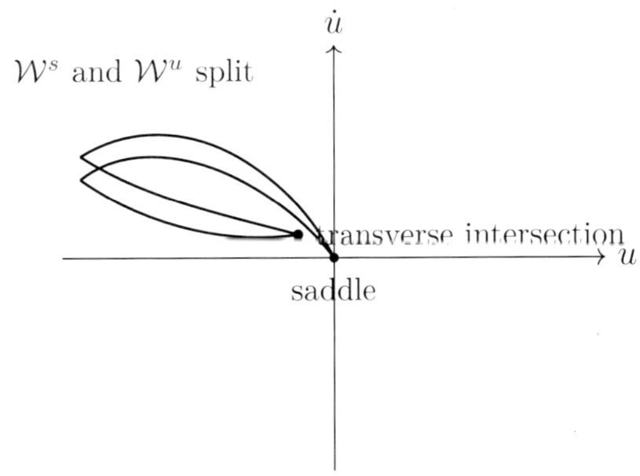

Figure 4.12: Perturbed case: separatrix splitting and transverse intersection imply chaos.

```
xh[t_]  := Sqrt[2 a/b] Sech[Sqrt[a] t];
dxh[t_] := D[xh[t], t];

(* damping integral *)
I1 = FullSimplify[ Integrate[ dxh[t]^2, {t, -Infinity, Infinity},
      Assumptions -> a > 0 && b > 0 ] ]

(* forcing integral: int dxh(t) Sin( t) dt *)
I2 = FullSimplify[ Integrate[ dxh[t] Sin[ t], {t, -Infinity, Infinity},
      Assumptions -> a > 0 && b > 0 &&  > 0 ] ]

(* Melnikov function: M(t0) = -2  I1 -  Sin( t0) I2 *)
```

For the mixed–parity Duffing–Helmholtz oscillator, the homoclinic orbit is generally not a simple sech function, so one proceeds in two standard ways:

1. **Exact implicit form (quadrature).** For the conservative system, energy conservation gives
$$\frac{1}{2}\dot{u}^2 + V(u) = E_{\text{sep}} = V(u_s), \quad \Rightarrow \quad \dot{u} = \pm\sqrt{2(E_{\text{sep}} - V(u))}.$$

This gives t as an elliptic integral of u, and the Melnikov integrals can be transformed into integrals over u (often reducible to elliptic integrals) and attacked symbolically in Mathematica.

2. **Direct numerical Melnikov integral.** Compute a numerical approximation of the homoclinic orbit $(u_h(t), v_h(t))$ for $\epsilon = \gamma = 0$, then evaluate (4.27) numerically (high precision) to find simple zeros.

4.3.9 How to locate the saddle and the separatrix in terms of (a_0, a_1, a_2, a_3)

The existence of saddle equilibria and separatrices is the *key prerequisite* for the onset of chaos in the forced–damped Duffing–Helmholtz oscillator. In this subsection we give a *practical, step–by–step algorithm* to locate saddles and separatrices directly in terms of the physical parameters (a_0, a_1, a_2, a_3).

Everything is done at the level of the *conservative system*

$$\ddot{u} + a_0 + a_1 u + a_2 u^2 + a_3 u^3 = 0, \tag{4.36}$$

since separatrices are objects of the conservative dynamics.

Step 1: Write the system in phase–plane form

Introduce $v = \dot{u}$. Equation (4.36) becomes the planar system

$$\dot{u} = v, \qquad \dot{v} = -(a_0 + a_1 u + a_2 u^2 + a_3 u^3). \tag{4.37}$$

Equilibria are points $(u_*, 0)$ where the restoring force vanishes.

Step 2: Locate equilibrium points

Equilibria satisfy the cubic equation

$$a_3 u^3 + a_2 u^2 + a_1 u + a_0 = 0. \tag{4.38}$$

Interpretation.

- One real root $\Rightarrow$ one equilibrium.
- Three real roots $\Rightarrow$ three equilibria.

The number of real roots is determined by the discriminant of the cubic (4.38). Only the *three–root case* can generate separatrices.

Step 3: Classify equilibria (center vs saddle)

Linearize (4.37) at $(u_*, 0)$. The Jacobian matrix is

$$J(u_*) = \begin{pmatrix} 0 & 1 \\ -(a_1 + 2a_2u_* + 3a_3u_*^2) & 0 \end{pmatrix}.$$

The eigenvalues satisfy

$$\lambda^2 = -(a_1 + 2a_2u_* + 3a_3u_*^2).$$

Hence:

- If
$$a_1 + 2a_2u_* + 3a_3u_*^2 > 0,$$
then $\lambda = \pm i\omega$ and $(u_*, 0)$ is a center.
- If
$$a_1 + 2a_2u_* + 3a_3u_*^2 < 0,$$
then $\lambda = \pm\sigma$ and $(u_*, 0)$ is a saddle.

Key rule (memorize this).

Separatrices can exist only around saddle equilibria.

Step 4: Construct the potential and identify separatrix energy

Equation (4.36) is Hamiltonian with potential

$$V(u) = a_0u + \frac{a_1}{2}u^2 + \frac{a_2}{3}u^3 + \frac{a_3}{4}u^4. \tag{4.39}$$

The energy first integral is

$$E = \frac{1}{2}v^2 + V(u). \tag{4.40}$$

Let u_s be a saddle. Then the separatrix corresponds to the energy level

$$E_{\text{sep}} = V(u_s). \tag{4.41}$$

All trajectories with $E = E_{\text{sep}}$ form the separatrix loop(s).

Step 5: Equation of the separatrix (quadrature)

On the separatrix,

$$\frac{1}{2}\dot{u}^2 + V(u) = V(u_s),$$

so

$$\dot{u} = \pm\sqrt{2(V(u_s) - V(u))}. \tag{4.42}$$

This gives the separatrix implicitly as

$$t - t_0 = \pm \int_{u_s}^{u} \frac{ds}{\sqrt{2(V(u_s) - V(s))}}.$$

This integral is elementary only in special cases (classical Duffing); in general it is an elliptic integral.

Geometric regimes (with pictures)

(A) One equilibrium: no separatrix. If (4.38) has one real root and it is a center, then all trajectories are closed and *no chaos is possible.*

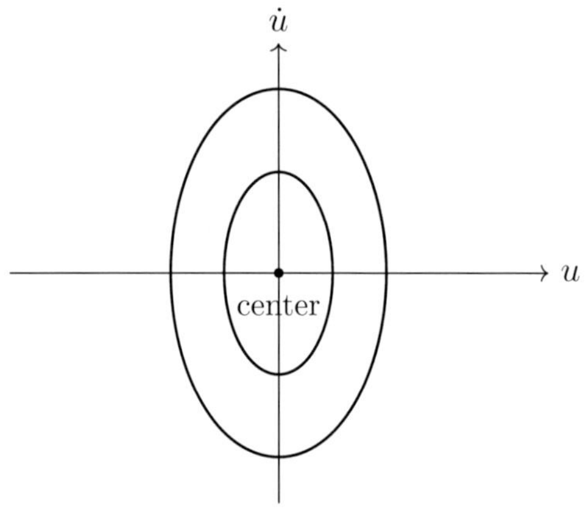

Figure 4.13: Single–well potential: only a center, no separatrix, no chaos.

(B) Three equilibria: one saddle between two centers. This is the *generic chaotic configuration.*

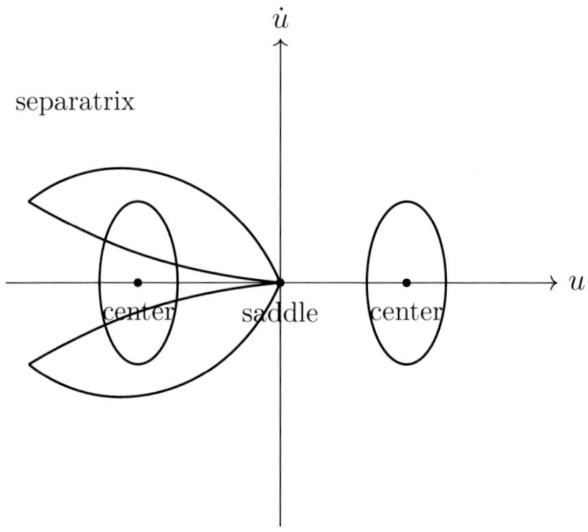

Figure 4.14: Double–well potential: one saddle and two centers. This configuration supports homoclinic orbits and chaos under forcing.

(C) Asymmetric wells (mixed parity case). If $a_0 \neq 0$ or $a_2 \neq 0$, the wells are asymmetric.

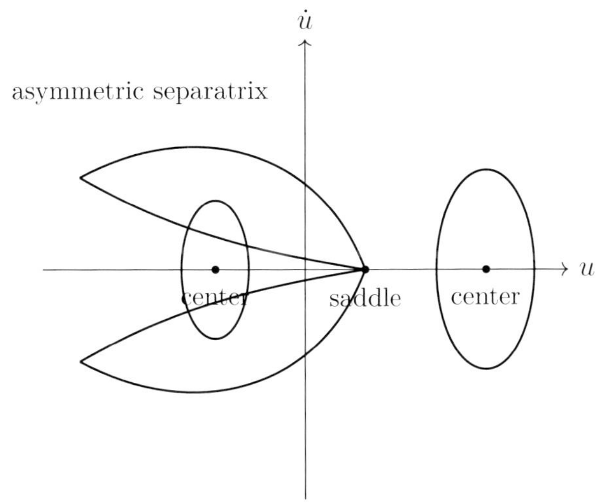

Figure 4.15: Asymmetric Duffing–Helmholtz oscillator: broken symmetry but the same separatrix–chaos mechanism applies.

Algorithm summary (practical checklist)

1. Solve the cubic (4.38).
2. Compute $a_1 + 2a_2u_* + 3a_3u_*^2$ for each root.
3. Identify at least one saddle.
4. Compute $V(u)$ from (4.39).
5. Set $E_{\text{sep}} = V(u_s)$.
6. Use (4.42) to characterize the separatrix.
7. Add damping and forcing $\Rightarrow$ apply Melnikov analysis.

Bottom line.

No saddle $\Rightarrow$ no separatrix $\Rightarrow$ no chaos. One saddle $\Rightarrow$ separatrix $\Rightarrow$ chaos is possible under forcing.

4.3.10 Parameter maps, explicit examples, and closure with the Melnikov chaos criterion

In this subsection we complete the analysis of the Duffing–Helmholtz oscillator by (i) constructing explicit parameter maps showing where saddles exist in terms of (a_0, a_1, a_2, a_3), (ii) presenting worked numerical examples with concrete values and phase portraits, and (iii) closing the analysis by connecting these results directly to the Melnikov chaos criterion derived earlier.

The goal is that *nothing remains abstract*: every condition is translated into inequalities on parameters, and every picture corresponds to an explicit choice of numbers.

Parameter maps: where do saddles exist?

Recall that equilibria are roots of the cubic

$$a_3u^3 + a_2u^2 + a_1u + a_0 = 0. \tag{4.43}$$

Step 1: number of equilibria. The cubic (4.43) has three real roots iff its discriminant $\mathcal{D}$ satisfies

$$\mathcal{D} > 0.$$

This inequality defines a three–dimensional region in parameter space (a_0, a_1, a_2, a_3). Only inside this region can separatrices and chaos occur.

Step 2: existence of a saddle. Let u_* be any real root. It is a saddle if

$$a_1 + 2a_2u_* + 3a_3u_*^2 < 0. \tag{4.44}$$

Thus, a saddle equilibrium exists if and only if

the cubic has three real roots and at least one of them satisfies (4.44).

Interpretation (for dummies).

- One real root ⇒ single well ⇒ no chaos.
- Three real roots ⇒ at least one hilltop.
- Hilltop ⇒ saddle ⇒ separatrix.

Canonical parameter slice: fixing $a_3 > 0$

To visualize the parameter map, fix $a_3 > 0$ (softening at large amplitudes) and consider the reduced parameter space (a_0, a_1, a_2).

Rule of thumb. Large $|a_0|$ destroys the double–well structure. Moderate a_0 combined with $a_1 < 0$ typically produces a saddle.

Worked Example 1: symmetric double–well

Let

$$a_0 = 0, \quad a_1 = -1, \quad a_2 = 0, \quad a_3 = 1.$$

Then

$$\ddot{u} - u + u^3 = 0,$$

the classical symmetric Duffing oscillator.

Equilibria.

$$u(u^2 - 1) = 0 \Rightarrow u_* = 0, \pm 1.$$

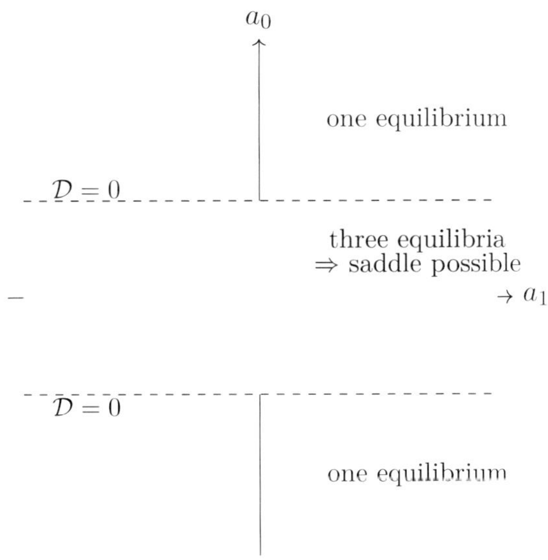

Figure 4.16: Typical parameter map in the (a_0, a_1)–plane (for fixed a_2 and $a_3 > 0$). The shaded region corresponds to three equilibria and the possible existence of a saddle.

Classification.

$$a_1 + 3a_3u^2 = -1 + 3u^2.$$

Hence:

$$u = 0 \Rightarrow -1 < 0 \Rightarrow \textit{saddle}, \qquad u = \pm 1 \Rightarrow 2 > 0 \Rightarrow \textit{centers}.$$

Phase portrait.

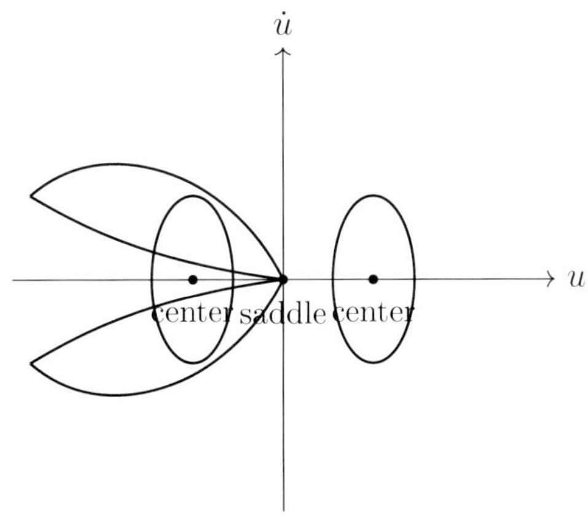

Figure 4.17: Symmetric double–well Duffing oscillator with a central saddle and a homoclinic separatrix.

This configuration supports homoclinic orbits and therefore chaos under forcing.

Worked Example 2: asymmetric Duffing–Helmholtz oscillator

Let

$$a_0 = 0.2, \quad a_1 = -1, \quad a_2 = 0.6, \quad a_3 = 1.$$

Equilibria. The cubic

$$u^3 + 0.6u^2 - u + 0.2 = 0$$

has three real roots (numerically):

$$u_* \approx -1.28, \quad 0.22, \quad 0.86.$$

Classification. Evaluate $a_1 + 2a_2u + 3a_3u^2$:

$$u \approx -1.28 \Rightarrow > 0 \Rightarrow \text{center},$$
$$u \approx 0.22 \Rightarrow < 0 \Rightarrow \text{saddle},$$
$$u \approx 0.86 \Rightarrow > 0 \Rightarrow \text{center}.$$

Phase portrait.

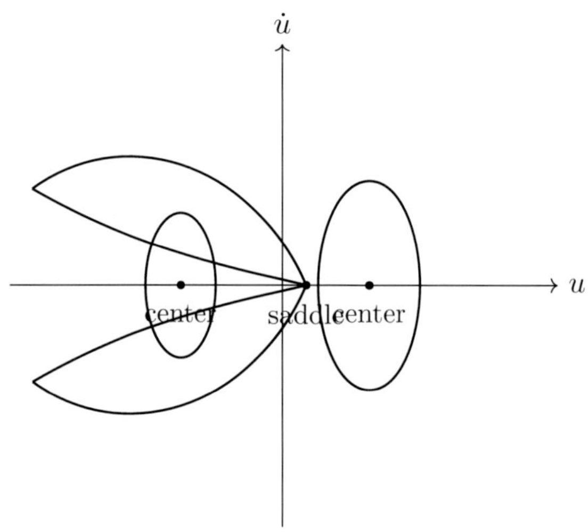

Figure 4.18: Asymmetric Duffing–Helmholtz oscillator. Despite broken symmetry, a saddle and separatrix still exist.

Closure with the Melnikov chaos threshold

Once a separatrix exists, the forced–damped system

$$\ddot{u} + 2\epsilon\dot{u} + a_0 + a_1u + a_2u^2 + a_3u^3 = \gamma\cos(\omega t)$$

falls under the Melnikov framework.

The Melnikov function

$$M(t_0) = \int_{-\infty}^{\infty} \dot{u}_h(t)\big[\gamma\cos(\omega(t+t_0)) - 2\epsilon\dot{u}_h(t)\big]dt$$

has simple zeros iff

$$\gamma > \gamma_{\text{crit}} = \frac{2\epsilon \displaystyle\int_{-\infty}^{\infty} \dot{u}_h^2(t)\, dt}{\left| \displaystyle\int_{-\infty}^{\infty} \dot{u}_h(t) \sin(\omega t)\, dt \right|}. \tag{4.45}$$

Final chain of logic (close the book).

Saddle $\Rightarrow$ separatrix $\Rightarrow$ homoclinic orbit $\Rightarrow$ Melnikov splitting $\Rightarrow$ chaos.

Everything is now explicit:

- saddle location $\Rightarrow$ algebra in (a_0, a_1, a_2, a_3),
- separatrix $\Rightarrow$ quadrature,
- chaos $\Rightarrow$ inequality (4.45).

4.4 Global signatures of chaos: bifurcations, Poincaré sections, Lyapunov exponents and a sufficient chaos theorem

We now complete the chaotic analysis of the generalized Duffing oscillator by presenting the principal global indicators of chaos. The goal of this section is twofold:

- to show how the local separatrix splitting detected by the Melnikov method manifests itself in observable global structures, and
- to provide practical diagnostic tools that allow chaos to be identified numerically and experimentally

Throughout this section we consider the forced and damped equation

$$\ddot{u} + 2\varepsilon\dot{u} + a_0 + a_1 u + a_2 u^2 + a_3 u^3 = \gamma\cos(\omega t), \tag{4.46}$$

where $\varepsilon > 0$ and $\gamma > 0$ are assumed small.

4.4.1 Bifurcation diagrams in the (γ, ω)–plane

A bifurcation diagram represents the qualitative changes in the long–time dynamics as a control parameter is varied.

Choice of parameters. In Duffing–type systems the natural control parameters are:

- the forcing amplitude γ,
- the forcing frequency ω.

For fixed $(a_0, a_1, a_2, a_3, \varepsilon)$, one records the values of $u(t)$ at times $t_n = 2\pi n/\omega$ after transients have decayed.

Interpretation.

- One point per period $\Rightarrow$ periodic motion.
- Two or four points $\Rightarrow$ period–doubling.
- Infinitely many points $\Rightarrow$ chaotic attractor.

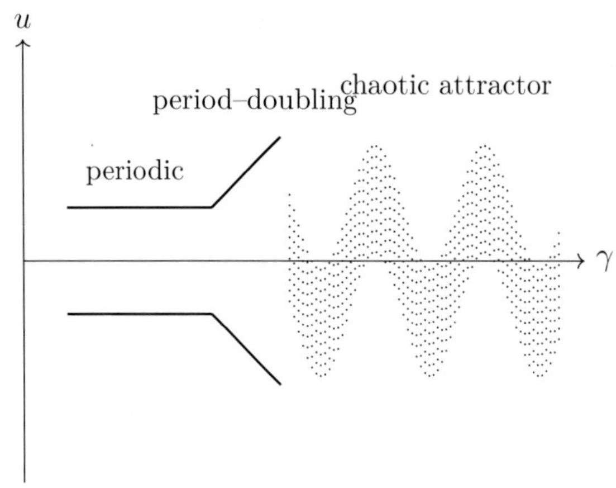

Figure 4.19: Schematic bifurcation diagram versus forcing amplitude γ for fixed ω.

Connection with Melnikov. The onset of the chaotic cloud occurs near the critical value γ_{crit} predicted by the Melnikov condition.

4.4.2 Poincaré sections: geometry of chaos

The most powerful geometric diagnostic of chaos is the Poincaré section.

Definition. Fix a stroboscopic time

$$t_n = \frac{2\pi n}{\omega}.$$

The Poincaré map sends

$$(u(t_n), \dot{u}(t_n)) \mapsto (u(t_{n+1}), \dot{u}(t_{n+1})).$$

Geometric meaning.

- Fixed point $\Rightarrow$ periodic orbit.
- Closed curve $\Rightarrow$ quasiperiodic motion.
- Fractal cloud $\Rightarrow$ chaos.

Why this proves chaos. A smooth invariant curve cannot fold infinitely many times. The observed fractal geometry implies sensitive dependence on initial conditions.

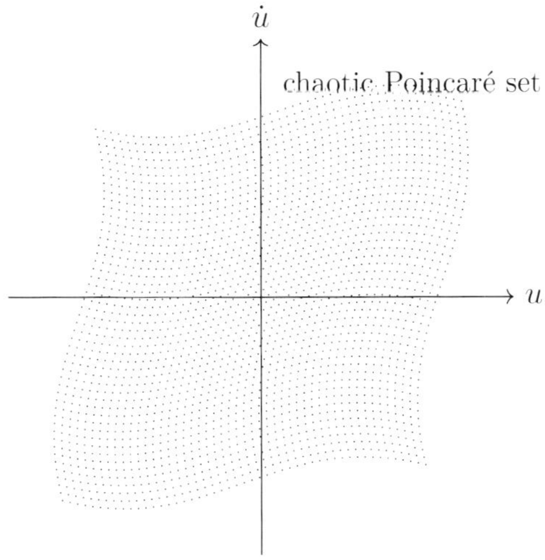

Figure 4.20: Typical Poincaré section of the forced generalized Duffing oscillator: a fractal cloud indicating chaos.

4.4.3 Lyapunov exponents and their link to Melnikov theory

The most quantitative chaos diagnostic is the Lyapunov exponent.

Definition. Let $\delta(t)$ be the distance between two initially nearby trajectories. The maximal Lyapunov exponent is

$$\lambda = \lim_{t\to\infty} \frac{1}{t} \ln \frac{\|\delta(t)\|}{\|\delta(0)\|}.$$

Interpretation.

- $\lambda < 0$: stable periodic motion.
- $\lambda = 0$: quasiperiodic motion.
- $\lambda > 0$: chaos.

Link with Melnikov. Melnikov theory predicts *transverse intersections* of stable and unstable manifolds. Such intersections imply a Smale horseshoe, which in turn guarantees:

$$\lambda > 0.$$

Thus, the Melnikov threshold

$$\gamma > \gamma_{\text{crit}}$$

is a *sufficient condition for a positive Lyapunov exponent.*

4.4.4 A sufficient condition for chaos

We now summarize the entire analysis in a single theorem.

Theorem 4.2 (Sufficient condition for chaos)**.** *Consider the forced generalized Duffing equation* (4.46)*. Assume that:*

1. *the unforced system admits a saddle with a homoclinic orbit,*
2. *the forcing and damping are weak ($0 < \varepsilon \ll 1$, $0 < \gamma \ll 1$),*
3. *the Melnikov function has a simple zero.*

Then:

- *the stable and unstable manifolds intersect transversely,*
- *the Poincaré map contains a Smale horseshoe,*
- *the system exhibits deterministic chaos,*
- *the maximal Lyapunov exponent is positive.*

Final message to the reader.

> Chaos in generalized Duffing oscillators is not mysterious. It is the inevitable consequence of:
>
> $$\text{saddle} \Rightarrow \text{separatrix} \Rightarrow \text{splitting} \Rightarrow \text{horseshoe.}$$

This completes the global chaos analysis of the generalized Duffing oscillator.

4.5 Summary

In this chapter we studied the generalized Duffing oscillator (or Duffing–Helmholtz oscillator, mixed–parity Duffing equation) governed by

$$\ddot{u} + a_0 + a_1 u + a_2 u^2 + a_3 u^3 = 0, \qquad u(0) = u_0,\ \dot{u}(0) = \dot{u}_0,$$

which provides a natural extension of the classical Duffing oscillator to systems with asymmetric restoring forces. The main conclusions of the chapter can be summarized as follows:

1. **Unified formulation and mechanical interpretation.** The generalized Duffing–Helmholtz equation was shown to arise from a polynomial restoring force containing constant, linear, quadratic, and cubic terms. This structure allows one to model a wide class of nonlinear oscillators with broken symmetry, shifted equilibria, and non-even potential functions, while preserving a clear mechanical interpretation in terms of stiffness coefficients and effective forces.

2. **Energy integral and potential-based organization of the dynamics.** In the conservative case, the equation admits a first integral of the form

$$\frac{1}{2}\dot{u}^2 + V(u) = E,$$

where the associated potential energy $V(u)$ is a quartic polynomial. This energy representation was identified as the central organizing principle for understanding phase–plane dynamics, allowing equilibria, turning points, bounded motions, and escape to be characterized geometrically.

3. **Equilibria and stability in terms of** (a_0, a_1, a_2, a_3). All equilibrium points were obtained explicitly as the real roots of a cubic polynomial. A complete linear stability analysis was carried out directly in terms of the parameters a_0, a_1, a_2, a_3, yielding explicit criteria for centers, saddles, and unstable equilibria. The role of the cubic discriminant in determining the number of equilibria and the qualitative structure of the phase portrait was made explicit.

4. **Global boundedness versus escape.** The influence of the leading coefficient a_3 on the global behavior of solutions was analyzed in detail. It was shown that the sign of a_3 determines whether the potential is confining or allows escape to infinity, thereby providing a simple and physically transparent criterion for global boundedness of the motion.

5. **Exact solutions via multiple constructive approaches.** Five complementary analytical approaches were developed to construct exact solutions of the generalized Duffing equation. These include reductions to cubic Duffing equations, cubic quintic Duffing equations, and representations in terms of Jacobi elliptic functions and Weierstrass elliptic functions. Explicit algebraic relations were derived linking the parameters of the generalized model to those of the reduced equations.

6. **Systematic reduction algorithms.** The chapter provided explicit reduction algorithms that map the generalized Duffing–Helmholtz oscillator onto simpler integrable models. These procedures allow one to pass constructively from the physical parameters and initial conditions to closed–form solutions and to reconstruct the original dynamics from the reduced system.

7. **Onset of chaos in the forced–damped case.** By introducing damping and periodic forcing, the generalized equation was shown to exhibit deterministic chaos. The geometric mechanism underlying chaos was identified as the transversal intersection of stable and unstable manifolds associated with homoclinic orbits emanating from saddle equilibria.

8. **Melnikov analysis and exact chaos thresholds.** The Melnikov method was applied to the forced–damped Duffing–Helmholtz oscillator, leading to explicit Melnikov functions and analytical chaos thresholds. Exact and semi–exact evaluation of the Melnikov integrals was carried out, with symbolic computations performed using Mathematica, thereby providing quantitative conditions for separatrix splitting.

9. **Parameter maps, phase portraits, and global diagnostics.** Detailed parameter maps in the space (a_0, a_1, a_2, a_3) were constructed, indicating the existence of saddles, separatrices, and chaotic regimes. These analytical predictions were complemented by

phase portraits, Poincaré sections, bifurcation diagrams, and Lyapunov exponents, yielding a coherent global picture of the dynamics.

10. **Foundational role within the book.** Overall, Chapter 4 establishes the generalized Duffing–Helmholtz oscillator as a unifying model bridging classical Duffing dynamics, higher–order nonlinearities, and forced chaotic motion. The analytical framework developed here provides the conceptual and technical foundation for the perturbative methods of Chapters 5 and 6, the numerical investigations of Chapter 7, and the fractional generalizations explored in Chapter 10.

4.6 Summary of chaotic dynamics in the generalized Duffing–Helmholtz oscillator

This section collects and synthesizes the main results concerning the emergence, structure, and characterization of chaotic dynamics in the generalized Duffing (Duffing–Helmholtz, mixed–parity) oscillator studied throughout Chapter 4.

4.6.1 Sufficient conditions for chaos

Theorem 4.3 (Melnikov criterion for chaos in the generalized Duffing–Helmholtz oscillator). *Consider the damped and periodically forced generalized Duffing–Helmholtz oscillator*

$$\ddot{u} + \delta\dot{u} + a_0 + a_1 u + a_2 u^2 + a_3 u^3 = \gamma \cos(\omega t), \tag{4.47}$$

where $\delta > 0$ and $\gamma > 0$ are sufficiently small. Assume that the unperturbed system ($\delta = \gamma = 0$) admits a saddle equilibrium u_s and a homoclinic orbit $\Gamma_h(t)$ associated with a separatrix of the conservative dynamics.

Let $M(t_0)$ denote the Melnikov function associated with $\Gamma_h(t)$. If there exist parameters (δ, γ, ω) such that

$$\max_{t_0} |M(t_0)| > 0 \quad \text{and} \quad M(t_0) \text{ has simple zeros,}$$

then the stable and unstable manifolds of the perturbed saddle intersect transversely. Consequently, system (4.47) exhibits Smale horseshoes and chaotic dynamics.

Corollary 4.4 (Generalization of classical Duffing chaos). *When $a_0 = a_2 = 0$, equation (4.47) reduces to the classical forced Duffing oscillator. Hence Theorem 4.3 extends the classical Duffing chaos criterion to a four–parameter family of asymmetric nonlinear oscillators. The quadratic and constant stiffness terms deform the potential wells, shift the saddle locations, and modify the homoclinic orbit, but they do not destroy the geometric mechanism responsible for chaos.*

Table 4.1: Qualitative dynamics of the generalized Duffing–Helmholtz oscillator

Parameter regime	Observed dynamics	Main tool
$a_3 > 0$, $\Delta_3 < 0$	Single potential well, periodic motion	Energy integral
$a_3 > 0$, $\Delta_3 > 0$	Double well, separatrix	Phase–plane analysis
$a_3 < 0$	Escape to infinity	Potential analysis
$\delta > 0$, $\gamma = 0$	Spiral decay to equilibrium	Linear stability
$\delta > 0$, $\gamma > 0$, no separatrix	Periodic/quasiperiodic motion	Averaging
$\delta > 0$, $\gamma > 0$, separatrix present	Homoclinic chaos	Melnikov method
Large γ	Global chaos	Numerical Poincaré sections

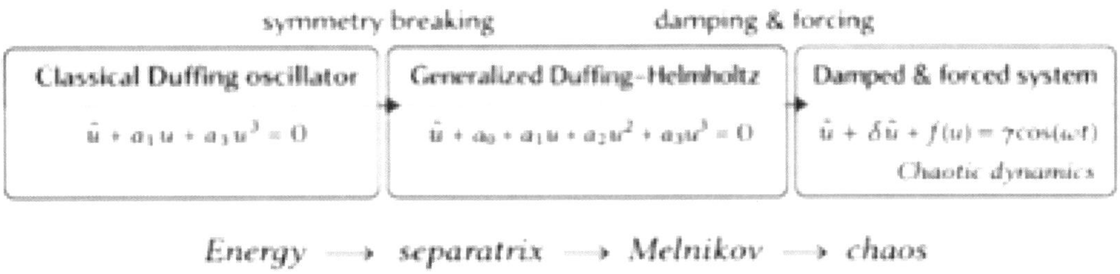

Figure 4.21: Conceptual route from the classical Duffing oscillator to chaotic dynamics in the generalized Duffing–Helmholtz system. Symmetry breaking introduces mixed–parity nonlinearities, while damping and external forcing lead to separatrix splitting and chaos, as detected by Melnikov analysis.

4.6.2 Global view: parameters, dynamics, and analytical tools

4.6.3 Conceptual pathway from Duffing to chaos

Figure 4.22 complements Fig. 4.21 by providing a global view of how parameter choices determine the qualitative dynamics of the system. In the conservative regime, the motion is organized by the potential landscape and the energy first integral, leading to periodic motion, separatrices, or escape to infinity. When damping and forcing are introduced, separatrix structures become the key mechanism for the onset of chaos: their transversal splitting, detected by the Melnikov method, signals homoclinic chaos, while larger forcing amplitudes give rise to global chaotic dynamics characterized numerically by Poincaré sections.

Remarks on novelty and scope

The analysis presented in this chapter goes beyond the classical Duffing framework in several essential aspects. First, the inclusion of constant and quadratic stiffness terms removes the artificial symmetry assumption inherent in the standard Duffing model, allowing the description of realistic mechanical systems with asymmetric restoring forces. Second, the location of saddle equilibria and separatrices is obtained explicitly in terms of the physical parameters (a_0, a_1, a_2, a_3), providing a direct connection between system design and nonlinear dynamics. Third, the Melnikov analysis is carried out using exact homoclinic solutions inherited from the conservative problem, yielding closed–form chaos thresholds. Finally, the results show that

Parameter regime	Observed dynamics	Main analytical tool
$a_3 > 0,\ \Delta_3 < 0$	Single potential well, periodic motion	Energy first integral
$a_3 > 0,\ \Delta_3 > 0$	Double well, separatrix structure	Phase–plane analysis
$a_3 < 0$	Escape to infinity	Potential analysis
$\delta > 0,\ \gamma = 0$	Spiral decay to equilibrium	Linear stability analysis
$\delta = 0,\ \gamma > 0$, no separatrix	Periodic / quasiperiodic motion	Averaging method
$\delta > 0,\ \gamma > 0$, separatrix present	Homoclinic chaos	Melnikov method
Large γ	Global chaotic dynamics	Numerical Poincaré sections

Figure 4.22: Qualitative classification of the dynamical regimes of the generalized Duffing–Helmholtz oscillator. The table relates parameter regimes to the observed dynamics and to the appropriate analytical tools, highlighting the central role of energy methods, separatrix structures, and Melnikov analysis in the transition to chaos.

chaotic dynamics in asymmetric oscillators arise from the same geometric mechanism as in the classical Duffing equation, while exhibiting new phase–space structures absent in symmetric models.

This completes the analysis of stability, integrability, and chaos for the generalized Duffing–Helmholtz oscillator and provides a solid foundation for the numerical and fractional extensions developed in subsequent chapters.

Chapter 5

Perturbative solution methods

It is well known that most nonlinear oscillators do not admit closed form solution. For this resaon we need to use methods that give approximate anlaytical solutions to them. These methods are called perturbative methods. Suppose we are given that

$$\ddot{x} + \omega_0^2 x = F(x, \dot{x}, t),\ x(0) = x_0 \text{ and } x'(0) = \dot{x}_0.$$

5.1 Lindstedt–Poincaré method for conservative oscillators

Let

$$\ddot{x} + \omega_0^2 x = F(x),\ x(0) = x_0 \text{ and } x'(0) - \dot{x}_0.$$

We consider the homotopy

$$H_p = \ddot{x} + \omega_0^2 x - pF(x)$$

We assume the solution in the ansatz form

$$x(t) = x_0 \cos(\omega t) + \frac{\dot{x}_0}{\omega} \sin(\omega t) + p u_1(\omega t) + p^2 u_2(\omega t) + \cdots,$$

where

$$\omega^2 = \omega_0^2 + p\omega_1 + p^2\omega_2 + \cdots$$

and

$$u_j(0) = u_j'(0) = 0 \text{ for all } j.$$

We choose the solutions $u_j(\omega t)$ in order to avoid the so called secularity terms. In practice, the first or second order approximation gives reasonable approximate analytical solution when $|x_0| \leq 1$ and $|\dot{x}_0| \leq 1$. Let us consider several examples.

5.2 The undamped and unforced Duffing oscillator

Let

$$\ddot{x} + \omega_0^2 x + \varepsilon x^3 = 0,\ x(0) = x_0 \text{ and } x'(0) = \dot{x}_0.$$

Let

$$x(t) = x_0 \cos(\omega t) + \frac{\dot{x}_0}{\omega} \sin(\omega t) + \varepsilon u_1(\omega t) + \varepsilon^2 u_2(\omega t).$$

and

$$\tau = \omega t,\ \omega = \sqrt{\omega_0^2 + \varepsilon\omega_1 + \varepsilon^2\omega_2},\ p = \varepsilon.$$

We have

$$\ddot{x} + \omega_0^2 x + \varepsilon x^3 = \varepsilon[\omega_0^2 u_1''(\tau) + \omega_0^2 u_1(\tau) + \left(\tfrac{x_0^3}{4} - \tfrac{3x_0\dot{x}_0^2}{4\omega_0^2}\right)\cos(3\tau) + \left(\tfrac{3x_0^2\dot{x}_0}{4\omega_0} - \tfrac{\dot{x}_0^3}{4\omega_0^3}\right)\sin(3\tau)$$
$$+\left(\tfrac{3\dot{x}_0^3}{4\omega_0^3} - \tfrac{\dot{x}_0\omega_1}{\omega_0} + \tfrac{3x_0^2\dot{x}_0}{4\omega_0}\right)\sin(\tau) + \left(\tfrac{3\dot{x}_0^2 x_0}{4\omega_0^2} - x_0\omega_1 + \tfrac{3x_0^3}{4}\right)\cos(\tau)]$$
$$+\varepsilon^2[\omega_1 u_1''(\tau) + \omega_0^2 u_2''(\tau) + \omega_0^2 u_2(\tau) + \tfrac{3\dot{x}_0^2 u_1(\tau)}{2\omega_0^2} + \tfrac{3\dot{x}_0 x_0}{\omega_0} u_1(\tau)\sin(2\tau)$$
$$-\tfrac{3\dot{x}_0^2}{2\omega_0^2} u_1(\tau)\cos(2\tau) + \tfrac{3}{2}x_0^2 u_1(\tau) + \tfrac{3}{2}x_0^2 u_1(\tau)\cos(2\tau) +$$
$$+\left(\tfrac{3\dot{x}_0^3\omega_1}{8\omega_0^5} - \tfrac{3x_0^2\dot{x}_0\omega_1}{8\omega_0^3}\right)\sin(3\tau) + \tfrac{3x_0\dot{x}_0^2\omega_1}{4\omega_0^4}\cos(3\tau) +$$
$$\left(-\tfrac{9\dot{x}_0^3\omega_1}{8\omega_0^5} + \tfrac{\dot{x}_0\omega_1^2}{2\omega_0^3} - \tfrac{3x_0^2\dot{x}_0\omega_1}{8\omega_0^3} - \tfrac{\dot{x}_0\omega_2}{\omega_0}\right)\sin(\tau) + \left(-\tfrac{3x_0\dot{x}_0^2\omega_1}{4\omega_0^4} - x_0\omega_2\right)\cos(\tau)]$$

In order to obtain the respective solutions, we equate to zero the coefficients of ε^j, $\cos\tau$ and $\sin\tau$. Solving the obtained system of algebraic-differential equations we get

$$\omega_1 = \frac{3\left(x_0^2\omega_0^2 + \dot{x}_0^2\right)}{4\omega_0^2},\ \omega_2 = -\frac{9\left(x_0^2\dot{x}_0^2\omega_0^2 + \dot{x}_0^4\right)}{16\omega_0^6}$$

$$u_1(\tau) = \frac{3\dot{x}_0\left(\dot{x}_0^2 - 3x_0^2\omega_0^2\right)}{32\omega_0^5}\sin(\tau) - \frac{x_0\left(x_0^2\omega_0^2 - 3\dot{x}_0^2\right)}{32\omega_0^4}\cos(\tau) - \frac{\dot{x}_0\left(\dot{x}_0^2 - 3x_0^2\omega_0^2\right)}{32\omega_0^5}\sin(3\tau) + \frac{x_0\left(x_0^2\omega_0^2 - 3\dot{x}_0^2\right)}{32\omega_0^4}\cos(3\tau).$$

$$u_2(\tau) = \tfrac{1}{256\omega_0^9}[6\tau x_0^5\omega_0^5 - 4\tau x_0^3\omega_0^3\left(\omega_0^2\omega_1 - 6\dot{x}_0^2\right) + 6\tau x_0\dot{x}_0^2\omega_0\left(2\omega_0^2\omega_1 - 5\dot{x}_0^2\right) + 35x_0^4\dot{x}_0\omega_0^4$$
$$-51\dot{x}_0^3\omega_1\omega_0^2 + x_0^2\left(81\dot{x}_0\omega_1\omega_0^4 + 11\dot{x}_0^3\omega_0^2\right) - 20\dot{x}_0^5]\sin(\tau)$$
$$-\tfrac{1}{256\omega_0^9}[30\tau x_0^4\dot{x}_0\omega_0^4 + 6\tau\dot{x}_0^3\left(2\omega_0^2\omega_1 - 5\dot{x}_0^2\right) + x_0^2\left(48\tau\dot{x}_0^3\omega_0^2 - 36\tau\dot{x}_0\omega_0^4\omega_1\right) + x_0^5\omega_0^5$$
$$+x_0^3\left(11\dot{x}_0^2\omega_0^3 - 9\omega_0^5\omega_1\right) + x_0\left(51\dot{x}_0^2\omega_0^3\omega_1 - 10\dot{x}_0^4\omega_0\right)]\cos(\tau)$$
$$-\tfrac{3\dot{x}_0}{1024\omega_0^9}\left(5x_0^4\omega_0^4 - 28\dot{x}_0^2\omega_1\omega_0^2 + x_0^2\left(52\omega_0^4\omega_1 - 22\dot{x}_0^2\omega_0^2\right) + 5\dot{x}_0^4\right)\sin(3\tau) +$$
$$\frac{3x_0\left(x_0^4\omega_0^4 + 68\dot{x}_0^2\omega_1\omega_0^2 - 6x_0^2\left(2\omega_0^4\omega_1 - 3\dot{x}_0^2\omega_0^2\right) - 15\dot{x}_0^4\right)}{1024\omega_0^8}\cos(3\tau)$$
$$+\tfrac{\dot{x}_0}{1024\omega_0^9}\left(5x_0^4\omega_0^4 - 10x_0^2\dot{x}_0^2\omega_0^2 + \dot{x}_0^4\right)\sin(5\tau) + \frac{x_0\left(x_0^4\omega_0^4 - 10x_0^2\dot{x}_0^2\omega_0^2 + 5\dot{x}_0^4\right)}{1024\omega_0^8}\cos(5\tau).$$

Example 9. Let

$$\ddot{x} + x + 2x^3 = 0 \wedge x(0) = 0.25 \wedge x'(0) = -0.1.$$

The solution is periodic with period $T = 5.97303$. The exact solution is given by

$$x_{\text{exact}}(t) = 0.267058\text{cn}(1.06894t + 0.359816, 0.0624167)$$

$$x_{\text{approx}}(t) = (0.249473 - 0.0000912452t)\cos(1.0522t)$$

$$+(0.0000955609t - 0.0925992)\sin(1.0522t) - 0.00077572\sin(3.15659t)$$

$$-5.23\text{e-}6\sin(5.26099t) + 0.000528595\cos(3.15659t) - 1.8\text{e-}6\cos(5.26099t)$$

See Figure 5.1.

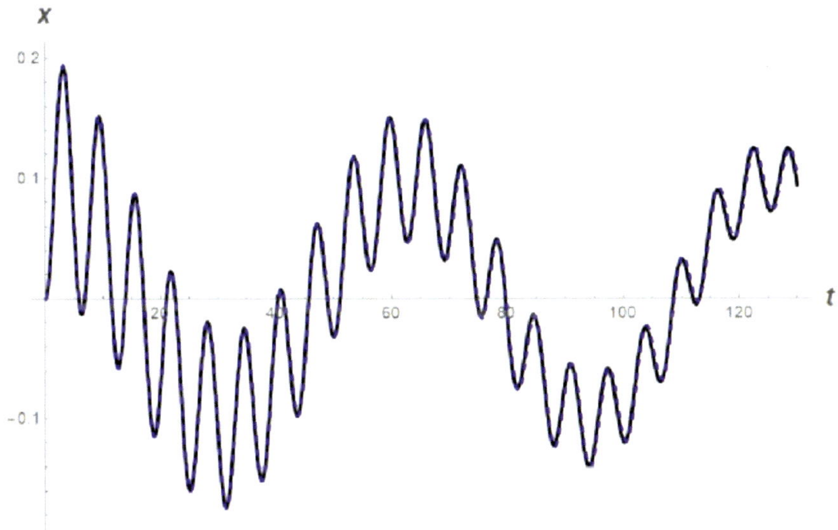

Figure 5.1: The errorequals $\epsilon = 0.00071157$

5.2.1 The quadratic oscillator

This is the oscillator

$$\ddot{x} + \omega_0^2 x + \varepsilon x^2 = 0,\ x(0) = x_0 \text{ and } x'(0) = \dot{x}_0.$$

We consider the homotopy

$$H_\varepsilon = \ddot{x} + \omega_0^2 x + \varepsilon x^2.$$

The solution is assumed in the ansatz form

$$x(t) = x_0 \cos(\omega t) + \frac{\dot{x}_0}{\omega}\sin(\omega t) + \varepsilon u_1(\omega t) + \varepsilon^2 u_2(\omega t).$$

and

$$\tau = \omega t,\ \omega = \sqrt{\omega_0^2 + \varepsilon\omega_1 + \varepsilon^2\omega_2},\ p = \varepsilon.$$

We have :

$H_\varepsilon = \varepsilon(\omega_0^2 u_1''(\tau) + \omega_0^2 u_1(\tau) + \frac{\dot{x}_0 x_0 \sin(2\tau)}{\omega_0} - \frac{\dot{x}_0\omega_1\sin(\tau)}{\omega_0} - x_0\omega_1\cos(\tau) - \frac{\dot{x}_0^2\cos(2\tau)}{2\omega_0^2} +$
$\frac{1}{2}x_0^2\cos(2\tau) + \frac{\dot{x}_0^2}{2\omega_0^2} + \frac{x_0^2}{2}) +$
$\varepsilon^2(\omega_0^2 u_2''(\tau) + \omega_1 u_1''(\tau) + \omega_0^2 u_2(\tau) + \sin(\tau)\left(\frac{2\dot{x}_0 u_1(\tau)}{\omega_0} + \frac{\dot{x}_0\omega_1^2}{2\omega_0^3} - \frac{\dot{x}_0\omega_2}{\omega_0}\right) + \cos(\tau)\left(2x_0 u_1(\tau) - x_0\omega_2\right) -$
$\frac{x_0\dot{x}_0\omega_1\sin(2\tau)}{2\omega_0^3} + \frac{\dot{x}_0^2\omega_1\cos(2\tau)}{2\omega_0^4} - \frac{\dot{x}_0^2\omega_1}{2\omega_0^4}) +$
$o(\varepsilon^3)$.

In order to obtain the respective solutions, we equate to zero the coefficients of ε^j, $\cos\tau$ and $\sin\tau$. Solving the obtained system of algebraic-differential equations we get

$\omega_1 = 0,\ \omega_2 = \frac{-x_0^2\omega_0^2 - \dot{x}_0^2}{\omega_0^4}$.

$u_1(\tau) = \frac{1}{6\omega_0^4}(-4x_0\dot{x}_0\omega_0\sin(\tau) + 2x_0\dot{x}_0\omega_0\sin(2\tau) + 2x_0^2\omega_0^2\cos(\tau) + x_0^2\omega_0^2\cos(2\tau) +$
$4\dot{x}_0^2\cos(\tau) - \dot{x}_0^2\cos(2\tau) - 3x_0^2\omega_0^2 - 3\dot{x}_0^2)$.

$u_2(\tau) = 0.$

The approximate solution will then be

$x_{\text{approx}}(t) = x_0 \cos(\omega t) + \frac{\dot{x}_0}{\omega} \sin(\omega t) -$
$\frac{1}{3\omega_0^4} \left[2\varepsilon \sin^2\left(\frac{\omega t}{2}\right)\left(2x_0\dot{x}_0\omega_0 \sin(\omega t) + x_0^2\omega_0^2(\cos(\omega t)+2) - \dot{x}_0^2(\cos(\omega t)-1)\right)\right].$

$\omega = \sqrt{\omega_0^2 - \frac{x_0^2\omega_0^2 + \dot{x}_0^2}{\omega_0^4}\varepsilon^2}.$

Example 10. Let

$$\ddot{x} - \frac{x^2}{4} + x = 0 \wedge x(0) = 0.4 \wedge x'(0) = 0.1$$

The approximate solution is

$x_{\text{approx}}(t) = 0.02125 + 0.107202 \sin(0.994673t) - 0.00333333 \sin(1.98935t) +$
$0.385 \cos(0.994673t) - 0.00625 \cos(1.98935t)$

Exact solution:

$x_{\text{exact}}(t) = -0.386887 + \frac{1}{4.71356\wp\left(t+2.88256; \frac{1}{12}, 0.00435301\right)+0.468781}.$

See Figure 5.2.

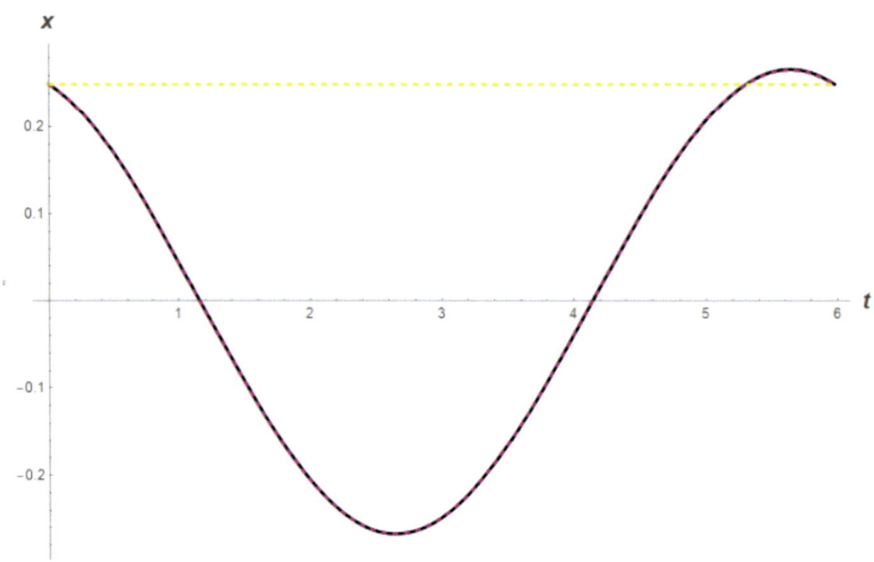

Figure 5.2

5.3 Krylov–Bogoliubov–Mitropolski (KBM) method

This method may be employed for solving both, conservative or non conservative oscillators. Suppose we are given that

$$\ddot{x} + \omega_0^2 x = F(x, \dot{x}, t)$$

We consider the homotopy

$$H_p = \ddot{x} + \omega_0^2 x - pF(x, \dot{x}, t)$$

The solution is assumed in the ansatz form

$$x(t) = a(t)\cos\psi(t) + \sum_{j=1}^{n} p^j v_j(a(t), \psi(t)),$$

where the functions $a = a(t)$ and $\psi = \psi(t)$ vary with time according to

$$\dot{a} = +\sum_{j=1}^{n} p^j A_j(a) \text{ and } \dot{\psi} = \omega_0 + \sum_{j=1}^{n} p^j \varphi_j(a).$$

The expression

$$x(t) = a(t)\cos\psi(t) + \sum_{j=1}^{n} v_j(a(t), \psi(t))$$

is called the $n-$th order approximation ($p = 1$). In practice, the first or second order approximation gives reasonable approximate analytical solution when $|x_0| \le 1$ and $|\dot{x}_0| \le 1$. Following formulas may be useful for computing the first or second order approximation (for the first order approximation we set $p^2 = p^3 = \cdots = 0$):

$$\begin{aligned}\ddot{x} = \quad & a\omega_0^2\cos(\psi) + p\left(\omega_0^2 {v_1}^{(0,2)}(a,\psi) - 2a\varphi_1\omega_0\cos(\psi) - 2A_1\omega_0\sin(\psi)\right) + \\ & p^2[2A_1\omega_0 {v_1}^{(1,1)}(a,\psi) + 2\varphi_1\omega_0 {v_1}^{(0,2)}(a,\psi) + \omega_0^2 {v_2}^{(0,2)}(a,\psi) + \\ & \sin(\psi)\left(-aA_1\dot{\varphi}_1 - 2A_1\varphi_1 - 2A_2\omega_0\right) + \cos(\psi)\left(-2a\varphi_2\omega_0 \quad a\varphi_1^2 + A_1\dot{A}_1\right)].\end{aligned}$$

$$\begin{aligned}\dot{x} = & -a\omega_0\sin(\psi) + p[\omega_0 {v_1}^{(0,1)}(a,\psi) - a\varphi_1\sin(\psi) + A_1\cos(\psi)] + \\ & p^2[A_1 {v_1}^{(1,0)}(a,\psi) + \varphi_1 {v_1}^{(0,1)}(a,\psi) + \omega_0 {v_2}^{(0,1)}(a,\psi) - a\varphi_2\sin(\psi) + A_2\cos(\psi)].\end{aligned}$$

$$\begin{aligned}x^2 = & \tfrac{1}{2}a^2\cos(2\psi) + \tfrac{a^2}{2} + \\ & 2ap\cos(\psi)v_1(a,\psi) + \\ & p^2[v_1(a,\psi)^2 + 2a\cos(\psi)v_2(a,\psi)].\end{aligned}$$

$$\begin{aligned}x^3 = & \tfrac{3}{4}a^3\cos(\psi) + \tfrac{1}{4}a^3\cos(3\psi) + \\ & p[\tfrac{3}{2}a^2 v_1(a,\psi) + \tfrac{3}{2}a^2\cos(2\psi)v_1(a,\psi)] + \\ & p^2[\tfrac{3}{2}a^2 v_2(a,\psi) + \tfrac{3}{2}a^2\cos(2\psi)v_2(a,\psi) + 3a\cos(\psi)v_1(a,\psi)^2].\end{aligned}$$

$$\begin{aligned}x^4 = & \tfrac{1}{2}a^4\cos(2\psi) + \tfrac{1}{8}a^4\cos(4\psi) + \tfrac{3a^4}{8} + \\ & p[3a^3\cos(\psi)v_1(a,\psi) + a^3\cos(3\psi)v_1(a,\psi)] + \\ & p^2[3a^3\cos(\psi)v_2(a,\psi) + a^3\cos(3\psi)v_2(a,\psi) + 3a^2 v_1(a,\psi)^2 + 3a^2\cos(2\psi)v_1(a,\psi)^2].\end{aligned}$$

$$x^5 = \frac{5}{8}a^5\cos(\psi) + \frac{5}{16}a^5\cos(3\psi) + \frac{1}{16}a^5\cos(5\psi) + \\ p[\frac{15}{8}a^4v_1(a,\psi) + \frac{5}{2}a^4\cos(2\psi)v_1(a,\psi) + \frac{5}{8}a^4\cos(4\psi)v_1(a,\psi)] + \\ p^2[\frac{15}{8}a^4v_2(a,\psi) + \frac{5}{2}a^4\cos(2\psi)v_2(a,\psi) + \frac{5}{8}a^4\cos(4\psi)v_2(a,\psi) + \\ \frac{15}{2}a^3\cos(\psi)v_1(a,\psi)^2 + \frac{5}{2}a^3\cos(3\psi)v_1(a,\psi)^2].$$

$$\dot{x}^2 = \frac{1}{2}a^2\omega_0^2 - \frac{1}{2}a^2\omega_0^2\cos(2\psi) + \\ p[-2a\omega_0^2\sin(\psi){v_1}^{(0,1)}(a,\psi) + a^2\varphi_1\omega_0(-\cos(2\psi)) + a^2\varphi_1\omega_0 - aA_1\omega_0\sin(2\psi)] + \\ p^2[\sin(\psi)\left(-2aA_1\omega_0{v_1}^{(1,0)}(a,\psi) - 4a\varphi_1\omega_0{v_1}^{(0,1)}(a,\psi) - 2a\omega_0^2{v_2}^{(0,1)}(a,\psi)\right) + \\ 2A_1\omega_0\cos(\psi){v_1}^{(0,1)}(a,\psi) + \omega_0^2{v_1}^{(0,1)}(a,\psi)^2 - a^2\varphi_2\omega_0\cos(2\psi) - \\ \frac{1}{2}a^2\varphi_1^2\cos(2\psi) + a^2\varphi_2\omega_0 + \frac{1}{2}a^2\varphi_1^2 - aA_1\varphi_1\sin(2\psi) - aA_2\omega_0\sin(2\psi) + \frac{1}{2}A_1^2\cos(2\psi) + \frac{A_1^2}{2}].$$

$$x\dot{x} = -\frac{1}{2}a^2\omega_0\sin(2\psi) + \\ p[-a\omega_0\sin(\psi)v_1(a,\psi) + a\omega_0\cos(\psi){v_1}^{(0,1)}(a,\psi) - \frac{1}{2}a^2\varphi_1\sin(2\psi) + \frac{1}{2}aA_1\cos(2\psi) + \frac{aA_1}{2}] + \\ p^2[\cos(\psi)\left(A_1v_1(a,\psi) + aA_1{v_1}^{(1,0)}(a,\psi) + a\varphi_1{v_1}^{(0,1)}(a,\psi) + a\omega_0{v_2}^{(0,1)}(a,\psi)\right) + \\ \sin(\psi)\left(-a\varphi_1v_1(a,\psi) - a\omega_0v_2(a,\psi)\right) + \omega_0v_1(a,\psi){v_1}^{(0,1)}(a,\psi) \\ -\frac{1}{2}a^2\varphi_2\sin(2\psi) + \frac{1}{2}aA_2\cos(2\psi) + \frac{aA_2}{2}].$$

$$x^2\dot{x} = -\frac{1}{4}a^3\omega_0\sin(\psi) - \frac{1}{4}a^3\omega_0\sin(3\psi) + \\ p[\frac{1}{2}a^2\omega_0{v_1}^{(0,1)}(a,\psi) - a^2\omega_0\sin(2\psi)v_1(a,\psi) + \frac{1}{2}a^2\omega_0\cos(2\psi){v_1}^{(0,1)}(a,\psi) - \frac{1}{4}a^3\varphi_1\sin(\psi) \\ -\frac{1}{4}a^3\varphi_1\sin(3\psi) + \frac{3}{4}a^2A_1\cos(\psi) + \frac{1}{4}a^2A_1\cos(3\psi)] + \\ p^2[\sin(\psi)\left(-a\omega_0v_1(a,\psi)^2 - \frac{1}{4}a^3\varphi_2\right) + \cos(\psi)\left(2a\omega_0v_1(a,\psi){v_1}^{(0,1)}(a,\psi) + \frac{3a^2A_2}{4}\right) + \\ \frac{1}{2}a^2A_1{v_1}^{(1,0)}(a,\psi) + \frac{1}{2}a^2A_1\cos(2\psi){v_1}^{(1,0)}(a,\psi) \\ +\frac{1}{2}a^2\varphi_1{v_1}^{(0,1)}(a,\psi) - a^2\varphi_1\sin(2\psi)v_1(a,\psi) + \frac{1}{2}a^2\varphi_1\cos(2\psi){v_1}^{(0,1)}(a,\psi) + \frac{1}{2}a^2\omega_0{v_2}^{(0,1)}(a,\psi) - \\ a^2\omega_0\sin(2\psi)v_2(a,\psi) + \frac{1}{2}a^2\omega_0\cos(2\psi){v_2}^{(0,1)}(a,\psi) \\ +aA_1v_1(a,\psi) + aA_1\cos(2\psi)v_1(a,\psi) - \frac{1}{4}a^3\varphi_2\sin(3\psi) + \frac{1}{4}a^2A_2\cos(3\psi)].$$

Let us consider some illustarive examples.

5.3.1 Dufifng oscillator with forcing and damping

Let

$$\ddot{x} + 2\varepsilon\dot{x} + \omega_0^2x + \beta x^3 = f(t),\ x(0) = x_0 \text{ and } x'(0) = \dot{x}_0.$$

We define the homotopy as

$$H_p = \ddot{x} + \omega_0^2x + p\left[2\varepsilon\dot{x} + \beta x^3 - f(t)\right]$$

We assume the solution in the ansatz form

$$x(t) = a(t)\cos\psi(t) + pv_1(a(t),\psi(t)) + p^2v_2(a(t),\psi(t)).$$

$$\dot{a} = pA_1(a) + p^2A_2(a) \text{ and } \dot{\psi} = \omega_0 + p\varphi_1(a) + p^2\varphi_2(a).$$

We have :

$H_p = p[\frac{1}{4}\left(4\omega_0^2 v_1(a,\psi) + 4\omega_0^2 {v_1}^{(0,2)}(a,\psi) - 4f(t)\right) + \frac{1}{4}\left(3a^3\beta - 8a\varphi_1\omega_0\right)\cos(\psi) +$
$\frac{1}{4}a^3\beta\cos(3\psi) + \frac{1}{4}\left(-8a\varepsilon\omega_0 - 8A_1\omega_0\right)\sin(\psi)] +$
$p^2[1/2(3a^2\beta v_1(a,\psi) + 3a^2\beta\cos(2\psi)v_1(a,\psi) + 4A_1\omega_0 {v_1}^{(1,1)}(a,\psi) + 4\varepsilon\omega_0 {v_1}^{(0,1)}(a,\psi) +$
$4\varphi_1\omega_0 {v_1}^{(0,2)}(a,\psi) + 2\omega_0^2 v_2(a,\psi) + 2\omega_0^2 {v_2}^{(0,2)}(a,\psi)) +$
$\frac{1}{2}\left(-2aA_1\dot{\varphi}_1 - 4a\varepsilon\varphi_1 - 4A_1\varphi_1 - 4A_2\omega_0\right)\sin(\psi) + \frac{1}{2}\left(-4a\varphi_2\omega_0 - 2a\varphi_1^2 + 4A_1\varepsilon + 2A_1\dot{A}_1\right)\cos(\psi)]$.

We equate to zero the coefficients of p^j, $\cos\psi$ and $\sin\psi$ to zero to obtain a system of algebraic-differential equations. We solve this system so that we did no obtain secularity terms. The solution is given by

$$A_1(a) = -a\varepsilon,\ A_2(a) = \frac{3a^3\beta\varepsilon}{8\omega_0^2}.$$

$$\varphi_1(a) = \frac{3a^2\beta}{8\omega_0},\ \varphi_2(a) = \frac{-15a^4\beta^2 - 128\varepsilon^2\omega_0^2}{256\omega_0^3}.$$

$$v_1(a,\psi) = \frac{a^3\beta\cos(3\psi) + 32f(t)}{32\omega_0^2}.$$

$$v_2(a,\psi) = \frac{a^2\beta}{1024\omega_0^4}\left(a^3\beta(\cos(5\psi) - 21\cos(3\psi)) + 48a\varepsilon\omega_0\sin(3\psi) + 512f(t)(\cos(2\psi) - 3)\right)$$

The odes for determining $a = a(t)$ and $\psi = \psi(t)$ are

$$\dot{a} = -a\varepsilon p + \frac{3a^3\beta\varepsilon}{8\omega_0^2}p^2 \text{ and } \dot{\psi} = \omega_0 + \frac{3a^2\beta}{8\omega_0}p + \left(-\frac{15a^4\beta^2}{256\omega_0^3} - \frac{\varepsilon^2}{2\omega_0}\right)p^2.$$

Now, we set $p = 1$. The solution to the odes are

$$a = \frac{2\sqrt{2}c_0\omega_0}{\sqrt{8\omega_0^2 e^{2\varepsilon t} - 3\beta c_0^2(e^{2\varepsilon t} - 1)}}.$$

$$\psi = c_1 + \left(\frac{19\omega_0}{12} - \frac{\varepsilon^2}{2\omega_0}\right)t + \frac{5\beta c_0^2}{64\varepsilon\omega_0} + \frac{5\beta c_0^2\omega_0}{8\varepsilon\left(3\beta c_0^2(e^{2\varepsilon t} - 1) - 8\omega_0^2 e^{2\varepsilon t}\right)} - \frac{7\omega_0}{24\varepsilon}\log\left(e^{2\varepsilon t} - \frac{3\beta c_0^2(e^{2\varepsilon t} - 1)}{8\omega_0^2}\right)$$

Finally, the second order approximation is given by

$$x(t) = a\cos(\psi) + \left(\frac{a^3\beta\cos(3\psi)}{32\omega_0^2} + \frac{f(t)}{\omega_0^2}\right)$$

$$+\left(-\frac{21a^5\beta^2\cos(3\psi)}{1024\omega_0^4} + \frac{a^5\beta^2\cos(5\psi)}{1024\omega_0^4} + \frac{3a^3\beta\varepsilon\sin(3\psi)}{64\omega_0^3} + \frac{a^2\beta f(t)\cos(2\psi)}{2\omega_0^4} - \frac{3a^2\beta f(t)}{2\omega_0^4}\right).$$

The constants c_0 and c_1 are obtained from the initial conditions $x(0) = x_0$ and $x'(0) = \dot{x}_0$.

Example 11. Let

$$\ddot{x} + 0.2\dot{x} + x + 3x^3 = 0 \wedge x(0) = 0.2 \wedge x'(0) = 0.1$$

Approximate solution:

$$x(t) = a\cos(\psi) - \frac{189a^5\cos(3\psi)}{1024} + \frac{9a^5\cos(5\psi)}{1024} + 0.0140625a^3\sin(3\psi) + \frac{3}{32}a^3\cos(3\psi).$$

$$a = a(t) = \frac{0.645684}{\sqrt{8e^{0.2t} - 0.469021(e^{0.2t}-1)}}.$$

$$\psi = \psi(t) = 1.57833t - 0.379965 + \frac{0.977128}{0.469021(e^{0.2t}-1) - 8e^{0.2t}} - 2.91667\log\left(e^{0.2t} - 0.0586277\left(e^{0.2t}-1\right)\right).$$

See Figure 5.3 for a comparison with Runge-Kutta numerical solution. The errorequals $\mathcal{E} = 0.00031$.

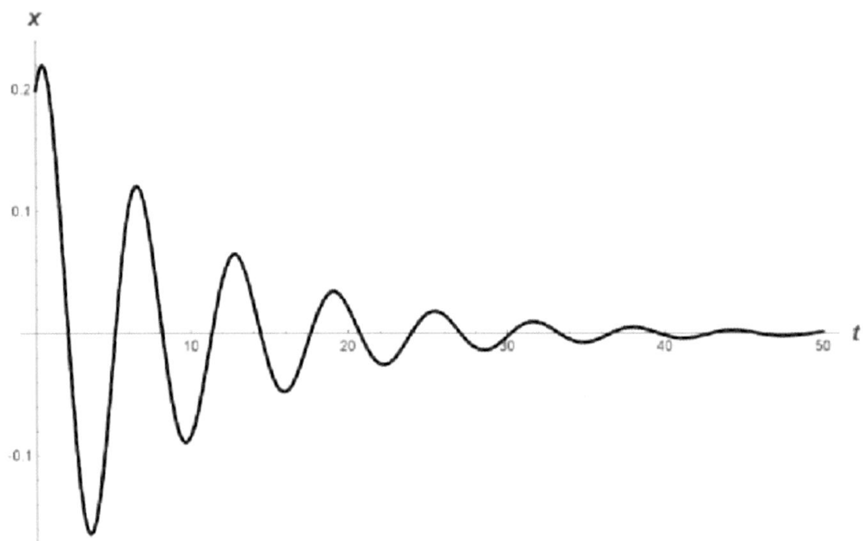

Figure 5.3

Example 12. Let

$$\ddot{x} + 0.02\dot{x} + 0.1x^3 + x - 0.01\cos\left(\frac{t}{10}\right) \wedge x(0) = 0 \wedge x'(0) = 0.$$

Approximate solution :

$$x(t) = -0.000205078a^5\cos(3\psi) + 9.77\mathrm{e} - 6a^5\cos(5\psi) + 0.000046875a^3\sin(3\psi) +$$

$$0.003125a^3\cos(3\psi) + 0.0005a^2\cos\left(\tfrac{t}{10}\right)\cos(2\psi) - 0.0015a^2\cos\left(\tfrac{t}{10}\right) + a\cos(\psi) + 0.01\cos\left(\tfrac{t}{10}\right).$$

$$a = a(t) = -\frac{0.0282854}{\sqrt{8e^{0.02t} - 0.0000300024(e^{0.02t}-1)}}.$$

$$\psi = \psi(t) = 1.58328t + \frac{0.00062505}{0.0000300024(e^{0.02t}-1) - 8e^{0.02t}} - 29.1667\log\left(e^{0.02t} - 3.75\mathrm{e} - 6\left(e^{0.02t}-1\right)\right) - 0.009$$

See Figure5.4 for a comparison with Runge-Kutta numerical solution. The errorequals $\mathcal{E} = 0.0001744$.

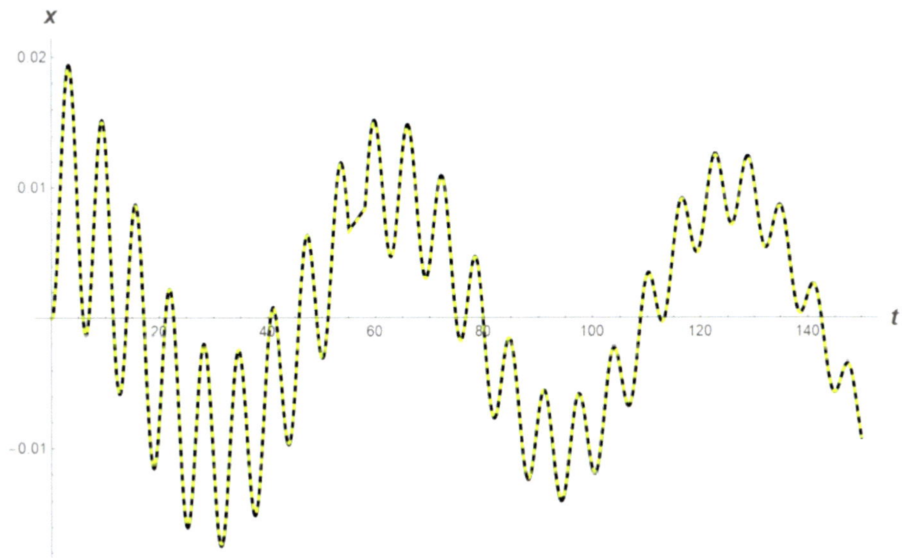

Figure 5.4

5.3.2 Double–Well Duffing oscillator

Let

$$\ddot{x} + 2\varepsilon\dot{x} - \alpha x + \beta x^3 = \gamma\cos\omega t,\ x(0) = x_0 \text{ and } x'(0) = \dot{x}_0, \text{ where } \alpha > 0 \text{ and } \beta > 0$$

Here where 2ε is the damping coefficient, and γ and $\omega$are the amplitude and frequency of the external driving force, respectively.The Duffing equation with negative linear stiffness describes the dynamics of a buckled beam as well as a plasma oscillator. Define

$$x(t) = \sqrt{\frac{\alpha}{\beta}} + u(t).$$

Then

$$\ddot{u} + 2\varepsilon\dot{u} + 2\alpha u + 3\sqrt{\alpha\beta}u^2 + \beta u^3 = \gamma\cos(\omega t).$$

Consider the homotopy

$$H_p = \ddot{u} + \omega_0^2 u + p\left[2\varepsilon\dot{u} + 3\sqrt{\alpha\beta}u^2 + \beta u^3 - f(t)\right],\ \omega_0 = \sqrt{2\alpha} \text{ and } f(t) = \gamma\cos(\omega t).$$

We assume the solution in nthe ansatz form

$$u(t) = a(t)\cos\psi(t)) + pu_1(a(t), \psi(t)) + p^2 u_2(a(t), \psi(t)).$$

so that

$$\dot{a} = pA_1(a) + p^2 A_2(a) \text{ and } \dot{\psi} = \omega_0 + p\varphi_1(a) + p^2\varphi_2(a).$$

We have :

$H_p = p[\frac{1}{4}\left(8\alpha v_1(a,\psi) + 8\alpha {v_1}^{(0,2)}(a,\psi) + 6\sqrt{\alpha\beta}a^2\cos(2\psi) + 6\sqrt{\alpha\beta}a^2 - 4f(t)\right) +$
$\frac{1}{4}\left(3a^3\beta - 8a\varphi_1\omega_0\right)\cos(\psi) + \frac{1}{4}a^3\beta\cos(3\psi) + \frac{1}{4}\left(-8a\varepsilon\omega_0 - 8A_1\omega_0\right)\sin(\psi)]+$
$p^2[\frac{1}{2}(3a^2\beta v_1(a,\psi) + 3a^2\beta\cos(2\psi)v_1(a,\psi) + 4A_1\omega_0 {v_1}^{(1,1)}(a,\psi) + 4\alpha v_2(a,\psi)+$
$4\alpha {v_2}^{(0,2)}(a,\psi) + 4\varepsilon\omega_0 {v_1}^{(0,1)}(a,\psi) + 4\varphi_1\omega_0 {v_1}^{(0,2)}(a,\psi))$
$+\frac{1}{2}\left(12a\sqrt{\alpha\beta}v_1(a,\psi) - 4a\varphi_2\omega_0 - 2a\varphi_1^2 + 4A_1\varepsilon + 2A_1\dot{A}_1\right)\cos(\psi)+$
$\frac{1}{2}\left(-2aA_1\dot{\varphi}_1 - 4a\varepsilon\varphi_1 - 4A_1\varphi_1 - 4A_2\omega_0\right)\sin(\psi)]+$
$o(p^3)$.

We equate to zero the coefficients of p^j,$\cos\psi$ and $\sin\psi$ to zero to obtain a system of algebraic-differential equations. We solve this system so that we did no obtain secularity terms. The solution is given by

$A_1(a) = -a\varepsilon,\ A_2(a) = \frac{3a^3\beta\varepsilon}{16\alpha}$.
$\varphi_1(a) = \frac{3a^2\beta}{8\sqrt{2}\sqrt{\alpha}},\ \varphi_2(a) = -\frac{1}{512\sqrt{2}\alpha^{3/2}}\left[15a^4\beta^2 + 960\alpha a^2\beta + 256\alpha\varepsilon^2 - 768\sqrt{\alpha}\sqrt{\beta}f(t)\right]$.
$v_1(a,\psi) = \frac{1}{64\alpha}\left[a^3\beta\cos(3\psi) + 16\sqrt{\alpha}a^2\sqrt{\beta}(\cos(2\psi)-3) + 32f(t)\right]$.
$v_2(a,\psi) = \frac{a^2}{12288\alpha^2}[3a^3\beta^2(\cos(5\psi) - 21\cos(3\psi)) + 96\sqrt{\alpha}a^2\beta^{3/2}(-31\cos(2\psi) + \cos(4\psi) + 60)+$
$144a\left(\sqrt{2}\sqrt{\alpha}\beta\varepsilon\sin(3\psi) + 4\alpha\beta\cos(3\psi)\right) + 512\left(4\sqrt{2}\alpha\sqrt{\beta}\varepsilon\sin(2\psi) + 3\beta f(t)(\cos(2\psi)-3)\right)]$.

The odes for determining $a = a(t)$ and $\psi = \psi(t)$ are

$$\dot{a} = -a\varepsilon + \frac{3\beta\varepsilon}{16\alpha}a^3.$$

$$\dot{\psi} = \sqrt{2\alpha} - \frac{\varepsilon^2}{2\sqrt{2\alpha}} - \frac{3\beta}{2\sqrt{2}\sqrt{\alpha}}a^2 - \frac{15\beta^2}{512\sqrt{2}\alpha^{3/2}}a^4 + \frac{3\sqrt{\beta}}{2\sqrt{2}\alpha}f(t).$$

First order approximation:

$x(t) = a\cos(\psi) + \frac{a^3\beta\cos(3\psi)}{64\alpha} + \frac{a^2\sqrt{\beta}\cos(2\psi)}{4\sqrt{\alpha}} - \frac{3a^2}{4}\sqrt{\frac{\beta}{\alpha}} + \frac{f(t)}{2\alpha}$.

Second order approximation :

$x(t) = a\cos(\psi) + \frac{f(t)}{2\alpha} + \frac{15a^4\beta^{3/2}}{32\alpha^{3/2}}+$
$\frac{a^2\sqrt{\beta}\varepsilon}{3\sqrt{2}\alpha}\sin(2\psi) + \left(-\frac{31a^4\beta^{3/2}}{128\alpha^{3/2}} + \frac{a^2\sqrt{\beta}}{4\sqrt{\alpha}} + \frac{a^2\beta f(t)}{8\alpha^2}\right)\cos(2\psi)+$
$\frac{3a^3\beta\varepsilon}{128\sqrt{2}\alpha^{3/2}}\sin(3\psi) - \frac{3a^2}{4}\sqrt{\frac{\beta}{\alpha}} - \frac{3a^2\beta}{8\alpha^2}f(t) + \left(\frac{a^3\beta}{16\alpha} - \frac{21a^5\beta^2}{4096\alpha^2}\right)\cos(3\psi) + \frac{a^4\beta^{3/2}}{128\alpha^{3/2}}\cos(4\psi)+$
$\frac{a^5\beta^2}{4096\alpha^2}\cos(5\psi)$.
$a = a(t) = \frac{4\sqrt{\alpha}c_0}{\sqrt{3\beta c_0^2 + e^{2\varepsilon t}\left(16\alpha - 3\beta c_0^2\right)}}$.
$\psi(t) = c_1 + \frac{1}{192\sqrt{2\alpha}}\left[15\beta c_0^2\left(\frac{16\alpha}{\varepsilon e^{2\varepsilon t}\left(3\beta c_0^2 - 16\alpha\right) - 3\beta c_0^2\varepsilon} + \frac{1}{\varepsilon}\right) - 32t\left(41\alpha + 3\varepsilon^2\right)\right]+$
$\frac{1}{12\sqrt{2}\alpha\varepsilon}\left[18\sqrt{\beta}\varepsilon\int_0^t f(\tau)\,d\tau + 53\alpha^{3/2}\phi(t)\right]$.
$\phi(t) = \log\left(\frac{3\beta c_0^2 + e^{2\varepsilon t}\left(16\alpha - 3\beta c_0^2\right)}{16\alpha}\right)$.
$\int_0^t f(\tau)\,d\tau = \frac{\gamma}{\omega}\sin(\omega t)$.

The constants c_0 and c_1 are determined from the initial conditions.

Example 13. Let

$$\ddot{x} + 0.1\dot{x} - x + x^3 = 0.005\cos(0.2t),\ x(0) = 1,\ x'(0) = 0.$$

Approximate solution:

$$x(t) = 1 - 0.00512695a^5\cos(3\psi) + 0.000244141a^5\cos(5\psi) - 0.242188a^4\cos(2\psi)$$

$$+0.0078125a^4\cos(4\psi) + 0.46875a^4 + 0.000828641a^3\sin(3\psi)$$

$$+0.0625a^3\cos(3\psi) + 0.0003125a^2\cos(0.2t - 2\psi) + 0.0003125a^2\cos(0.2t + 2\psi)$$

$$-0.001875a^2\cos(0.2t) + 0.0117851a^2\sin(2\psi)$$

$$+0.25a^2\cos(2\psi) - 0.75a^2 + 1.a\cos(\psi) + 0.0025\cos(0.2t).$$

$$a = -\frac{0.00999378}{\sqrt{0.0000187267 + 16.e^{0.1t}}}.$$

$$\psi = -4.83278t + \frac{1}{-144996.e^{0.1t}\ \ 0.169706} + 62.4611\log\left(0.999999e^{0.1t} + 1.17\text{e-}6\right)$$

$$+0.0265165\log\left(0.999999e^{0.1t} + 1.17\text{e-}6\right)\sin(0.2t) - 0.0355924$$

See Figure 5.5 for a comparison with Runge-Kutta numerical solution. The errorequals $\mathcal{E} = 0.00009$.

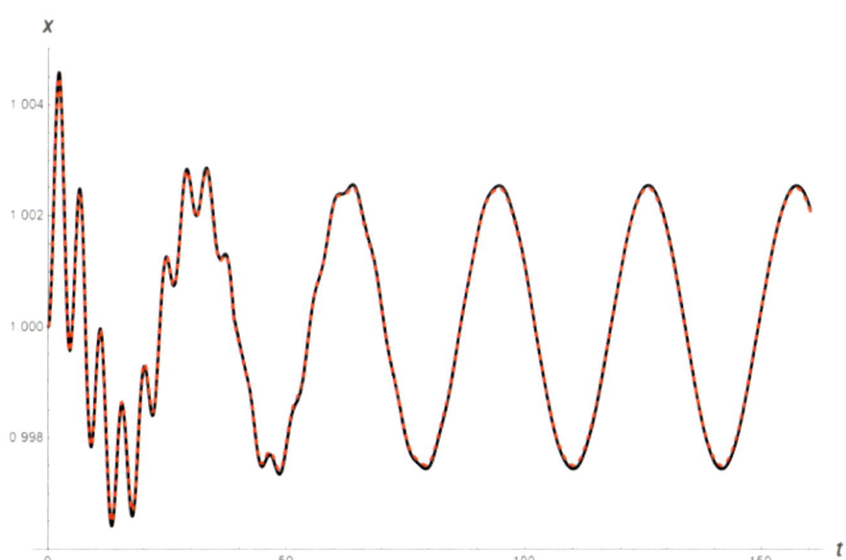

Figure 5.5

5.3.3 Duffing oscillator with forcing and damping

Let

$$\ddot{x} + 2\varepsilon\dot{x} + \omega_0^2 x + \beta x^3 + \gamma x^5 = f(t),\ x(0) = x_0 \text{ and } x'(0) = \dot{x}_0.$$

We define the homotopy as

$$H_p = \ddot{x} + \omega_0^2 x + p\left[2\varepsilon\dot{x} + \beta x^3 + \gamma x^5 - f(t)\right]$$

We assume the solution in the ansatz form

$$x(t) = a(t)\cos\psi(t) + pv_1(a(t),\psi(t)) + p^2 v_2(a(t),\psi(t)).$$

$$\dot{a} = pA_1(a) + p^2 A_2(a) \text{ and } \dot{\psi} = \omega_0 + p\varphi_1(a) + p^2\varphi_2(a).$$

We have :

$H_p = p[\frac{1}{16}\left(16\omega_0^2 v_1(a,\psi) + 16\omega_0^2 {v_1}^{(0,2)}(a,\psi) - 16f(t)\right) + \frac{1}{16}a^5\gamma\cos(5\psi)+$
$\frac{1}{16}\cos(\psi)\,(10a^5\gamma + 12a^3\beta - 32a\varphi_1\omega_0)$
$+\frac{1}{16}\cos(3\psi)\,(5a^5\gamma + 4a^3\beta) + \frac{1}{16}\sin(\psi)\,(-32a\varepsilon\omega_0 - 32A_1\omega_0)]+$
$p^2[\frac{1}{8}(15a^4\gamma v_1(a,\psi) + 20a^4\gamma\cos(2\psi)v_1(a,\psi) + 5a^4\gamma\cos(4\psi)v_1(a,\psi) + 12a^2\beta v_1(a,\psi)+$
$12a^2\beta\cos(2\psi)v_1(a,\psi) + 16A_1\omega_0 {v_1}^{(1,1)}(a,\psi)$
$+16\varepsilon\omega_0 {v_1}^{(0,1)}(a,\psi) + 16\varphi_1\omega_0 {v_1}^{(0,2)}(a,\psi) + 8\omega_0^2 v_2(a,\psi) + 8\omega_0^2 {v_2}^{(0,2)}(a,\psi)\ )+$
$\frac{1}{8}\sin(\psi)\,(-8aA_1\dot{\varphi}_1 - 16a\varepsilon\varphi_1 - 16A_1\varphi_1 - 16A_2\omega_0)+$
$\frac{1}{8}\cos(\psi)\left(-16a\varphi_2\omega_0 - 8a\varphi_1^2 + 16A_1\varepsilon + 8A_1\dot{A}_1\right)]+$
$o(p^3)$.

We equate to zero the coefficients of p^j,$\cos\psi$ and $\sin\psi$ to zero to obtain a system of algebraic-differential equations. We solve this system so that we did no obtain secularity terms. The solution is given by

$A_1(a) = -a\varepsilon,\ A_2(a) = \frac{5a^5\gamma\varepsilon + 3a^3\beta\varepsilon}{8\omega_0^2}$.

$\varphi_1(a) = \frac{5a^4\gamma + 6a^2\beta}{16\omega_0},\ \varphi_2(a) = \frac{-55a^8\gamma^2 - 240a^6\beta\gamma - 180a^4\beta^2 - 1536\varepsilon^2\omega_0^2}{3072\omega_0^3}$.

$v_1(a,\psi) = \frac{a^5\gamma(15\cos(3\psi) + \cos(5\psi)) + 12a^3\beta\cos(3\psi) + 384f(t)}{384\omega_0^2}$.

$v_2(a,\psi) = \frac{a^2}{294912\omega_0^4}[a^7\gamma^2(-5280\cos(3\psi) + 160\cos(5\psi) + 95\cos(7\psi)+$
$3\cos(9\psi)) + 72a^5\beta\gamma(-164\cos(3\psi) + 4\cos(5\psi) + \cos(7\psi))+$
$32a^3\left(-189\beta^2\cos(3\psi) + 9\beta^2\cos(5\psi) + 40\gamma\varepsilon\omega_0(27\sin(3\psi) + \sin(5\psi))\right)+$
$12288a^2\gamma f(t)(20\cos(2\psi) + \cos(4\psi) - 45) + 13824a\beta\varepsilon\omega_0\sin(3\psi) + 147456\beta f(t)(\cos(2\psi) - 3)]$.

The odes for determining $a = a(t)$ and $\psi = \psi(t)$ are

$$\dot{a} = -a\varepsilon + \frac{3\beta\varepsilon}{8\omega_0^2}a^3 + \frac{5\gamma\varepsilon}{8\omega_0^2}a^5.$$

$$\dot{\psi} = \omega_0 - \frac{\varepsilon^2}{2\omega_0} + \frac{3\beta}{8\omega_0}a^2 + \left(\frac{5\gamma}{16\omega_0} - \frac{15\beta^2}{256\omega_0^3}\right)a^4 - \frac{5\beta\gamma}{64\omega_0^3}a^6 - \frac{55\gamma^2}{3072\omega_0^3}a^8.$$

First order approximation:

$x(t) = a\cos(\psi) + \frac{f(t)}{\omega_0^2} + \left(\frac{5a^5\gamma}{128\omega_0^2} + \frac{a^3\beta}{32\omega_0^2}\right)\cos(3\psi) + \frac{a^5\gamma}{384\omega_0^2}\cos(5\psi).$

Second order approximation:

$x(t) = a\cos(\psi) + \frac{f(t)}{\omega_0^2} - \frac{15a^4\gamma f(t)}{8\omega_0^4} - \frac{3a^2\beta f(t)}{2\omega_0^4} +$
$\frac{a^2 f(t)\left(5a^2\gamma+3\beta\right)}{6\omega_0^4}\cos(2\psi)+$
$\frac{3a^3\varepsilon\left(5a^2\gamma+2\beta\right)}{128\omega_0^3}\sin(3\psi) + \frac{\left(24\omega_0^2\left(5a^5\gamma+4a^3\beta\right)-a^5\left(55a^4\gamma^2+123a^2\beta\gamma+63\beta^2\right)\right)}{3072\omega_0^4}\cos(3\psi)+$
$\frac{a^4\gamma f(t)}{24\omega_0^4}\cos(4\psi)+$
$\frac{5a^5\gamma\varepsilon}{1152\omega_0^3}\sin(5\psi)+\frac{a^5}{9216\omega_0^4}\left(5a^4\gamma^2 + 9a^2\beta\gamma + 9\beta^2 + 24\gamma\omega_0^2\right)\cos(5\psi)+\frac{a^7\gamma\left(95a^2\gamma+72\beta\right)}{294912\omega_0^4}\cos(7\psi)+\frac{a^9\gamma^2}{983040\omega_0^4}\cos(9\psi).$

. The ode for finding $a(t)$ is difficult to solve in closed form. We may obatin approximate solution to it as follows. We have

$-a\varepsilon + \frac{3\beta\varepsilon}{8\omega_0^2}a^3 + \frac{5\gamma\varepsilon}{8\omega_0^2}a^5 \approx -a\rho + \nu a^3,$

where

$\rho = -\varepsilon\left(\frac{5\gamma M^4}{8\omega_0^2}\sin^2\left(\frac{\pi}{8}\right) + 1\right),\ \nu = \frac{\varepsilon}{8\omega_0^2}\left(3\beta + 5\gamma M^2\left(1 + \sin^2\left(\frac{\pi}{8}\right)\right)\right)$

and

$M = \sqrt{\frac{-3\beta+\sqrt{9\beta^2+160\gamma\omega_0^2}}{10\gamma}}.$

Then

$$a(t) \approx \frac{c_0\sqrt{\rho}}{\sqrt{c_0^2\nu + e^{2\rho t}\left(\rho - c_0^2\nu\right)}}.$$

The solution for $\psi = \psi(t)$ is easily obtained taking into account that

$\int_0^t a^2(\tau)d\tau = \frac{1}{2\nu}\left[2\rho t - \log\left(\frac{c_0^2\nu + e^{2\rho t}\left(\rho-c_0^2\nu\right)}{\rho}\right)\right].$

$\int_0^t a^4(\tau)d\tau = \frac{\rho}{2\nu^2}\left[c_0^2\nu\left(\frac{1}{c_0^2\nu+e^{2\rho t}\left(\rho-c_0^2\nu\right)} - \frac{1}{\rho}\right) - \log\left(\frac{c_0^2\nu+e^{2\rho t}\left(\rho-c_0^2\nu\right)}{\rho}\right) + 2\rho t\right].$

$\int_0^t a^6(\tau)d\tau = \frac{\rho^2}{4\nu^3}[c_0^2\nu\left(c_0^2\nu\left(\frac{1}{\left(c_0^2\nu+e^{2\rho t}\left(\rho-c_0^2\nu\right)\right)^2} - \frac{1}{\rho^2}\right) + \frac{2}{c_0^2\nu+e^{2\rho t}\left(\rho-c_0^2\nu\right)} - \frac{2}{\rho}\right) - 2\log\left(\frac{c_0^2\nu+e^{2\rho t}\left(\rho-c_0^2\nu\right)}{\rho}\right) + 4\rho t].$

$\int_0^t a^8(\tau)d\tau = \frac{\rho^3}{12\nu^4}[c_0^2\nu(2c_0^4\nu^2\left(\frac{1}{\left(c_0^2\nu+e^{2\rho t}\left(\rho-c_0^2\nu\right)\right)^3} - \frac{1}{\rho^3}\right) + 3c_0^2\nu\left(\frac{1}{\left(c_0^2\nu+e^{2\rho t}\left(\rho-c_0^2\nu\right)\right)^2} - \frac{1}{\rho^2}\right) + \frac{6}{c_0^2\nu+e^{2\rho t}\left(\rho-c_0^2\nu\right)} - \frac{6}{\rho}) - 6\log\left(\frac{c_0^2\nu+e^{2\rho t}\left(\rho-c_0^2\nu\right)}{\rho}\right) + 12\rho t]$

Then

$\psi(t) = \left(\omega_0 - \frac{\varepsilon^2}{2\omega_0}\right)t + c_1 + \frac{3\beta}{8\omega_0}\int_0^t a^2(\tau)d\tau + \left(\frac{5\gamma}{16\omega_0} - \frac{15\beta^2}{256\omega_0^3}\right)\int_0^t a^4(\tau)d\tau - \frac{5\beta\gamma}{64\omega_0^3}\int_0^t a^6(\tau)d\tau - \frac{55\gamma^2}{3072\omega_0^3}\int_0^t a^8(\tau)d\tau.$

The constants c_0 and c_1 are determined fron the initial conditions.

In the particular case when $\varepsilon = 0$ the solution is given by

$x(t) = c_0\cos(\psi)+\left(\frac{\beta f(t)}{2\omega_0^4}c_0^2 + \frac{5\gamma f(t)}{6\omega_0^4}c_0^4\right)\cos(2\psi)+\left(-\frac{21\beta^2 c_0^5}{1024\omega_0^4} - \frac{41\beta\gamma c_0^7}{1024\omega_0^4} + \frac{\beta c_0^3}{32\omega_0^2} - \frac{55\gamma^2 c_0^9}{3072\omega_0^4} + \frac{5\gamma c_0^5}{128\omega_0^2}\right)\cos(3\psi)+$
$\frac{\gamma f(t)}{24\omega_0^4}c_0^4\cos(4\psi) + \left(\frac{\beta^2 c_0^5}{1024\omega_0^4} + \frac{\beta\gamma c_0^7}{1024\omega_0^4} + \frac{5\gamma^2 c_0^9}{9216\omega_0^4} + \frac{\gamma c_0^5}{384\omega_0^2}\right)\cos(5\psi) + \left(\frac{\beta\gamma c_0^7}{4096\omega_0^4} + \frac{95\gamma^2 c_0^9}{294912\omega_0^4}\right)\cos(7\psi)+$

$\frac{\gamma^2 c_0^9}{98304\omega_0^4}\cos(9\psi) - \frac{3\beta f(t)}{2\omega_0^4}c_0^2 - \frac{15\gamma f(t)}{8\omega_0^4}c_0^4 + \frac{f(t)}{\omega_0^2}.$

$\psi(t) = \left(c_0^4\left(\frac{5\gamma}{16\omega_0} - \frac{15\beta^2}{256\omega_0^3}\right) - \frac{5\beta\gamma c_0^6}{64\omega_0^3} + \frac{3\beta c_0^2}{8\omega_0} - \frac{55\gamma^2 c_0^8}{3072\omega_0^3} + \omega_0\right)t + c_1$

If $f(t) = \varepsilon = 0$ we get $a(t) \equiv A$ and then we obtain the following frequency-amplitude formulation :

$$\omega = \omega_0 + \frac{3\beta}{8\omega_0}A^2 + \left(\frac{5\gamma}{16\omega_0} - \frac{15\beta^2}{256\omega_0^3}\right)A^4 - \frac{5\beta\gamma}{64\omega_0^3}A^6 - \frac{55\gamma^2}{3072\omega_0^3}A^8.$$

5.3.4 Duffing–Helmholtz oscillator with damping and forcing

Let us consider the i.v.p.

$$\ddot{x} + 2\varepsilon\dot{x} + q_0 + q_1 x + q_2 x^2 + q_3 x^3 = f(t),\ x(0) = x_0 \text{ and } x'(0) = \dot{x}_0.$$

Define

$$x(t) = d + u(t),$$

where

$$q_0 + q_1 d + q_2 d^2 + q_3 d^3 = 0.$$

Then

$$\ddot{u} + 2\varepsilon\dot{u} + \left(3q_3 d^2 + 2q_2 d + q_1\right)u + (3q_3 d + q_2)\,u^2 + q_3 u^3 = f(t),\ u(0) = x_0 - d \text{ and } u'(0) = \dot{x}_0.$$

The homotopy to be considered is

$$H_p = \ddot{u} + \omega_0^2 u + p\left[2\varepsilon\dot{u} + ru^2 + su^3 - f(t)\right],$$

where

$$\omega_0^2 = 3q_3 d^2 + 2q_2 d + q_1 \text{ and } r = 3a_3 d + a_2 \text{ and } s = a_3$$

We assume the solution in the ansatz form

$$u(t) = a(t)\cos\psi(t) + pv_1(a(t), \psi(t)) + p^2 v_2(a(t), \psi(t)).$$

$$\dot{a} = pA_1(a) + p^2 A_2(a) \text{ and } \dot{\psi} = \omega_0 + p\varphi_1(a) + p^2\varphi_2(a).$$

We have :

$H_p = p[\frac{1}{4}\left(4\omega_0^2 v_1(a,\psi) + 4\omega_0^2 {v_1}^{(0,2)}(a,\psi) + 2a^2 r\cos(2\psi) + 2a^2 r - 4f(t)\right) +$
$\frac{1}{4}\cos(\psi)\left(3a^3 s - 8a\varphi_1\omega_0\right) + \frac{1}{4}a^3 s\cos(3\psi) + \frac{1}{4}\sin(\psi)\left(-8a\varepsilon\omega_0 - 8A_1\omega_0\right)]+$
$p^2[\frac{1}{2}(3a^2 s v_1(a,\psi) + 3a^2 s\cos(2\psi)v_1(a,\psi) + 4A_1\omega_0 {v_1}^{(1,1)}(a,\psi) + 4\varepsilon\omega_0 {v_1}^{(0,1)}(a,\psi)+$
$4\varphi_1\omega_0 {v_1}^{(0,2)}(a,\psi) + 2\omega_0^2 v_2(a,\psi) + 2\omega_0^2 {v_2}^{(0,2)}(a,\psi))+$
$\frac{1}{2}\cos(\psi)\left(4arv_1(a,\psi) - 4a\varphi_2\omega_0 - 2a\varphi_1^2 + 4A_1\varepsilon + 2A_1\dot{A}_1\right) +$
$\frac{1}{2}\sin(\psi)\left(-2aA_1\dot{\varphi}_1 - 4a\varepsilon\varphi_1 - 4A_1\varphi_1 - 4A_2\omega_0\right)].$

We equate to zero the coefficients of p^j,cos ψ and sin ψ to zero to obtain a system of algebraic-differential equations. We solve this system so that we did no obtain secularity terms. The solution is given by

$A_1(a) = -a\varepsilon,\ A_2(a) = \frac{3a^3\varepsilon s}{8\omega_0^2}.$

$\varphi_1(a) = \frac{3a^2 s}{8\omega_0},\ \varphi_2(a) = \frac{-45a^4s^2-320a^2r^2-384\varepsilon^2\omega_0^2+768rf(t)}{768\omega_0^3}.$

$v_1(a,\psi) = \frac{3a^3 s\cos(3\psi)+16a^2r(\cos(2\psi)-3)+96f(t)}{96\omega_0^2}.$

$v_2(a,\psi) = \frac{a^2}{9216\omega_0^4}[9a^3s^2(\cos(5\psi) - 21\cos(3\psi)) + 96a^2rs(-31\cos(2\psi) + \cos(4\psi) + 60)+$
$48a\left(4r^2\cos(3\psi) + 9\varepsilon s\omega_0\sin(3\psi)\right) + 512\left(9sf(t)(\cos(2\psi) - 3) + 4\varepsilon r\omega_0\sin(2\psi)\right)]$The odes for $a = a(t)$ and $\psi = \psi(t)$ are

$$\dot{a} = -\varepsilon a + \frac{3\varepsilon s}{8\omega_0^2}a^3 \ \text{ and } \dot{\psi} = \omega_0 - \frac{\varepsilon^2}{2\omega_0} + \left(\frac{3s}{8\omega_0} - \frac{5r^2}{12\omega_0^3}\right)a^2 + \frac{15s^2}{256\omega_0^3}a^4 + \frac{rf(t)}{\omega_0^3}.$$

First order approximation

$x(t) = a\cos(\psi) + \frac{a^2r}{6\omega_0^2}\cos(2\psi) + \frac{a^3s}{32\omega_0^2}\cos(3\psi) + d - \frac{a^2r}{2\omega_0^2} + \frac{f(t)}{\omega_0^2}.$

Second order approximation:

$x(t) = a\cos(\psi)+$
$\frac{2a^2\varepsilon r}{9\omega_0^3}\sin(2\psi) + \left(-\frac{31a^4rs}{96\omega_0^4} + \frac{a^2sf(t)}{2\omega_0^4} + \frac{a^2r}{6\omega_0^2}\right)\cos(2\psi)+$
$\frac{3a^3\varepsilon s}{64\omega_0^3}\sin(3\psi) + \left(-\frac{21a^5s^2}{1024\omega_0^4} + \frac{a^3r^2}{48\omega_0^4} + \frac{a^3s}{32\omega_0^2}\right)\cos(3\psi)+$
$\frac{a^4rs}{96\omega_0^4}\cos(4\psi) + \frac{a^5s^2}{1024\omega_0^4}\cos(5\psi) + d + \frac{5a^4rs}{8\omega_0^4} - \frac{a^2r}{2\omega_0^2} + \left(\frac{1}{\omega_0^2} - \frac{3a^2s}{2\omega_0^4}\right)f(t).$

$a = a(t) = \frac{2\sqrt{2}c_0\omega_0}{\sqrt{3c_0^2s(1-e^{2\varepsilon t})+8\omega_0^2e^{2\varepsilon t}}}.$

$\psi = \psi(t) = \left(\omega_0 - \frac{\varepsilon^2}{2\omega_0}\right)t + c_1 + \left(\frac{3s}{8\omega_0} - \frac{5r^2}{12\omega_0^3}\right)\int_0^t a^2(\tau)d\tau + \frac{15s^2}{256\omega_0^3}\int_0^t a^4(\tau)d\tau + \frac{r}{\omega_0^3}\int_0^t f(\tau)d\tau.$

$..\int_0^t a^2(\tau)d\tau = \frac{4\omega_0^2}{3\varepsilon s}\left(2\varepsilon t - \log\left(-\frac{3c_0^2s\left(e^{2\varepsilon t}-1\right)}{8\omega_0^2} - e^{2\varepsilon t}\right)\right).$

$\int_0^t a^4(\tau)d\tau = \frac{4\omega_0^4}{9\varepsilon s^2}\left(3c_0^2s\left(-\frac{8}{3c_0^2s(e^{2\varepsilon t}-1)-8\omega_0^2e^{2\varepsilon t}} - \frac{1}{\omega_0^2}\right) - 8\log\left(e^{2\varepsilon t} - \frac{3c_0^2s\left(e^{2\varepsilon t}-1\right)}{8\omega_0^2}\right) + 16\varepsilon t\right)$

The constants c_0 and c_1 are obtained from the initial conditions.

5.4 The multiple time scales method (MSM)

This is another useful method for solving either conservative or nonconservative oscillators. Suppose we are given that

$$\ddot{x} + \omega_0^2 x = F(x, \dot{x}, t)$$

We consider the homotopy

$$H_p = \ddot{x} + \omega_0^2 x - pF(x, \dot{x}, t)$$

and the solution is assumed in tha ansatz form

$$x(t) = a(pt)\cos(\omega_0 t + \phi(pt)) + pv_1(t, pt) : \text{First order approximation}$$

or

$$x(t) = a(pt, p^2t)\cos(\omega_0 t + \phi(pt, p^2t)) + pv_1(t, pt, p^2t) + p^2 v_2(t, pt, p^2t) : \text{Second order approximation}$$

Following rules may be useful for computing .

For ther first order approximation:

$\dot{x} = -\omega_0 a(\tau)\sin(\theta)$.

$\ddot{x} = pv_1^{(2,0)}(t,\tau) - 2p\omega_0\sin(\theta)a'(\tau) + \cos(\theta)\left(\omega_0^2(-a(\tau)) - 2p\omega_0 a(\tau)\phi'(\tau)\right)$.

$x^2 = \frac{1}{2}\left(a(\tau)^2\cos(2\theta) + a(\tau)^2\right)$.

$x^3 = \frac{3}{4}a(\tau)^3\cos(\theta) + \frac{1}{4}a(\tau)^3\cos(3\theta)$.

$x^4 = \frac{1}{8}\left(4a(\tau)^4\cos(2\theta) + a(\tau)^4\cos(4\theta) + 3a(\tau)^4\right)$.

$x^5 = \frac{5}{8}a(\tau)^5\cos(\theta) + \frac{1}{16}\left(5a(\tau)^5\cos(3\theta) + a(\tau)^5\cos(5\theta)\right)$.

$x\dot{x} = -\frac{1}{2}\omega_0 a(\tau)^2\sin(2\theta)$.

$x\dot{x}^2 = \frac{1}{4}\omega_0^2 a(\tau)^3\cos(\theta) - \frac{1}{4}\omega_0^2 a(\tau)^3\cos(3\theta)$.

Here, $\theta = \omega_0 t + \phi(\tau)$, $\tau = pt$.

For the second order approximation :

$\dot{x} = pv_1^{(1,0,0)}(t,\tau,\eta) + p\cos(\theta)a^{(1,0)}(\tau,\eta) + \sin(\theta)\left(\omega_0(-a(\tau,\eta)) - pa(\tau,\eta)\phi^{(1,0)}(\tau,\eta)\right)$.

$\ddot{x} = 2p^2 v_1^{(1,1,0)}(t,\tau,\eta) + p^2 v_2^{(2,0,0)}(t,\tau,\eta) + pv_1^{(2,0,0)}(t,\tau,\eta) +$
$[-2p^2\omega_0 a^{(0,1)}(\tau,\eta) - 2p^2 a^{(1,0)}(\tau,\eta)\phi^{(1,0)}(\tau,\eta) - 2p\omega_0 a^{(1,0)}(\tau,\eta) -$
$p^2 a(\tau,\eta)\phi^{(2,0)}(\tau,\eta)]\sin(\theta) +$
$[p^2 a^{(2,0)}(\tau,\eta) - \omega_0^2 a(\tau,\eta) - 2p^2\omega_0 a(\tau,\eta)\phi^{(0,1)}(\tau,\eta) +$
$p^2(-a(\tau,\eta))\phi^{(1,0)}(\tau,\eta)^2 - 2p\omega_0 a(\tau,\eta)\phi^{(1,0)}(\tau,\eta)]\cos(\theta)$.

$x^2 = 2p\cos(\theta)a(\tau,\eta)v_1(t,\tau,\eta) + \frac{1}{2}\left(\cos(2\theta)a(\tau,\eta)^2 + a(\tau,\eta)^2\right)$.

$x^3 = \frac{1}{4}\left(6p\cos(2\theta)a(\tau,\eta)^2 v_1(t,\tau,\eta) + 6pa(\tau,\eta)^2 v_1(t,\tau,\eta) + \cos(3\theta)a(\tau,\eta)^3\right) +$
$\frac{3}{4}\cos(\theta)a(\tau,\eta)^3$.

$x^4 = 3p\cos(\theta)a(\tau,\eta)^3 v_1(t,\tau,\eta) + \frac{1}{8}(8p\cos(3\theta)a(\tau,\eta)^3 v_1(t,\tau,\eta) +$
$4\cos(2\theta)a(\tau,\eta)^4 + \cos(4\theta)a(\tau,\eta)^4 + 3a(\tau,\eta)^4)$.

$x^5 = \frac{1}{16}[40p\cos(2\theta)a(\tau,\eta)^4 v_1(t,\tau,\eta) + 10p\cos(4\theta)a(\tau,\eta)^4 v_1(t,\tau,\eta) +$
$30pa(\tau,\eta)^4 v_1(t,\tau,\eta) + 5\cos(3\theta)a(\tau,\eta)^5 +$
$\cos(5\theta)a(\tau,\eta)^5 + \frac{5}{8}\cos(\theta)a(\tau,\eta)^5]$.

$\dot{x} = -p\omega_0\sin(\theta)a(\tau,\eta)v_1(t,\tau,\eta) + p\cos(\theta)a(\tau,\eta)v_1^{(1,0,0)}(t,\tau,\eta) +$
$\frac{1}{2}(p\cos(2\theta)a^{(1,0)}(\tau,\eta)a(\tau,\eta) + pa^{(1,0)}(\tau,\eta)a(\tau,\eta) -$
$\omega_0 a(\tau,\eta)^2\sin(2\theta) - p\sin(2\theta)a(\tau,\eta)^2\phi^{(1,0)}(\tau,\eta))$.

$x\dot{x}^2 = \frac{1}{4}[-4p\omega_0\sin(2\theta)a(\tau,\eta)^2 v_1^{(1,0,0)}(t,\tau,\eta) - 2p\omega_0^2\cos(2\theta)a(\tau,\eta)^2 v_1(t,\tau,\eta) +$
$2p\omega_0^2 a(\tau,\eta)^2 v_1(t,\tau,\eta) - 2p\omega_0\sin(3\theta)a^{(1,0)}(\tau,\eta)a(\tau,\eta)^2 -$
$\omega_0^2\cos(3\theta)a(\tau,\eta)^3 - 2p\omega_0\cos(3\theta)a(\tau,\eta)^3\phi^{(1,0)}(\tau,\eta) - \frac{1}{2}p\omega_0\sin(\theta)a^{(1,0)}(\tau,\eta)a(\tau,\eta)^2 +$
$\frac{1}{4}\cos(\theta)\left(\omega_0^2 a(\tau,\eta)^3 + 2p\omega_0 a(\tau,\eta)^3\phi^{(1,0)}(\tau,\eta)\right)]$.

Here, $\theta = \omega_0 t + \phi(\tau,\eta)$, $\tau = pt$, $\eta = p^2 t$.

Let us see how to apply this method,.

5.4.1 Unforced and undamped Duffing equation

Let

$$\ddot{x}+\omega_0^2x+\beta x^3=0,\ x(0)=x_0 \text{ and } x'(0)=\dot{x}_0.$$

Define

$$H_p=\ddot{x}+\omega_0^2x+p\beta x^3$$

and assume that

$$x(t)=a(pt,p^2t)\cos(\omega_0t+\phi(pt,p^2t))+pv_1(t,pt,p^2t)+p^2v_2(t,pt,p^2t).$$

Then

$H_p=p[-2\omega_0\sin(\theta)a^{(1,0)}(\tau,\eta)+\frac{1}{4}\cos(\theta)\left(3\beta a(\tau,\eta)^3-8\omega_0a(\tau,\eta)\phi^{(1,0)}(\tau,\eta)\right)+$
$\frac{1}{4}\left(\beta\cos(3\theta)a(\tau,\eta)^3+4\omega_0^2v_1(t,\tau,\eta)+4{v_1}^{(2,0,0)}(t,\tau,\eta)\right)]+$
$p^2[\frac{1}{2}\sin(\theta)\left(-4\omega_0a^{(0,1)}(\tau,\eta)-4a^{(1,0)}(\tau,\eta)\phi^{(1,0)}(\tau,\eta)-2a(\tau,\eta)\phi^{(2,0)}(\tau,\eta)\right)+$
$\frac{1}{2}\cos(\theta)\left(2a^{(2,0)}(\tau,\eta)-4\omega_0a(\tau,\eta)\phi^{(0,1)}(\tau,\eta)-2a(\tau,\eta)\phi^{(1,0)}(\tau,\eta)^2\right)+$
$\frac{1}{2}(3\beta\cos(2\theta)a(\tau,\eta)^2v_1(t,\tau,\eta)+3\beta a(\tau,\eta)^2v_1(t,\tau,\eta)+2\omega_0^2v_2(t,\tau,\eta)+$
$4{v_1}^{(1,1,0)}(t,\tau,\eta)+2{v_2}^{(2,0,0)}(t,\tau,\eta))]+o(p^3)$.

Now, we equate to zero the coefficients of p^j, $\cos\theta$ and $\sin\theta$ to obtain a system of pdes. Solving it gives

$$a(t,t)=c_0,\ \phi(t,t)=\frac{1}{8\omega_0}\left[3\beta c_0^2t+8c_1\omega_0\right].$$

$$v_1(t,t,t)=\frac{\beta c_0^3}{32\omega_0^2}\cos\left(\frac{3\left(3\beta c_0^2t+8c_1\omega_0\right)}{8\omega_0}+3\omega_0t\right).$$

$$v_2(t,t,t)=\frac{1}{1024\omega_0^4}\left[\beta^2c_0^5\cos\left(\frac{5\left(3\beta c_0^2t+8c_1\omega_0\right)}{8\omega_0}+5\omega_0t\right)-21\beta^2c_0^5\cos\left(\frac{3\left(3\beta c_0^2t+8c_1\omega_0\right)}{8\omega_0}+3\omega_0t\right)\right].$$

The solution is

$$x(t)=c_0\cos\left(\frac{3\beta c_0^2t+8c_1\omega_0}{8\omega_0}+\omega_0t\right)$$

$$+\frac{1}{1024\omega_0^4}\left[\beta^2c_0^5\cos\left(\frac{5\left(3\beta c_0^2t+8c_1\omega_0\right)}{8\omega_0}+5\omega_0t\right)-21\beta^2c_0^5\cos\left(\frac{3\left(3\beta c_0^2t+8c_1\omega_0\right)}{8\omega_0}+3\omega_0t\right)\right]$$

$$+\frac{\beta c_0^3}{32\omega_0^2}\cos\left(\frac{3\left(3\beta c_0^2t+8c_1\omega_0\right)}{8\omega_0}+3\omega_0t\right).$$

The constants c_0 and c_1 are obtained from the initial conditions.

Example 14. Let

$$\ddot{x}+x+2x^3=0,\ x(0)=0.2 \text{ and } x'(0)=0.3.$$

Then

$$x_{\text{approx}}(t) = 0.0000178602\cos(4.67823 - 5.43459t) + 0.00209073\cos(2.80694 - 3.26076t) + 0.340428\cos(0.935647 - 1.08692t)$$

See Figure 5.6.

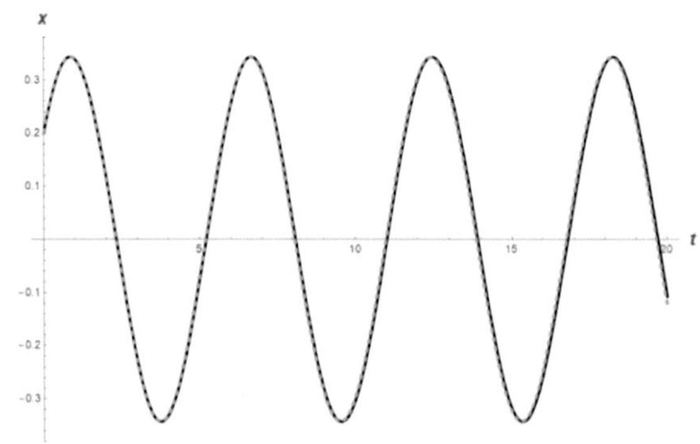

Figure 5.6

5.4.2 The oscillator $\ddot{x} + \omega_0 x + \beta x^2 = 0$

We will solve the following i.v.p.

$$\ddot{x} + \omega_0 x + \beta x^2 = 0,\ x(0) = x_0 \text{ and } x'(0) = \dot{x}_0$$

The homotopy :

$$H_p = \ddot{x} + \omega_0 x + p\beta x^2$$

Assume that

$$x(t) = a(pt, p^2 t)\cos(\omega_0 t + \phi(pt, p^2 t)) + pv_1(t, pt, p^2 t) + p^2 v_2(t, pt, p^2 t).$$

It is easy to see that

$$v_1(t, \tau, \eta) = \frac{\beta c_0^2 \cos(2\omega t_0 + 2\phi) - 3\beta c_0^2}{6\omega_0^2},\ v_2(t, \tau, \eta) = 0,\ \phi(\tau, \eta) = c_1 \text{ and } a(\tau, \eta) = c_0.$$

The solution is given by

$$x(t) = c_0 \cos(c_1 + \omega_0 t) + \frac{1}{6\omega_0^2}\left[\beta c_0^2 \cos(2c_1 + 2\omega_0 t) - 3\beta c_0^2\right]$$

The constants c_0 and c_1 are found from the initila conditions.

5.4.3 The undamped and unforced cubic-quintic Duffing equation

Let us consider the i.v.p.

$$\ddot{x} + \omega_0^2 x + \beta x^3 + \gamma x^5 = 0,\ x(0) = x_0 \text{ and } x'(0) = \dot{x}_0.$$

We consider the homotopy

$$H_p = \ddot{x} + \omega_0^2 x + p\left[\beta x^3 + \gamma x^5\right].$$

Assume that

$$x(t) = a(pt, p^2 t)\cos(\omega_0 t + \phi(pt, p^2 t)) + p v_1(t, pt, p^2 t) + p^2 v_2(t, pt, p^2 t).$$

Then

$H_p = p[-2\omega_0 \sin(\theta) a^{(1,0)}(\tau,\eta) + \frac{1}{4}\cos(\theta)\left(3\beta a(\tau,\eta)^3 - 8\omega_0 a(\tau,\eta)\phi^{(1,0)}(\tau,\eta)\right) +$
$\frac{1}{4}\left(\beta\cos(3\theta)a(\tau,\eta)^3 + 4\omega_0^2 v_1(t,\tau,\eta) + 4{v_1}^{(2,0,0)}(t,\tau,\eta)\right)]+$
$p^2[\frac{1}{2}\sin(\theta)\left(-4\omega_0 a^{(0,1)}(\tau,\eta) - 4a^{(1,0)}(\tau,\eta)\phi^{(1,0)}(\tau,\eta) - 2a(\tau,\eta)\phi^{(2,0)}(\tau,\eta)\right) +$
$\frac{1}{2}\cos(\theta)\left(2a^{(2,0)}(\tau,\eta) - 4\omega_0 a(\tau,\eta)\phi^{(0,1)}(\tau,\eta) - 2a(\tau,\eta)\phi^{(1,0)}(\tau,\eta)^2\right)$
$+\frac{1}{2}(3\beta\cos(2\theta)a(\tau,\eta)^2 v_1(t,\tau,\eta) + 3\beta a(\tau,\eta)^2 v_1(t,\tau,\eta) + 2\omega_0^2 v_2(t,\tau,\eta)+$
$4{v_1}^{(1,1,0)}(t,\tau,\eta) + 2{v_2}^{(2,0,0)}(t,\tau,\eta))]$

Now, we equate to zero the coefficients of p^j, $\cos\theta$ and $\sin\theta$ to obtain a system of pdes. Solving it gives

$a(t,t) = c_0$, $\phi(t,t) = \frac{1}{8\omega_0}\left[3\beta c_0^2 t + 8c_1\omega_0\right]$.

$v_1(t,t,t) = \frac{\beta c_0^3}{32\omega_0^2}\cos\left(\frac{3(3\beta c_0^2 t + 8c_1\omega_0)}{8\omega_0} + 3\omega_0 t\right)$.

$v_2(t,t,t) = \frac{1}{1024\omega_0^4}\left[\beta^2 c_0^5\cos\left(\frac{5(3\beta c_0^2 t + 8c_1\omega_0)}{8\omega_0} + 5\omega_0 t\right) - 21\beta^2 c_0^5\cos\left(\frac{3(3\beta c_0^2 t + 8c_1\omega_0)}{8\omega_0} + 3\omega_0 t\right)\right]$.

The solution is

$x(t) = c_0\cos\left(\frac{3\beta c_0^2 t + 8c_1\omega_0}{8\omega_0} + \omega_0 t\right) +$
$\frac{1}{1024\omega_0^4}\left[\beta^2 c_0^5\cos\left(\frac{5(3\beta c_0^2 t + 8c_1\omega_0)}{8\omega_0} + 5\omega_0 t\right) - 21\beta^2 c_0^5\cos\left(\frac{3(3\beta c_0^2 t + 8c_1\omega_0)}{8\omega_0} + 3\omega_0 t\right)\right] +$
$\frac{\beta c_0^3}{32\omega_0^2}\cos\left(\frac{3(3\beta c_0^2 t + 8c_1\omega_0)}{8\omega_0} + 3\omega_0 t\right)$.

The constants c_0 and c_1 are obtained from the initial conditions.

Example 15 . Let

$$\ddot{x} + x - x^3 + x^5 = 0,\ x(0) = 0.5 \text{ and } x'(0) = 0.$$

The approximate solutuon is

$$x_{\text{approx}}(t) = 0.502534\cos(0.925228t) - 0.00262719\cos(2.77568t) + 0.0000949649\cos(4.62614t)$$
$$-1.32e^{-6}\cos(6.47659t) + 2.1e^{-8}\cos(8.32705t)$$

See Figure 5.7 for a comparison with the exact solution.

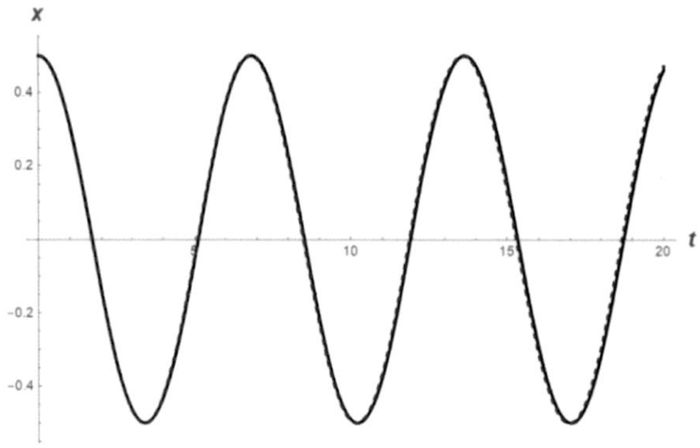

Figure 5.7

5.4.4 Damped and forced Duffing equation

We are given that

$$\ddot{x} + 2\varepsilon\dot{x} + \omega_0^2 x + \beta x^3 = f(t),\ x(0) = x_0 \text{ and } x'(0) = \dot{x}_0.$$

It is not difficult to see that the first order approximation is given by

$x(t) = c_0 \exp(-\varepsilon t)\cos\left(\frac{3\beta c_0^2}{8\varepsilon\omega_0}\exp(-\varepsilon t)\sinh(\varepsilon t) + c_1 + \omega t_0\right) + \frac{\beta c_0^3}{32\omega_0^2}e^{-3\varepsilon t}\cos\left(3\omega t_0 + 3\phi\right) + \frac{1}{\omega_0}\int_0^t f(\zeta)\sin\left(\omega_0(t-\zeta)\right)d\zeta,$

where $\phi = \phi(t) = \frac{3\beta c_0^2}{8\varepsilon\omega_0}\exp(-\varepsilon t)\sinh(\varepsilon t) + c_1$.

Our aim is to obtain the second order approximation. To this end, we will take

$$f(t) = \gamma\cos\omega t$$

and we consider the homotopy

$$H_p = \ddot{x} + \omega_0^2 x + p\left[2\varepsilon\dot{x} + \beta x^3 - f(t)\right].$$

Then

$H_p = p[\frac{1}{4}\sin(\theta)\left(-8\omega_0 a^{(1,0)}(\tau,\eta) - 8a\varepsilon\omega_0\right) + \frac{1}{4}\cos(\theta)\left(3a^3\beta - 8a\omega_0\phi^{(1,0)}(\tau,\eta)\right) + \frac{1}{4}\left(4\omega_0^2 v_1(t,\tau,\eta) + 4{v_1}^{(2,0,0)}(t,\tau,\eta) + a^3\beta\cos(3\theta) - 4\gamma\cos(\omega t)\right)] + p^2[\frac{1}{2}\sin(\theta)\left(-4\omega_0 a^{(0,1)}(\tau,\eta) - 4a^{(1,0)}(\tau,\eta)\phi^{(1,0)}(\tau,\eta) - 4a\varepsilon\phi^{(1,0)}(\tau,\eta) - 2a\phi^{(2,0)}(\tau,\eta)\right) + \frac{1}{2}\cos(\theta)\left(4\varepsilon a^{(1,0)}(\tau,\eta) + 2a^{(2,0)}(\tau,\eta) - 4a\omega_0\phi^{(0,1)}(\tau,\eta) - 2a\phi^{(1,0)}(\tau,\eta)^2\right) + \frac{1}{2}(3a^2\beta\cos(2\theta)v_1(t,\tau,\eta) + 3a^2\beta v_1(t,\tau,\eta) + 4\varepsilon{v_1}^{(1,0,0)}(t,\tau,\eta) + 2\omega_0^2 v_2(t,\tau,\eta) + 4{v_1}^{(1,1,0)}(t,\tau,\eta) + 2{v_2}^{(2,0,0)}(t,\tau,\eta))] + o(p^3).$

Now, we equate to zero the coefficients of p^j, $\cos\theta$ and $\sin\theta$ to obtain a system of pdes. Solving it gives

$v_1(t,\tau,\eta) = -\frac{1}{4\omega_0^2(\omega^2-\omega_0^2)}\left[\beta\left(\omega^2-\omega_0^2\right)a(\tau,\eta)^3\cos\left(3\left(\omega t_0+\phi\right)\right) + 4\gamma\omega_0^2\cos(\omega t)\right].$

$v_2(t,\tau,\eta) = \frac{1}{128(\omega-\omega_0)^2\omega_0^4(\omega+\omega_0)^2\left(\omega^2-9\omega_0^2\right)}\times$

$[-24\beta\omega_0 a(\tau,\eta)^2(4\gamma\omega_0^3(2\left(\omega^2-9\omega_0^2\right)\cos(\omega t) + \left(\omega^2-4\omega_0\omega+3\omega_0^2\right)\cos\left(\omega t+2\omega t_0+2\phi\right) +$

$\left(\omega^2+4\omega_0\omega+3\omega_0^2\right)\cos\left(-\omega t+2\omega t_0+2\phi\right)|)-3\left(\omega^2-9\omega_0^2\right)\left(\omega^2-\omega_0^2\right){}^2 a^{(1,0)}(\tau,\eta)\sin\left(3\left(\omega t_0+\phi\right)\right))+$
$\beta^2\left(\omega^2-9\omega_0^2\right)\left(\omega^2-\omega_0^2\right){}^2 a(\tau,\eta)^5\left(-6\cos\left(3\left(\omega t_0+\phi\right)\right)-\cos\left(5\left(\omega t_0+\phi\right)\right)\right)+$
$24\beta\varepsilon\omega_0\left(\omega^2-9\omega_0^2\right)\left(\omega^2-\omega_0^2\right){}^2 a(\tau,\eta)^3\sin\left(3\left(\omega t_0+\phi\right)\right)+$
$256\gamma\varepsilon\omega\omega_0^4\left(\omega^2-9\omega_0^2\right)\sin(\omega t)].$
$a(\tau,\eta)=d_0(\eta)e^{-\varepsilon\tau},$
$\phi(\tau,\eta)=\frac{3\beta d_0(\eta)^2 e^{-\varepsilon\tau}\sinh(\varepsilon\tau)}{8\varepsilon\omega_0}+d_1(\eta).$
$d_0(\eta)=\frac{2\omega_0 e^{\frac{\varepsilon\tau}{2}}}{\sqrt{\frac{4\omega_0^2 e^{\varepsilon\tau}}{c_0^2}-3\beta\varepsilon\eta e^{-\varepsilon\tau}}}$, $d_1(\eta)=\frac{\eta\left(9\beta^2c_0^4\left(2-e^{-2\varepsilon\tau}\right)-48\beta c_0^2\varepsilon^3\eta+64\varepsilon^2\omega_0^2e^{2\varepsilon\tau}\right)}{32\left(3\beta c_0^2\varepsilon\eta\omega_0-4\omega_0^3e^{2\varepsilon\tau}\right)}+c_1.$

The constants c_0 and c_1 are found from the initial conditions.

Example 16. Let

$$\ddot{x}+0.25x^3+x+0.02\dot{x}=0.1\cos(0.1t)\wedge x(0)=0.\wedge x'(0)=0$$

The solution is

$$x(t)=a(t)\cos(t+\phi(t))+v_1(t)+v_2(t).$$

$$\phi(t)=3.17888+\frac{9.08\text{e-}7t-0.0476477}{1.e^{0.02t}-0.0000190591t}+\frac{9.53\text{e-}10t\left(e^{0.02t}(1.t-952.953)-52468.5e^{0.04t}+476.477\right)}{e^{0.04t}-0.0000190591e^{0.02t}t}.$$

$$a(t)=\frac{2e^{-0.005t}}{\sqrt{393.514e^{0.01t}-0.0075e^{-0.01t}t}}.$$

$$v_1=0.252525\left(0.4\cos(0.1t)-\frac{1.98e^{-0.015t}\cos(3(\phi(t)+t))}{(393.514e^{0.01t}-0.0075e^{-0.01t}t)^{3/2}}\right).$$

$$v_2=+0.000204061\sin(0.1t)$$

$$+a(t)^2\left(\begin{array}{c}0.140625a'^3(-0.00292969\cos(3(\phi(t)+t))-0.000488281\cos(5(\phi(t)+t)))\\ \\ +0.00725647\cos(2\phi(t)+1.9t)+0.00555407\cos(2\phi(t)+2.1t)-0.0382614\cos(0.1t)\end{array}\right).$$

$c_0=-0.100821$, $c_1=3.13124$

See Figure 5.8.

5.4.5 The quadratic oscillatlor $\ddot{x}+2\varepsilon\dot{x}+\omega_0^2x+\beta x^2=f(t)$

Let

$$\ddot{x}+2\varepsilon\dot{x}+\omega_0^2x+\beta x^2=f(t), x(0)=x_0 \text{ and } x'(0)=\dot{x}_0.$$

The homotopy reads

$$H_p=\ddot{x}+\omega_0^2x+p\left[2\varepsilon\dot{x}+\beta x^2-f(t)\right]$$

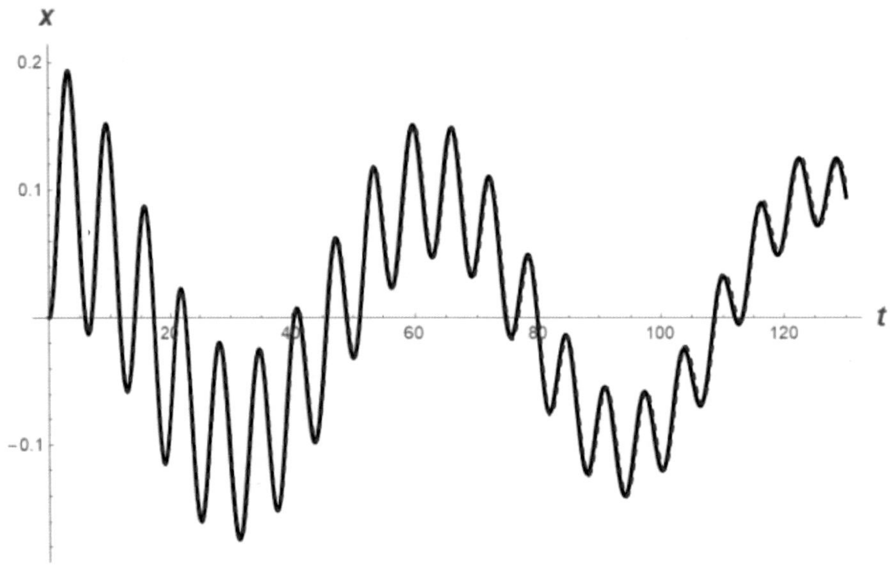

Figure 5.8

It is easy to see that the first order approximation is given by

$$x_{\text{first}}(t) = \frac{\beta c_0^2 e^{-2\varepsilon t}\left(\cos\left(2\left(c_1+\omega t_0\right)\right)-3\right)}{6\omega_0^2} + c_0 e^{-\varepsilon t}\cos\left(c_1+\omega t_0\right) + \frac{1}{\omega_0}\int_0^t f(\zeta)\sin\left(\omega_0(t-\zeta)\right)\,d\zeta.$$

For the choice $f(t)=\gamma\cos\omega t$ we have the following second order approximation:

$$x_{\text{second}}(t) = c_0 e^{-\varepsilon t}\cos\left(c_1+\omega t_0\right)$$

$$+\frac{\beta c_0^2}{18\omega_0^3} e^{-2\varepsilon t}\left(4\varepsilon\sin\left(2\left(c_1+\omega t_0\right)\right)+3\omega_0\left(\cos\left(2\left(c_1+\omega t_0\right)\right)-3\right)\right)$$

$$+\frac{1}{(\omega^2-\omega_0^2)^2}\left[2\gamma\varepsilon\omega\sin(\omega t)+\gamma\left(\omega_0^2-\omega^2\right)\cos(\omega t)\right].$$

The constants c_0 and c_1 are obtained from the initial conditions.

5.5 Summary

In Chapter 5 we studied perturbative solution methods for the (damped and/or forced) Duffing oscillator and related nonlinear systems. The main conclusions are the following:

1. **Systematic formulation of perturbative frameworks.** The chapter presented a systematic formulation of several classical perturbative schemes for weakly nonlinear oscillators, including the Lindstedt–Poincaré method, the Krylov–Bogoliubov–Mitropolsky (KBM) approach, the method of multiple scales and, when appropriate, harmonic balance. Each method was derived from first principles, with explicit ordering in terms of a small parameter and clear solvability conditions that remove secular terms.

2. **Amplitude–frequency relations for the Duffing oscillator.** Applying these methods to the Duffing equation led to explicit amplitude–frequency relations for weak nonlinearity, both in the undamped and weakly damped cases. The corrections to the linear natural frequency were obtained in closed form and expressed in terms of the physical parameters of the system. These formulas make the dependence of the period on the oscillation amplitude fully transparent and provide simple design rules for engineering applications.

3. **Primary resonance and response curves under external forcing.** For the externally forced Duffing oscillator, the method of multiple scales and the KBM approach yielded modulation equations for the slowly varying amplitude and phase in the neighbourhood of primary resonance. From these equations, steady–state response curves were derived, showing the characteristic bending of the resonance peak, multi–valued regions and jump (hysteresis) phenomena. Stability of the steady solutions was analyzed using the modulation system and confirmed by comparison with numerical simulations.

4. **Higher-order resonances and qualitative behaviour.** The perturbative framework was extended to treat superharmonic, subharmonic and (when relevant) combination resonances. Although the algebra becomes more involved, the resulting amplitude equations again provide simple criteria for the existence, multiplicity and stability of periodic responses. These results clarify how a rich variety of qualitative behaviours can emerge from relatively simple nonlinear models under periodic forcing.

5. **Accuracy and limitations of perturbation methods.** The perturbative solutions were systematically compared with high–accuracy numerical integrations (e.g. high–order Runge–Kutta schemes). In the weakly nonlinear regime the agreement is excellent, with small errors in both amplitude and phase. However, the study also showed that the accuracy deteriorates as the nonlinearity and damping increase, or as the forcing approaches thresholds associated with strong asymmetry and incipient complex dynamics. This quantifies the practical domain of validity of each perturbation method.

6. **Basis for subsequent averaging and global analyses.** Overall, Chapter 5 demonstrates that perturbative methods provide powerful and relatively simple tools to analyse weakly nonlinear Duffing type oscillators, especially near resonance. At the same time, it highlights the need for complementary techniques—such as higher-order averaging, numerical continuation and global phase-space analysis—to describe large-amplitude motions and strongly nonlinear regimes. The perturbative results obtained here serve as a bridge between the exact conservative solutions of previous chapters and the more global analytical and numerical studies developed in the following chapters.

Chapter 6

Elliptic Averaging Methods

As one of the classical methods for analyzing nonlinear oscillations, the averaging method is particularly useful for weakly nonlinear problems. The averaging method has had a profound influence in many fields of physics, mechanics and engineering. The averaging method is a perturbation technique devised for the asymptotic analysis of non-linear dynamical systems. In its modern form, the method is originally due to Krylov and Bogoliubov in the field of non-linear mechanics.

The study of nonlinear oscillators remains a central theme in applied mathematics, mechanics, and physics due to their ability to model a wide spectrum of complex phenomena ranging from structural vibrations and electrical circuits to nonlinear waves and biological rhythms. In contrast with linear systems, nonlinear oscillators cannot generally be solved by closed-form elementary functions. Even approximate solutions often require sophisticated methods to capture essential dynamical features such as amplitude–frequency dependence, resonance shifts, bifurcations, and chaotic transitions.

Among the available techniques, *averaging methods* play a distinguished role in simplifying the analysis of nonlinear oscillators. The basic idea is to replace a rapidly varying system with an averaged system that governs the slow evolution of amplitude and phase. For weak nonlinearities, classical perturbative approaches such as the Krylov–Bogoliubov averaging method or the method of multiple scales provide effective approximations. However, in the *strongly nonlinear regime*, where the nonlinearity dominates the dynamics, these classical perturbative methods lose accuracy or even fail altogether.

To overcome this limitation, *trigonometric averaging* and *elliptic averaging* have been developed as robust alternatives. Trigonometric averaging employs Fourier-based approximations to represent oscillatory solutions in terms of trigonometric functions, allowing one to extract frequency–amplitude relations and approximate periodic responses with relatively simple expressions. On the other hand, elliptic averaging exploits the fact that many strongly nonlinear oscillators admit exact solutions in terms of *Jacobi elliptic functions*. By averaging over periods of elliptic functions, one can capture highly nonlinear behaviors, including large-amplitude oscillations, with significantly better accuracy than purely trigonometric expansions.

The combination of trigonometric and elliptic averaging methods thus provides a unified framework for analyzing strongly nonlinear oscillators. These methods are not only mathematically appealing but also applicable to a wide range of physical problems, such as the Duffing oscillator, the cubic–quintic nonlinear oscillator, nonlinear pendula, and fractional-order extensions. By bridging the gap between exact elliptic solutions and approximate trigonometric representations, averaging methods offer both analytical tractability and physical insight, making them indispensable tools for modern nonlinear dynamics.

The method of averaging (also called averaging theory) considers systems containing time-scales separation: a fast oscillation versus a slow drift. It suggests that we perform an averaging over a given amount of time in order to iron out the fast oscillations and observe the qualitative behavior from the resulting dynamics. The approximated solution holds under finite time inversely proportional to the parameter denoting the slow time scale. It turns out to be a customary problem where there exists the trade off between how good is the approximated solution balanced by how much time it holds to be close to the original solution. We will consider nonlinear oscillators having the form

$$\ddot{x}+F(x)=\varepsilon f(x,\dot{x}).$$

The basic idea of the averaging method is to approximate the original system by another system called the averaged system.

6.1 General Idea of Averaging Methods

Let $u(t)$ be the exact solution (or a good approximate analytical solution) to the unperturbed oscillator

$$\ddot{x}+F(x)=0,\ x(0)=x_0 \text{ and } x'(0)=\dot{x}_0.$$

Let now $\varepsilon\neq 0$ and consider the i.v.p.

$$\ddot{x}+F(x)=\varepsilon f(x,\dot{x}),\ x(0)=x_0 \text{ and } x'(0)=\dot{x}_0. \tag{6.1}$$

The solution to the i.v.p. (6.1) is assumed in the ansatz form

$$x(t)=a(t)u(wt+\phi(t)),$$

so that

$$x'(t)\approx\xi(t):=wa(t)u'(wt+\phi(t)),$$

where w is the main frequency of the solution $u=u(t)$.

The next step consists on solving the following system for $\dot{a}=a'(t)$ and $\dot{\phi}=\phi(t)$:

$$\begin{cases} \frac{d}{dt}\left[a(t)u(wt+\phi(t))\right]=\xi(t) \\ \xi'(t)+F(x(t))=\varepsilon f(x(t),\xi(t)) \end{cases} \tag{6.2}$$

Let $\theta = wt + \phi(t)$. Solving the system (6.2) gives

$\dot{a}w\left(F(u(\theta))u(\theta) + u'(\theta)^2\right) =$
$u'(\theta)\left(\varepsilon f\left(au(\theta), awu'(\theta)\right) - F(au(\theta)) + aw^2 F(u(\theta))\right)$, and
$aw\dot{\phi}\left(F(u(\theta))u(\theta) + u'(\theta)^2\right) =$
$u(\theta)\left(-\varepsilon f\left(au(\theta), awu'(\theta)\right) + F(au(\theta)) - aw^2 F(u(\theta))\right)$

Let T be the main period of the solution $u = u(t)$. Integrating the above differential equations over $0 \leq \theta \leq T$ we obtain the following averaged differential equations :

$\dot{a} = \frac{\varepsilon \int_0^T u'(\theta) f(au(\theta), awu'(\theta))\, d\theta + \left(aw^2 - 1\right) \int_0^T u'(\theta) F(au(\theta))\, d\theta}{w \int_0^T \left(F(u(\theta))u(\theta) + u'(\theta)^2\right) d\theta}$, and

$a\dot{\phi} = \frac{-\varepsilon \int_0^T u(\theta) f(au(\theta), awu'(\theta))\, d\theta - \left(aw^2 - 1\right) \int_0^T u(\theta) F(u(\theta))\, d\theta}{w(F(u(\theta))u(\theta) + u'(\theta)^2)}$.

Finally, solving the averaged differential equations for the initial conditions $a(0) = c_0$ and $\phi(0) = c_1$ we obtain approximate analytical solutionnto the i.v.p. (6.1). The constants c_0 and c_1 are obtained from the initial conditions $x(0) = x_0$ and $x'(0) = \dot{x}_0$.

6.2 Trigonometric Averaging Method

For weakly non-linear oscillators governed by the weakly nonlinear oscillator $\ddot{x} + \omega_0^2 x = \varepsilon f(x, \dot{x})$ we seek a solution in the ansatz form

$$x(t) = a(t)\cos(\omega_0 t + \phi(t)),$$

so that

$$x'(t) \approx -\omega_0 a(t)\sin(\omega_0 t + \phi(t)) := \xi(t).$$

The next step consists on solving the following system for $\dot{a} = a'(t)$ and $\dot{\phi} = \phi(t)$:

$$\begin{cases} \frac{d}{dt}\left[a(t)\cos(\omega_0 t + \phi(t))\right] = -\omega_0 a(t)\sin(\omega_0 t + \phi(t)) \\ \xi'(t) + \omega_0^2 x(t) = \varepsilon f(x(t), \xi(t)) \end{cases}$$

The differential equations are expresed as

$a'(t) = -\frac{\varepsilon}{\omega_0} f\left(a\cos\left(\omega_0 t + \phi\right), -a\omega_0 \sin\left(\omega_0 t + \phi\right)\right)\sin\left(\omega_0 t + \phi\right)$, and
$\phi'(t) = -\frac{\varepsilon}{a\omega_0} f\left(a\cos\left(\omega_0 t + \phi\right), -a\omega_0 \sin\left(\omega_0 t + \phi\right)\right)\cos\left(\omega_0 t + \phi\right)$

Now, we take the average over $0 \leq t \leq T = \frac{2\pi}{\omega_0}$. The averaged differential equations are

$$\begin{aligned} \dot{a} &= -\frac{\varepsilon}{\omega_0 T} \int_0^T f\left(a\cos\left(\omega_0 t + \phi\right), -a\omega_0 \sin\left(\omega_0 t + \phi\right)\right)\sin\left(\omega_0 t + \phi\right) dt \\ \dot{\phi} &= -\frac{\varepsilon}{a\omega_0 T} \int_0^T f\left(a\cos\left(\omega_0 t + \phi\right), -a\omega_0 \sin\left(\omega_0 t + \phi\right)\right)\cos\left(\omega_0 t + \phi\right) dt \end{aligned}$$

Finally, we solve the system of averaged differential equations for the initial conditions $a(0) = c_0$ and $\phi(0) = c_1$. The constants c_0 and c_1 are evaluated using the initial conditions $x(0) = x_0$ and $x'(0) = \dot{x}_0$.

Assuming that

$f(x, \dot{x}) = \eta_{0,0} + x^5\eta_{5,0} + \dot{x}x^5\eta_{5,1} + \dot{x}^2x^5\eta_{5,2} + \dot{x}^3x^5\eta_{5,3} + \dot{x}^4x^5\eta_{5,4}$
$+\dot{x}^5x^5\eta_{5,5} + x^4\eta_{4,0} + \dot{x}x^4\eta_{4,1} + \dot{x}^2x^4\eta_{4,2} + \dot{x}^3x^4\eta_{4,3} + \dot{x}^4x^4\eta_{4,4}$
$+\dot{x}^5x^4\eta_{4,5} + x^3\eta_{3,0} + \dot{x}x^3\eta_{3,1} + \dot{x}^2x^3\eta_{3,2} + \dot{x}^3x^3\eta_{3,3} + \dot{x}^4x^3\eta_{3,4}$

$+\dot{x}^5x^3\eta_{3,5} + x^2\eta_{2,0} + \dot{x}x^2\eta_{2,1} + \dot{x}^2x^2\eta_{2,2} + \dot{x}^3x^2\eta_{2,3} + \dot{x}^4x^2\eta_{2,4}$
$+\dot{x}^5x^2\eta_{2,5} + \dot{x}x\eta_{1,1} + \dot{x}^2x\eta_{1,2} + \dot{x}^3x\eta_{1,3} + \dot{x}^4x\eta_{1,4} + \dot{x}^5x\eta_{1,5} + \dot{x}\eta_{0,1}$
$+\dot{x}^2\eta_{0,2} + \dot{x}^3\eta_{0,3} + \dot{x}^4\eta_{0,4} + \dot{x}^5\eta_{0,5}$

will give the following avearged odes

$$\dot{a} = \tfrac{1}{2}\varepsilon\eta_{0,1}a + \tfrac{1}{2\pi\omega_0}\left[\tfrac{3}{4}\pi\varepsilon\omega_0^3\eta_{0,3} + \tfrac{1}{4}\pi\varepsilon\omega_0\eta_{2,1}\right]a^3$$

$$+\tfrac{1}{16}\varepsilon\left(5\omega_0^4\eta_{0,5} + \omega_0^2\eta_{2,3} + \eta_{4,1}\right)a^5$$

$$+\tfrac{1}{128}\varepsilon\left(5\omega_0^4\eta_{2,5} + 3\omega_0^2\eta_{4,3}\right)a^7 + \tfrac{3}{256}\varepsilon\omega_0^4\eta_{4,5}a^9.$$

$$\dot{\phi} = -\frac{\varepsilon\left(32\omega_0^2\eta_{1,2}+96\eta_{3,0}\right)}{256\omega_0}a^2 - \frac{\varepsilon\left(\omega_0^4\eta_{1,4}+\omega_0^2\eta_{3,2}+5\eta_{5,0}\right)}{16\omega_0}a^4$$

$$-\frac{\varepsilon\left(3\omega_0^4\eta_{3,4}+5\omega_0^2\eta_{5,2}\right)}{128\omega_0}a^6 - \tfrac{3}{256}\varepsilon\omega_0^3\eta_{5,4}a^8.$$

Example 17. Let us consider a pendulum with quadratic damping

$$\ddot{x} + 2\varepsilon\dot{x}\,|\dot{x}| + \omega_0^2\sin x = 0,\ x(0) = x_0 \text{ and } x'(0) = x_0.$$

We use the approximation

$$\sin x \approx x - x^3/6 + x^5/120.$$

The averaged odes are

$$\begin{cases} a'(t) = -\frac{8\varepsilon\omega_0}{3\pi}a(t)\,|a(t)| \\ \phi'(t) = \frac{\omega_0}{384}a(t)^2\left(a(t)^2 - 24\right) \end{cases}$$

Solving these odes gives

$$\begin{aligned} a(t) &= \frac{3\pi c_0}{8c_0\varepsilon t\omega_0\epsilon+3\pi} \\ \phi(t) &= \frac{\pi c_0^2 t\omega_0\left(64c_0^2\left(c_0^2-72\right)\varepsilon^2t^2\omega_0^2+72\pi c_0\left(c_0^2-48\right)\varepsilon t\omega_0\epsilon+27\pi^2\left(c_0^2-24\right)\right)}{384(8c_0\varepsilon t\omega_0\epsilon+3\pi)^3} + c_1 \\ \epsilon &= 1 \text{ if } c_0 > 0 \text{ and } \epsilon = -1 \text{ if } c_0 < 0. \end{aligned}$$

Let us consider the parameter values

$$\varepsilon = 0.05,\ \omega_0 = 1,\ x_0 = \pi/3 \text{ and } \dot{x}_0 = 0.$$

The solution with

$$\{c_0 = 1.04838,\ c_1 = -0.04758\}$$

reads (see Figure 6.1).

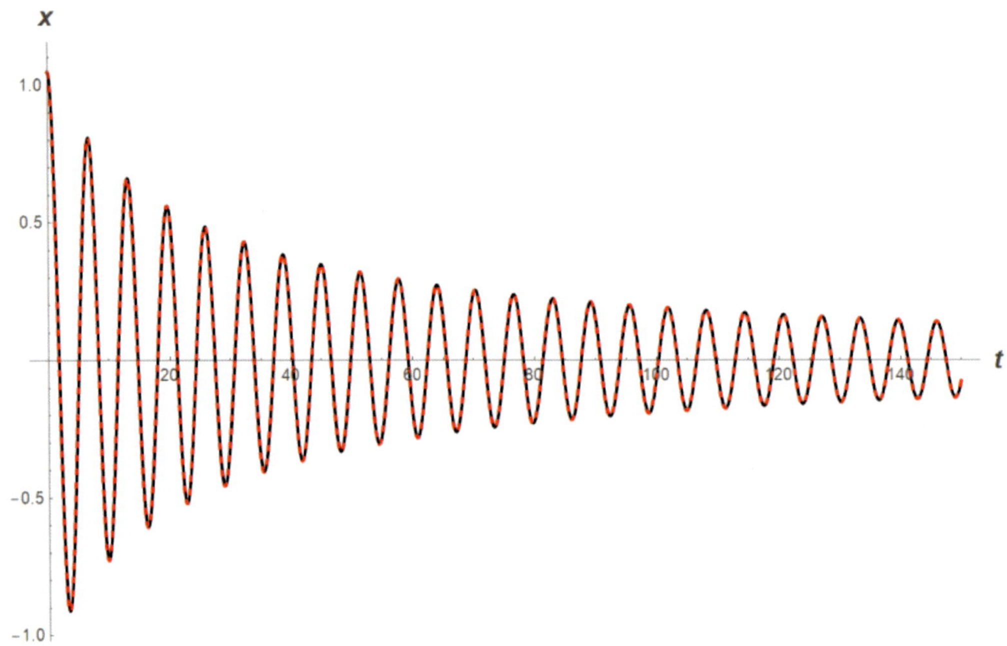

Figure 6.1: Quadratically dampeded pendulum Error = 0.0085

6.3 Elliptic Averaging Method

We will consider the following oscillator:

$$\ddot{x} + \alpha x + \beta x^3 = \varepsilon f(x, \dot{x}) = 0,\ x(0) = x_0 \text{ and } x'(0) = x_0.$$

Let

$$\ddot{x} + \alpha x + \beta x^3 = \varepsilon f(x, \dot{x})$$

Observe that when $\varepsilon = 0$ the oscillator admits the exact solution

$$x(t) = a(t)\mathrm{cn}\left(\sqrt{\alpha + c_0^2\beta}t + \phi(t), \frac{c_0^2\beta}{2(\alpha + c_0^2\beta)}\right),$$

where

$$a(t) \equiv c_0 \text{ and } \phi(t) \equiv c_1$$

Let $\varepsilon \neq 0$. The solution is assumed in the ansatz form

$$x(t) = a(t)\mathrm{cn}(wt + \phi(t), m),$$

where

$$w = \sqrt{\alpha + c_0^2\beta} \text{ and } m = \frac{c_0^2\beta}{2w^2},$$

so so that

$$x'(t) \approx -wa(t)\mathrm{sn}(wt + \phi(t), m)\mathrm{dn}(wt + \phi(t), m) := \eta(t).$$

The next step consists on solving the following system for $\dot{a} = a'(t)$ and $\dot{\phi} = \phi(t)$:

$$\begin{cases} \frac{d}{dt}\left[a(t)\text{cn}(wt+\phi(t),m)\right] = \eta(t). \\ \eta'(t) + \alpha x(t) = \varepsilon f(x(t), \eta(t)). \end{cases}$$

Let

$$a(t) = a,\ \phi(t) = \phi,\ \psi = wt + \phi.$$
$$\text{cn}(\psi, m) = \text{cn}(\psi) = \text{cn},\ \text{sn}(\psi, m) = \text{sn}(\psi) = \text{sn}.$$
$$\text{dn}(\psi, m) = \text{dn}(\psi) = \text{dn}.$$

The differential equations are expresed as

$(1-(1-\text{cn}^4)\,m)\,w\dot{a} = -\varepsilon f\,(a\text{cn}, -aw\text{dnsn})\text{dnsn}+a\text{cndnsn}\,(a^2\text{cn}^2\beta + 2\text{sn}^2 mw^2)$, and

$(1-(1-\text{cn}^4)\,m)\,aw\dot{\phi} = -\text{cn}\varepsilon f\,(a\text{cn}, -aw\text{dnsn}) + a^3\text{cn}^4\beta + 2a\text{cn}^2\text{sn}^2 mw^2.$

Next, we integrate these equations over $0 \leq \psi \leq T = 4K(m)$ to obtain the following averaged differential equations :

$$\begin{cases} \dot{a} = -\frac{\varepsilon}{wt\mathcal{C}}\int_0^T f\,(a\text{cn}(\psi), -aw\text{sn}(\psi)\text{dn}(\psi))\,\text{sn}(\psi)\text{dn}(\psi)d\psi \\ \dot{\phi} = -\frac{\varepsilon}{Taw\mathcal{C}}\int_0^T f\,(a\text{cn}, -aw\text{sn}(\psi)\text{dn}(\psi))\text{cn}(\psi)d\psi - \frac{\varepsilon}{wt\mathcal{C}}\mathcal{D}(a), \\ \mathcal{D}(a) = (2\beta(2m-1)\upsilon_4 + 4\beta(m-1)(3m-2)K(m))a^2 \\ +\,(16(m-1)mw^2K(m) - 2(m-2)mw^2\upsilon_4)\,. \\ a(0) = c_0 \text{ and } \phi(0) = c_1. \end{cases}$$

The constants c_0 and c_1 are evaluated using the initial conditions $x(0) = x_0$ and $x'(0) = \dot{x}_0$. Here,

$$\mathcal{C} = \frac{4}{3m}(2m-1)\upsilon_2 - 8(m-1)K(m),\ \upsilon_2 := E(\text{am}(2K(m)|m)|m).$$

Finally, we solve the system of averaged differential equations.

Assume that

$$f(x,\dot{x}) = \textstyle\sum_{k=0}^{2}\sum_{l=0}^{2}\eta_{k,l}x^k\dot{x}^l$$

Then

$\int_0^T f\,(a\text{cn}(\psi), -aw\text{sn}(\psi)\text{dn}(\psi))\text{sn}(\psi)\text{dn}(\psi)d\psi =$
$-\frac{(4(m-1)(m(m(8m-9)-3)+8)K_m+(8-m(m(16(m-2)m+9)+7))\upsilon_4)}{315m^3}a^5w^3\eta_{2,3}-$
$\frac{(4(m-1)(m(8m-5)-2)K_m-2(2m(4m^2-6m+1)+1)\upsilon_4)}{35m^2}a^3w^3\eta_{0,3}-$
$\frac{(2((m-1)m+1)\upsilon_4-4(m-2)(m-1)K_m)}{15m^2}a^3w\eta_{2,1} - \frac{((2m-1)\upsilon_4-4(m-1)K_m)}{3m}aw\eta_{0,1}.$
$\int_0^T f\,(a\text{cn}, -aw\text{sn}(\psi)\text{dn}(\psi))\text{cn}(\psi)d\psi =$
$\frac{((2m-1)(3(m-1)m+8)\upsilon_4-4(m-1)(3(m-5)m+8)K_m)}{105m^3}a^5w^2\eta_{3,2}+$
$\frac{2(2(m-1)(3m-2)K_m+2(2m-1)\upsilon_4)}{3m^2}a^3\eta_{3,0} + \frac{(2((m-1)m+1)\upsilon_4-4(m-2)(m-1)K_m)}{15m^2}a^3w^2\eta_{1,2}$
$+\frac{2(2(m-1)K_m+\upsilon_4)}{m}a\eta_{1,0}.$

Here,

$$\upsilon_2 = E(\text{am}(2K(m)|m)|m),\ \upsilon_4 = E(\text{am}(4K(m)|m)|m) \text{ and } K_m = K(m).$$

Following chebyshev approximation formulas for $-1 \leq m \leq 0.9$ will be useful for computing :

$$\upsilon_2 \approx \pi - 0.781423m - 0.134716m^2 - 0.0680332m^3 - 0.0735199m^4 - 0.0372182m^5$$

$$\upsilon_4 \approx 2\pi - 1.56285m - 0.269432m^2 - 0.136066m^3 - 0.14704m^4 - 0.0744364m^5$$

$$K_m = K(m) \approx \frac{\pi}{2} + 0.386997m + 0.200579m^2 + 0.212501m^3 + 0.355304m^4$$
$$-0.00911757m^5 - 0.665362m^6 - 0.109512m^7 + 0.792598m^8 + 0.461789m^9$$

Following approximations are valid for $0 \leq m \leq 0.9$:

$$J_2 = J_2(m) := \int_0^{4K(m)} \mathrm{cn}(t,m)^2\, dt =$$
$$\pi + 0.115799m^5 - 0.0113096m^4$$
$$+0.0931852m^3 + 0.145214m^2 + 0.392794m.$$

$$J_4 = J_4(m) := \int_0^{4K(m)} \mathrm{cn}(t,m)^4\, dt =$$
$$0.0207302m^5 + 0.00231775m^4 + 0.0256466m^3$$
$$+0.0548977m^2 + 0.196365m + 2.35619$$

$$J_6 = J_6(m) := \int_0^{4K(m)} \mathrm{cn}(t,m)^6\, dt =$$
$$0.0056423m^5 + 0.00186108m^4 + 0.0102345m^3$$
$$+0.027532m^2 + 0.122722m + 1.9635$$

$$J_8 = J_8(m) := \int_0^{4K(m)} \mathrm{cn}(t,m)^8\, dt =$$
$$0.00233569m^5 + 0.000558651m^4 + 0.00528521m^3$$
$$+0.0160106m^2 + 0.0859121m + 1.71806$$

$$J_{10} = J_{10}(m) := \int_0^{4K(m)} \mathrm{cn}(t,m)^{10}\, dt =$$
$$0.000946137m^5 + 0.000439666m^4 + 0.00287903m^3$$
$$+0.0103188m^2 + 0.0644306m + 1.54625$$

$$J_{12} = J_{12}(m) := \int_0^{4K(m)} \mathrm{cn}(t,m)^{12}\, dt =$$
$$-0.000161887m^5 + 0.00119489m^4 + 0.00122077m^3$$
$$+0.007236m^2 + 0.0506068m + 1.4174$$

$$J_{14} = J_{14}(m) := \int_0^{4K(m)} \mathrm{cn}(t,m)^{14}\, dt =$$
$$0.00020617m^5 + 0.000223092m^4 + 0.00109415m^3$$
$$+0.00513822m^2 + 0.0411299m + 1.31616$$

$$J_{16} = J_{16}(m) := \int_0^{4K(m)} \mathrm{cn}(t,m)^{16}\, dt =$$
$$0.000102628m^5 + 0.000164155m^4 + 0.000731799m^3$$
$$+0.0038575m^2 + 0.0342745m + 1.2339$$

$$J_{18} = J_{18}(m) := \int_0^{4K(m)} \mathrm{cn}(t,m)^{18}\, dt =$$
$$0.00101128m^5 - 0.00145513m^4 + 0.0015097m^3$$
$$+0.00267898m^2 + 0.0291772m + 1.16534$$

$J_{20} = J_{20}(m) := \int_0^{4K(m)} \text{cn}(t,m)^{20}\, dt =$
$-0.0107562m^5 + 0.0178996m^4 - 0.0109795m^3$
$+0.0058391m^2 + 0.0246495m + 1.10711.$

$J_{22} = J_{22}(m) := \int_0^{4K(m)} cn(t,m)^{22}\, dt =$
$0.00024039m^5 - 0.000321693m^4 + 0.00053609m^3$
$+0.00182329m^2 + 0.0220287m + 1.05676.$

$J_{24} = J_{24}(m) := \int_0^{4K(m)} \text{cn}(t,m)^{24}\, dt =$
$0.268897m^5 - 0.448074m^4 + 0.288805m^3$
$-0.0879195m^2 + 0.0327975m + 1.01197.$

The integrals in (6.3) may be quite difficult to evaluate. We may use the following chebyshevapproximations in order to facilitate the integration process :

$$\text{sn}(\psi)\text{dn}(\psi) \approx \sigma_0 \text{sn}(\psi) + \sigma_1 \text{sn}^3(\psi) \text{ for } 0 < m \leq 0.8,$$

where

$$\left\{ \begin{array}{l} \sigma_0 = 0.323792m^5 - 0.370372m^4 + 0.216124m^3 + 0.0032398m^2 + 0.00222037m + 1. \\ \sigma_1 - 0.621633m^5 + 0.684869m^4 - 0.433874m^3 - 0.089613m^2 - 0.504113m \end{array} \right\}$$

$\text{cn}(\psi) \approx \cos(\nu\psi)$, $\nu = 1 - 15/61m - 3/28m^2$ for $0 < m \leq 0.25$.

$\text{sn}(\psi) \approx \sin(\nu\psi)$, $\nu = 1 - 15/61m - 3/28m^2$ for $0 < m \leq 0.25$.
For $0 \leq m \leq 0.8$ we have the following accurate approximations :
$\text{cn}(t|m) = (-0.254169m^5 + 0.295035m^4 - 0.168789m^3 - 0.00533401m^2 - 0.0642244m + 1)\cos(\nu t)$
$+ (0.254169m^5 - 0.295035m^4 + 0.168789m^3 + 0.00533401m^2 + 0.0642244m)\cos(3\nu t).$
$\text{sn}(t|m) = (0.135515m^5 - 0.148848m^4 + 0.0887866m^3 + 0.0144684m^2 + 0.0631576m + 1)\sin(\nu t)$
$+ (0.173546m^5 - 0.188158m^4 + 0.118718m^3 + 0.0100297m^2 + 0.0642225m)\sin(3\nu t),$
where
$\nu = -0.341479m^5 + 0.382056m^4 - 0.235018m^3 - 0.0414976m^2 - 0.252254m + 1.$

Thus,

$$\text{cn} = \text{cn}(\psi) = \text{cn}(\psi, m) \approx C_0 \cos(\nu\psi) + C_1 \cos(3\nu\psi)$$

and

$$\text{sn} = \text{sn}(\psi) = \text{sn}(\psi, m) \approx S_0 \sin(\nu\psi) + S_1 \sin(3\nu\psi),$$

$$T = T_{\text{period}} \approx \frac{2\pi}{\nu}.$$

where

$$C_0 = S_0 = 1 \text{ and } C_1 = S_1 = 1 \text{ and } \nu = 1 - 15/61m - 3/28m^2 \text{ for } 0 \leq m \leq 0.25$$

In the case when $0.25 < m \leq 0.8$,

$$C_0 = -0.254169m^5 + 0.295035m^4 - 0.168789m^3 - 0.00533401m^2 - 0.0642244m + 1.$$

$$C_1 = 1 - C_0$$

$$S_0 = 0.135515m^5 - 0.148848m^4 + 0.0887866m^3 + 0.0144684m^2 + 0.0631576m + 1.$$

$$S_1 = 0.173546m^5 - 0.188158m^4 + 0.118718m^3 + 0.0100297m^2 + 0.0642225m.$$

$$\nu = -0.341479m^5 + 0.382056m^4 - 0.235018m^3 - 0.0414976m^2 - 0.252254m + 1.$$

Observe also that

$$\text{sn}(\psi)\text{dn}(\psi) \approx \sigma_0 \left(S_0 \sin(\nu\psi) + S_1 \sin(3\nu\psi)\right) + \sigma_1 \left(S_0 \sin(\nu\psi) + S_1 \sin(3\nu\psi)\right)^3.$$

Thus, the elliptic integrals in are reduced to trigonometric integrals.

Example 18. Let

$$\ddot{x} + 2\varepsilon\dot{x} + \alpha x + \beta x^3 = 0,\ x(0) = x_0 \text{ and } x'(0) = \dot{x}_0$$

In this case,

$$f(x, \dot{x}) = -2\dot{x}$$

The solution is assumed in the ansatz form

$$x(t) = a(t)\text{cn}(\omega t + \phi(t), m),$$

where

$$\omega = \sqrt{\alpha + c_0^2\beta} \text{ and } m = \frac{c_0^2\beta}{2\omega^2}.$$

We have :

$$\begin{gathered}
\left(\left(\text{cn}^4 - 1\right) m\omega + \omega\right) \dot{a} = \left(a^2\beta - 2m\omega^2\right) a\text{cn}^3\text{dnsn} + a\left(\alpha + 2m\omega^2 - \omega^2\right) \text{cndnsn} - 2a\varepsilon\omega\text{dn}^2\text{sn}^2 \\
\left(\left(\text{cn}^4 - 1\right) m\omega + \omega\right) \dot{\phi} = \left(a^2\beta - 2m\omega^2\right) \text{cn}^4 + \left(\alpha + 2m\omega^2 - \omega^2\right) \text{cn}^2 - 2\omega\varepsilon\text{cndnsn}. \\
\text{sn}=\text{sn}(\psi, m),\ \text{cn}=\text{cn}(\psi, m),\ \text{dn}=\text{dn}(\psi, m), \psi = \omega t + \phi(t).
\end{gathered}$$

In order to obtain the averaged odes we integrate the above odes on $0 \le \psi \le T = 4K(m)$ to get the following averaged odes :

$$\dot{a} = -\varepsilon a\ ,\ \dot{\phi} = r_0 + r_2 a^2.$$

Solving these odes we obtain the approximate solution in the form

$$a = a(t) = c_0 \exp(-\varepsilon t),$$

$$\phi = \phi(t) = \frac{c_0^2 r_2}{\varepsilon} \exp(-\varepsilon t)\sinh(\varepsilon t) + c_1 + r_0 t,$$

where

$r_0 = [3mE(\mathrm{am}(4K(m)|m)|m) + (1-2m)m\omega^2 E(\mathrm{am}(4K(m)|m)|m) + 12\alpha(m-1)mK(m) + 4(m-1)m\omega^2 K(m)]/$

$[4m\omega((2m-1)E(\mathrm{am}(2K(m)|m)|m) - 2(m-1)K(m))]$

and

$r_2 = [4\beta mE(\mathrm{am}(4K(m)|m)|m) - 2\beta E(\mathrm{am}(4K(m)|m)|m) - 8\beta(m-1)K(m) + 12\beta(m-1)mK(m)]/$

$[4m\omega((2m-1)E(\mathrm{am}(2K(m)|m)|m) - 2(m-1)K(m))]$.

The constants c_0 and c_1 are evaluated using the initial conditions $x(0) = x_0$ and $x'(0) = \dot{x}_0$.

On the other hand, using trigonometric averaging we obtain the following approximte solutuion:

$$x(t) = c_0 \exp(-\varepsilon t) \cos\left(\sqrt{\alpha} t + \frac{3\beta c_0^2}{8\sqrt{\alpha}\varepsilon} \sinh(\varepsilon t) \exp(-\varepsilon t) + c_1\right).$$

For a numerical example, let us consider the parameter values

$$\varepsilon = 0.05,\ \alpha = \beta = 1,\ x_0 = 0.25 \text{ and } \dot{x}_0 = 0.$$

In this case,

$$r_0 = -0.0231031,\ r_2 = 0.368781$$
$$c_0 = 0.250294,\ c_1 = -0.048468$$

The solution is given by

$$x(t) = 0.250294e^{-0.05t}\mathrm{cn}\left(-1.00774t - 0.462061e^{-0.05t}\sinh(0.05t) + 0.048468 \middle| 0.0294769\right)$$

See Figure 6.2.

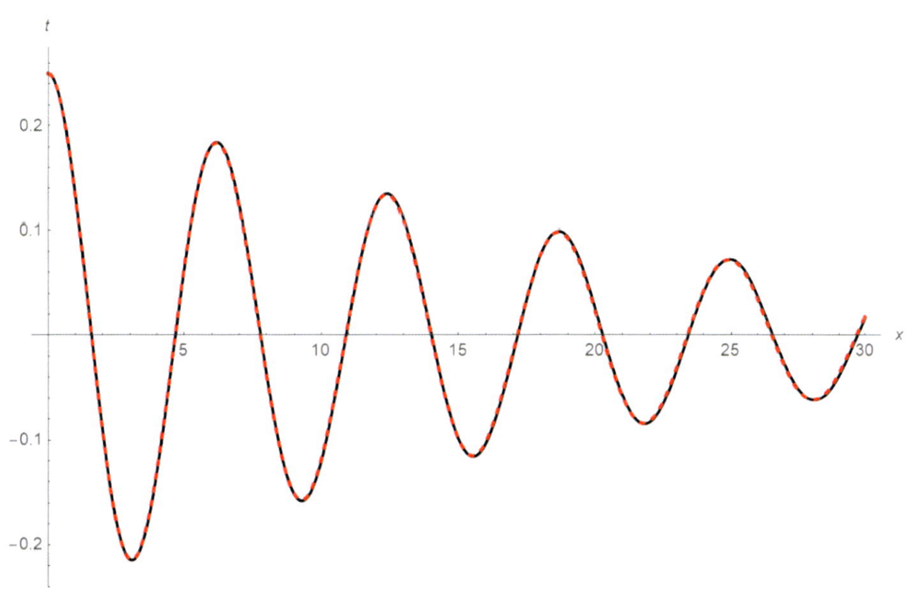

Figure 6.2: Comparison with numerical solution

Example 19. Let

$$\ddot{x} + 2\varepsilon\dot{x} + \alpha x + \beta x^3 = f\cos(\Omega t),\ x(0) = x_0 \ \text{ and } x'(0) = \dot{x}_0$$

The solution is assumed in ther ansatz form

$$x(t) = y(t) + r\cos(\Omega t) + s\cos(3\Omega t),$$

where

$$\ddot{y} + 2\varepsilon\ \dot{y} + \alpha y + \beta y^3 = 0,\ y(0) = x_0 - r - s \text{ and } \ y'(0) = \dot{x}_0.$$

and r and s are some constants to be defined. We have:

$$\begin{aligned}&\ddot{x} + 2\varepsilon\dot{x} + \alpha x + \beta x^3 - f\cos(\Omega t) = \\ &\tfrac{1}{4}\left(-4f + 3\beta r^3 + 3\beta r^2 s + 4\alpha r + 6\beta r s^2 - 4r\Omega^2\right)\cos(\Omega t) \\ &+\tfrac{1}{4}\left(\beta r^3 + 6\beta r^2 s + 3\beta s^3 + 4\alpha s - 36 s\Omega^2\right)\cos(3\Omega t) + \text{h.o.t}\end{aligned}$$

The constants r and s are the least in magnitude solutions to the algebraic system

$$\begin{cases} -4f + 3\beta r^3 + 3\beta r^2 s + 4\alpha r + 6\beta r s^2 - 4r\Omega^2 = 0. \\ \beta r^3 + 6\beta r^2 s + 3\beta s^3 + 4\alpha s - 36 s\Omega^2 = 0. \end{cases}$$

For example, let us consider the i.v.p.

$$\ddot{x} + \frac{1}{10}\dot{x} + x + x^3 = 1/2\cos(\pi t),\ x(0) = 0 \ \text{ and } x'(0) = 0$$

The algebraic system reads

$$\begin{aligned}&6r^3 + 6r^2 s + 12 r s^2 - 4\pi^2 r + 4r - 2 = 0. \\ &2r^3 + 12 r^2 s + 6 s^3 - 36\pi^2 s + 4s = 0.\end{aligned}$$

A solution is

$$\{r = -0.0564026, s = -1.022 \times 10^{-6}\ \}.$$

So,

$$x(t) = y(t) - 0.0564026\cos(\pi t) - 1.022 \times 10^{-6}\cos(3\pi t),$$

where

$$\ddot{y} + \frac{1}{10}\ \dot{y} + y + y^3 = 0,\ y(0) = -r - s \text{ and } \ y'(0) = 0.$$

Example 20. Let us consider the following delayed Dufing oscillator

$$\ddot{x} + 2\delta\dot{x} + px + qx^3 = \gamma x(t-\tau),\ x(0) = x_0 \ \text{ and } x'(0) = \dot{x}_0.$$

Assuming that τ is small, we use the approximation

$$x(t-\tau) \approx x(t) - \tau x'(t) + \frac{1}{2}\tau^2 x''(t),$$

and then we obtain the i.v.p.

$$\ddot{x} + 2\varepsilon\dot{x} + \alpha x + \beta x^3 = 0,\ x(0) = x_0 \text{ and } x'(0) = \dot{x}_0,$$

where

$$\varepsilon = \frac{\gamma\tau + 2\delta}{2 - \gamma\tau^2},\ \alpha = \frac{2(p-\gamma)}{2 - \gamma\tau^2},\ \beta = \frac{2q}{2 - \gamma\tau^2}.$$

Assume now that the delay is not small. Let us consider a more general problem (Raleygh-Duffing oscillator)

$$\ddot{x} + 2\delta\dot{x} + \mu\dot{x}^3 + px(t) + qx(t)^3 = \gamma x(t-\tau) + \zeta x'(t-\tau),\ x(0) = x_0 \text{ and } x'(0) = \dot{x}_0.$$

Assume that

$$x(t) = a(t)\mathrm{cn}(\omega t + \phi(t), m),$$

where

$$\omega = \sqrt{p + c_0^2 q} \text{ and } m = \frac{c_0^2 q}{2\omega^2}.$$

Using the approximations

$$a(t-\tau) \approx a(t),\ \phi(t-\tau) \approx \phi(t)$$

yields

$$\begin{cases} a\dot{\phi}\mathrm{dn}(\theta)\mathrm{sn}(\theta) - \dot{a}\mathrm{cn}(\theta) = 0. \\ \\ a^3\left(q\mathrm{cn}(\theta)^3 - \mu\omega^3\mathrm{dn}(\theta)^3\mathrm{sn}(\theta)^3\right) \\ +a[-\gamma\mathrm{cn}(\theta-\omega\tau) + \mathrm{cn}(\theta)\left(-\omega\mathrm{dn}(\theta)^2\left(\omega+\dot{\phi}\right) + m\omega\mathrm{sn}(\theta)^2\left(\omega+\dot{\phi}\right) + p\right) \\ -2\delta\omega\mathrm{dn}(\theta)\mathrm{sn}(\theta) + \zeta\omega\mathrm{dn}(\theta-\omega\tau)\mathrm{sn}(\theta-\omega\tau)] - \dot{a}\omega\mathrm{dn}(\theta)\mathrm{sn}(\theta) = 0. \end{cases}$$

Since it is difficult to obtain the averaged elliptic equations, we will use our trigonometric approximations as follows :

$$\mathrm{sn}(\psi)\mathrm{dn}(\psi) \approx \sigma_0\mathrm{sn}(\psi) + \sigma_1\mathrm{sn}^3(\psi),\ \mathrm{sn}(\psi) \approx \sin(\nu\psi) \text{ and } \mathrm{cn}(\psi) \approx \cos(\nu\psi).$$

After some algebra we obtain the following averaged odes :

$$\begin{aligned} \dot{a} &= \tfrac{\mu\omega^2}{128(5m-8)}\left(384\sigma_0^4 + 1280\sigma_1\sigma_0^3 + 1680\sigma_1^2\sigma_0^2 + 1008\sigma_1^3\sigma_0 + 231\sigma_1^4\right)a^3 \\ &+ \tfrac{1}{4(5m-8)\omega}\begin{pmatrix} 4\gamma(4\sigma_0 + 3\sigma_1)\sin(\nu\omega\tau) + 4\delta\left(8\sigma_0^2 + 12\sigma_1\sigma_0 + 5\sigma_1^2\right)\omega \\ -\zeta\sigma_1^2\omega\cos(3\nu\omega\tau) - \zeta(4\sigma_0 + 3\sigma_1)^2\omega\cos(\nu\omega\tau) \end{pmatrix} a. \end{aligned}$$

$$\dot{\phi} = \tfrac{1}{(5m-8)\omega}\left[4\gamma\cos(\nu\omega\tau) + \zeta(4\sigma_0 + 3\sigma_1)\omega\sin(\nu\omega\tau) - 2(m-2)\omega^2 - 4p\right] + \tfrac{3q}{(8-5m)\omega}a^2$$

$$\nu = -0.341479m^5 + 0.382056m^4 - 0.235018m^3 - 0.0414976m^2 - 0.252254m + 1.$$

These odes may be solved easily.

On the other hand, the averaged odes using trigonometric averaging are

$$\dot{a} = \tfrac{1}{8}\left(-8\delta - \frac{4\gamma\sin\left(\sqrt{p}\tau\right)}{\sqrt{p}} + 4\zeta\cos\left(\sqrt{p}\tau\right)\right)a - \tfrac{3}{8}\mu p a^3$$
$$\dot{\phi} = \tfrac{1}{8}\left(-\frac{4\gamma\cos\left(\sqrt{p}\tau\right)}{\sqrt{p}} - 4\zeta\sin\left(\sqrt{p}\tau\right)\right) + \tfrac{3q}{8\sqrt{p}}a^2$$

$$x_{\text{trig}}(t) = a(t)\cos(\sqrt{p} + \phi(t)),$$

$$a(t) = \frac{c_0\sqrt{s_1}e^{s_1 t}}{\sqrt{s_1 - c_0^2 s_3\left(e^{2s_1 t} - 1\right)}}$$

$$\phi(t) = \frac{r_2}{2s_3}\log\left(\frac{s_1}{s_1 - c_0^2 s_3\left(e^{2s_1 t} - 1\right)}\right) + c_1 + \sqrt{p}t + r_0 t.$$

$$r_0 = \frac{1}{8}\left(-\frac{4\gamma}{\sqrt{p}}\cos\left(\sqrt{p}\tau\right) - 4\zeta\sin\left(\sqrt{p}\tau\right)\right),\; r_2 = \frac{3q}{8\sqrt{p}}.$$

$$s_1 = \frac{1}{8}\left(-8\delta - \frac{4\gamma}{\sqrt{p}}\sin\left(\sqrt{p}\tau\right) + 4\zeta\cos\left(\sqrt{p}\tau\right)\right),\; s_3 = -\frac{3}{8}\mu p.$$

Example 21. Let us consider the Van der pol-Duffing oscillator

$$\ddot{x} - \varepsilon(1 - x^2)\dot{x} + \alpha x + \beta x^3 = 0,\; x(0) = x_0 \text{ and } x'(0) = \dot{x}_0$$

The solution is assumed as

$$x(t) = a(t)\ \mathrm{cn}(wt + \phi(t), m).$$

Then

$$\dot{a} = \frac{\varepsilon}{2}a - va^3 \text{ and } \dot{\phi} = w_0 + w_2 a^2,$$

where

$$w_0 \approx \frac{0.434591\alpha m^3}{w} - \frac{0.0150551\alpha m^2}{w} + \frac{0.274495\alpha m}{w} + \frac{0.499772\alpha}{w} - \frac{w}{2},$$
$$w_2 \approx \frac{0.266705\beta m^3}{w} - \frac{0.000935658\beta m^2}{w} + \frac{0.186649\beta m}{w} + \frac{0.374862\beta}{w},$$
$$v \approx \tfrac{\varepsilon}{8} + 0.0335772\varepsilon m^3 - 0.00282389\varepsilon m^2 + 0.0175692\varepsilon m,$$
$$m = \frac{\beta c_0^2}{2(\alpha+\beta c_0^2)} \text{ and } w = \sqrt{\alpha + \beta c_0^2}.$$

The solutions are

$$a(t) = \frac{c_0\sqrt{\varepsilon}}{\sqrt{2c_0^2 v(e^{\varepsilon t}-1)+\varepsilon}}\exp\left(\tfrac{\varepsilon t}{2}\right),$$

and

$$\phi(t) = \tfrac{w_2}{2v}\log\left(1 + \tfrac{2c_0^2 v}{\varepsilon}\left(\exp\left(\varepsilon t\right) - 1\right)\right) + c_1 + wt_0.$$

The limit cycle is obtained from the condition $\lim_{t\to\infty} a(t) = L$ and it is found from the equation

$$L = \sqrt{\frac{\varepsilon}{2v}},\; c_0 = L.$$

This gives the following octic for L :

$$-\frac{3.6423\alpha^3}{\beta^3} + \frac{0.910576\alpha^3 - 10.9269\alpha^2\beta}{\beta^3}L^2 + \frac{2.79572\alpha^2\beta - 10.9269\alpha\beta^2}{\beta^3}L^4$$
$$+\left(\frac{2.85457\alpha}{\beta} - 3.6423\right)L^6 + L^8 = 0,\; \beta \neq 0.$$

As $\beta \to 0$ we get $L = 2$.

Observe that the limit cycle depends on β : $L = L(\alpha, \beta)$. Using the trignomentric averaging method we do not obtain this dependence and so the elliptic averaging method is stronger. On the other hand,

$$\lim_{\beta \to \infty} L(1, \beta) = L(1, \infty) = 1.9084820724025244.$$

We obtained the following expression for $\alpha = 1$ and $\beta \geq 0$:

$$L(\beta) = 2 - \frac{0.0915195(\beta - 0.345917)\beta(\beta + 2.72086)}{(\beta - 0.345428)(\beta + 0.372614)(\beta + 2.76672)}.$$

Example 22. Let $\alpha = 1$, $\beta = 5$, $\varepsilon = 0.1$. Consider the i.v.p.

$$\ddot{x} - 0.1(1 - x^2)\dot{x} + x + 5x^3 = 0,\ x(0) = 1.7 \text{ and } x'(0) = 0.$$

The approximate solution is

$$x(t) = \frac{0.537589e^{0.0490782t}}{\sqrt{0.0213728 + 0.0786272e^{0.0981564t}}} \times$$
$$\mathrm{cn}(\kappa \left[-2.1356t - 23.3032 \log\left(0.271824 + e^{0.0981564t}\right) + 5.60053\right], 0.467638),$$

where $\kappa = 0.981564$.

The limit cycle is obtained from the octic

$L^8 - 3.07139L^6 - 2.07355L^4 - 0.429792L^2 - 0.0291384 = 0,$

and its value equals $L = 1.9154772358335026$.

See Figure 6.3.

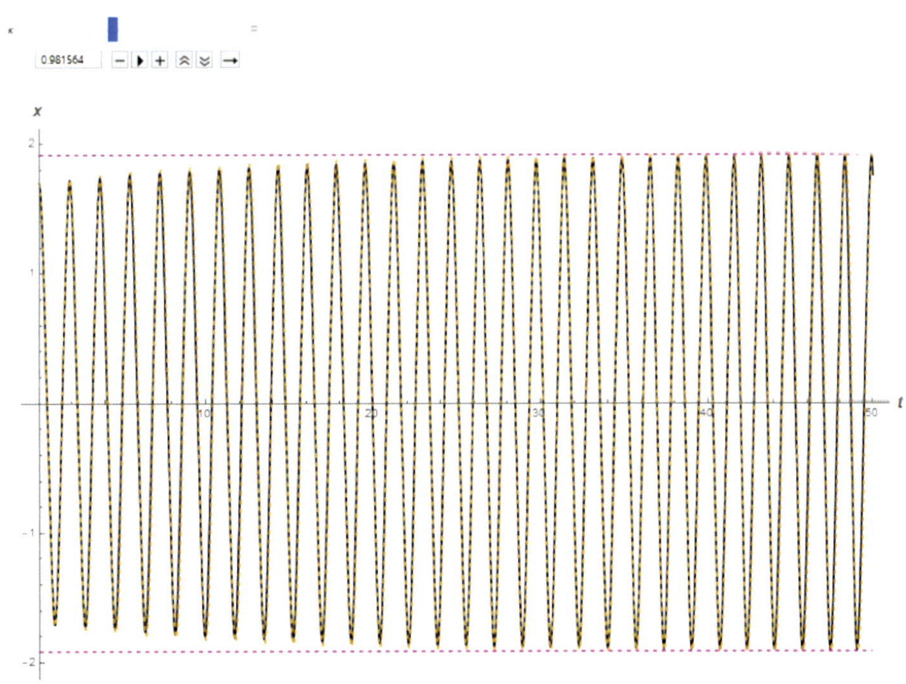

Figure 6.3: The analytical solution vs the Runge-Kutta solution.

Example 23. Consider the i.v.p.

$$\ddot{x} + \alpha x + \beta x^3 - \varepsilon\left(p\dot{x} + qx^2\dot{x} - r\dot{x}^3\right) = 0,\ x(0) = x_0 \ \text{ and } x'(0) = \dot{x}_0.$$

The solution is assumed as

$$x(t) = a(t)\ \mathrm{cn}(wt + \phi(t), m).$$

The averaged equatios are

$$\dot{a} = \frac{\varepsilon}{2}a - va^3 \ \text{ and } \dot{\phi} = w_0 + w_2 a^2,$$

where

$$v = 0.375001\varepsilon rw^2 \ - 0.125001\varepsilon q + -m\left(0.0155202\varepsilon q + 0.328192\varepsilon rw^2\right)$$
$$+ - m^2\left(0.012312\varepsilon q - 0.00979715\varepsilon rw^2\right) \ + -m^3\left(0.0000308249\varepsilon q - 0.000846768\varepsilon rw^2\right)$$
$$+ - m^4\left(0.0198577\varepsilon q - 0.0132176\varepsilon rw^2\right)\ ,$$
$$w_0 = \frac{0.500011\alpha}{w} - 0.5w + \frac{0.248692\alpha m}{w} + \frac{0.17586\alpha m^2}{w} + \frac{0.0100331\alpha m^3}{w} + \frac{0.253494\alpha m^4}{w}\ ,$$
$$w_2 \ = \frac{0.375007\beta}{w} + \frac{0.171091\beta m}{w} + \frac{0.1143\beta m^2}{w} + \frac{0.00987896\beta m^3}{w} + \frac{0.154206\beta m^4}{w}\ .$$

The solutions are

$$a(t) = \frac{c_0\sqrt{\varepsilon}}{\sqrt{2c_0^2 v(e^{\varepsilon t}-1)+\varepsilon}}\exp\left(\frac{\varepsilon t}{2}\right)\ ,$$

and

$$\phi(t) = \frac{w_2}{2v}\log\left(1 + \frac{2c_0^2 v}{\varepsilon}\left(\exp\left(\varepsilon t\right) - 1\right)\right) \ + c_1 \ + wt_0\ .$$

The limit cycle is obtained from the condition $\lim_{t\to\infty} a(t) = L$ and it is found from the equation

$$\frac{9.33333\alpha^4}{\beta^5 r} \ + \left(\frac{1.75\alpha^5}{\beta^5} - \frac{0.583338\alpha^4 q}{\beta^5 r} + \frac{37.3333\alpha^3}{\beta^4 r}\right)\ L^2 \ + \left(\frac{7.98423\alpha^4}{\beta^4} - \frac{2.36956\alpha^3 q}{\beta^4 r} + \frac{56.\alpha^2}{\beta^3 r}\right)\ L^4$$
$$+ \left(\frac{14.4483\alpha^3}{\beta^3} - \frac{3.62303\alpha^2 q}{\beta^3 r} + \frac{37.3333\alpha}{\beta^2 r}\right)\ L^6 + \left(\frac{12.9401\alpha^2}{\beta^2} - \frac{2.47074\alpha q}{\beta^2 r} + \frac{9.33333}{\beta r}\right)\ L^8$$
$$+ \left(\frac{5.72602\alpha}{\beta} - \frac{0.639725q}{\beta r}\right)\ L^{10} \ + L^{12} \ = 0.$$

Letting $r = -s/\beta$ and $\beta \to \infty$ gives

$$L = \sqrt{\sqrt{\frac{0.102312q^2}{s^2} + \frac{9.33332}{s}} - \frac{0.319863q}{s}}\ .$$

Example 24. Let

$$\ddot{x} - \varepsilon(1 - |x|^{2v})\,|\dot{x}|^{\mu}\,\mathrm{sign}(\dot{x}) + \alpha x + \beta x^3 = 0,\ x(0) = x_0 \text{ and } x'(0) = \dot{x}_0.$$

We use the following chebyshevapproximations:

$$|x|^{2v} \approx bx^2 + cx^4 + dx^6,\ |x| \leq M,$$

$$b = \frac{0.285714}{M^2}\left(1.27633M^{2v} + 16.\left(2.068530.433884^{2v}M^{2v} - 0.4096980.781831^{2v}M^{2v}\right)\right).$$

$$c = \frac{1.14286}{M^4}\left(-21.81030.433884^{2v}M^{2v} + 10.3440.781831^{2v}M^{2v} - 2.21695M^{2v}\right).$$

$$d = \frac{4.57143}{M^6}\left(3.384040.433884^{2v}M^{2v} - 2.176290.781831^{2v}M^{2v} + 0.693218M^{2v}\right).$$

and

$$|\dot{x}|^{\mu}\,\mathrm{sign}(\dot{x}) \approx p\dot{x} + q\dot{x}^3 + r\dot{x}^5,\ |\dot{x}| \leq R,$$

$$p = \tfrac{2R^{\mu-1}}{3+\sqrt{3}}^{-\frac{3\mu}{2}-\frac{1}{2}} \left(2^{\mu}\left(\left(9-5\sqrt{3}\right)2^{\frac{\mu+1}{2}}+4-4\sqrt{3}\right)+32\left(\sqrt{3}-1\right)^{\mu}\right).$$

$$q = \tfrac{R^{\mu-3}}{3+\sqrt{3}} \left(-3\ 2^{-\frac{3}{2}(\mu-3)}\left(\sqrt{3}-1\right)^{\mu}+\left(3+5\sqrt{3}\right)2^{\frac{3}{2}-\frac{\mu}{2}}-30+14\sqrt{3}\right).$$

$$r = \tfrac{8R^{\mu-5}}{3+\sqrt{3}} \left(2^{\frac{5}{2}-\frac{3\mu}{2}}\left(\sqrt{3}-1\right)^{\mu}-\left(1+\sqrt{3}\right)2^{\frac{1}{2}-\frac{\mu}{2}}+3-\sqrt{3}\right).$$

So, we solve the following i.v.p. :

$$\ddot{x}-\varepsilon(1-bx^2-cx^4-dx^6)(p\dot{x}+q\dot{x}^3+r\dot{x}^5)+\alpha x+\beta x^3=0,\ x(0)=x_0 \text{ and } x'(0)=\dot{x}_0.$$

Let

$$f(x,\dot{x})=(1-bx^2-cx^4-dx^6)(p\dot{x}+q\dot{x}^3+r\dot{x}^5).$$

Observe that

$$\int_0^T f\left(a\mathrm{cn}, -aw\mathrm{sn}(\psi)\mathrm{dn}(\psi)\right)\mathrm{cn}(\psi)d\psi=0,$$

and so

$$\dot{\phi}=-\frac{\varepsilon}{wt\mathcal{C}}\mathcal{D}(a).$$

On the other hand,

$$\dot{a}=-\frac{\varepsilon}{wt\mathcal{C}}\int_0^T f\left(a\mathrm{cn}, -aw\mathrm{sn}(\psi)\mathrm{dn}(\psi)\right)\mathrm{cn}(\psi)\mathrm{sn}(\psi)d\psi=-\frac{\varepsilon}{wt\mathcal{C}}aP(a),$$

where

$P(a) = (dr\omega^5J_{12}m^6 - 3dr\omega^5J_{14}m^6 + 3dr\omega^5J_{16}m^6 - dr\omega^5J_{18}m^6 - 3dr\omega^5J_{10}m^5$
$+6dr\omega^5J_{12}m^5 - 3dr\omega^5J_{14}m^5 + 3dr\omega^5J_8m^4 - 6dr\omega^5J_{12}m^4 + 3dr\omega^5J_{14}m^4 - dr\omega^5J_6m^3$
$-6dr\omega^5J_8m^3+6dr\omega^5J_{10}m^3+3dr\omega^5J_6m^2+3dr\omega^5J_8m^2-3dr\omega^5J_{10}m^2-3dr\omega^5J_6m+dr\omega^5J_6)a^{10}$
$+(cr\omega^5J_{10}m^6 - 3cr\omega^5J_{12}m^6 + 3cr\omega^5J_{14}m^6 - cr\omega^5J_{16}m^6 - 3cr\omega^5J_8m^5 + 6cr\omega^5J_{10}m^5$
$-3cr\omega^5J_{12}m^5 + 3cr\omega^5J_6m^4 - 6cr\omega^5J_{10}m^4 + dq\omega^3J_{10}m^4 + 3cr\omega^5J_{12}m^4 - 2dq\omega^3J_{12}m^4$
$+dq\omega^3J_{14}m^4 - cr\omega^5J_4m^3 - 6cr\omega^5J_6m^3 + 6cr\omega^5J_8m^3 - 2dq\omega^3J_8m^3 + 2dq\omega^3J_{10}m^3$
$+3cr\omega^5J_4m^2 + 3cr\omega^5J_6m^2 + dq\omega^3J_6m^2 - 3cr\omega^5J_8m^2 + 2dq\omega^3J_8m^2 - 2dq\omega^3J_{10}m^2$
$-3cr\omega^5J_4m - 2dq\omega^3J_6m + cr\omega^5J_4 + dq\omega^3J_6)a^8$
$+(bJ_8m^6r\omega^5 - 3bJ_{10}m^6r\omega^5 + 3bJ_{12}m^6r\omega^5 - bJ_{14}m^6r\omega^5 - 3bJ_6m^5r\omega^5 + 6bJ_8m^5r\omega^5$
$-3bJ_{10}m^5r\omega^5 + 3bJ_4m^4r\omega^5 - 6bJ_8m^4r\omega^5 + 3bJ_{10}m^4r\omega^5 - bJ_2m^3r\omega^5 - 6bJ_4m^3r\omega^5$
$+6bJ_6m^3r\omega^5 + 3bJ_2m^2r\omega^5 + 3bJ_4m^2r\omega^5 - 3bJ_6m^2r\omega^5 - 3bJ_2mr\omega^5 + bJ_2r\omega^5$
$+cJ_8m^4q\omega^3 - 2cJ_{10}m^4q\omega^3 + cJ_{12}m^4q\omega^3 - 2cJ_6m^3q\omega^3 + 2cJ_8m^3q\omega^3 + cJ_4m^2q\omega^3$
$+2cJ_6m^2q\omega^3-2cJ_8m^2q\omega^3-2cJ_4mq\omega^3+cJ_4q\omega^3+dJ_8m^2p\omega-dJ_{10}m^2p\omega-dJ_6mp\omega+dJ_6p\omega)a^6$
$+(-r\omega^5J_6m^6+3r\omega^5J_8m^6-3r\omega^5J_{10}m^6+r\omega^5J_{12}m^6+3r\omega^5J_4m^5-6r\omega^5J_6m^5+3r\omega^5J_8m^5-$
$3r\omega^5J_2m^4$
$+6r\omega^5J_6m^4 + bq\omega^3J_6m^4 - 3r\omega^5J_8m^4 - 2bq\omega^3J_8m^4 + bq\omega^3J_{10}m^4 + r\omega^5m^3 + 6r\omega^5J_2m^3 -$
$6r\omega^5J_4m^3$
$-2bq\omega^3J_4m^3 + 2bq\omega^3J_6m^3 - 3r\omega^5m^2 - 3r\omega^5J_2m^2 + bq\omega^3J_2m^2 + 3r\omega^5J_4m^2 + 2bq\omega^3J_4m^2 -$
$2bq\omega^3J_6m^2$

$+cp\omega J_6 m^2 - cp\omega J_8 m^2 + 3r\omega^5 m - 2bq\omega^3 J_2 m - cp\omega J_4 m - r\omega^5 + bq\omega^3 J_2 + cp\omega J_4)a^4$
$+(-q\omega^3 J_4 m^4 + 2q\omega^3 J_6 m^4 - q\omega^3 J_8 m^4 + 2q\omega^3 J_2 m^3 - 2q\omega^3 J_4 m^3 - q\omega^3 m^2 - 2q\omega^3 J_2 m^2 + 2q\omega^3 J_4 m^2$
$+bp\omega J_4 m^2 - bp\omega J_6 m^2 + 2q\omega^3 m - bp\omega J_2 m - q\omega^3 + bp\omega J_2)a^2$
$+(-p\omega J_2 m^2 + p\omega J_4 m^2 + p\omega m - p\omega),$
$\omega = \sqrt{\alpha + c_0^2\beta},\ m = \frac{c_0^2\beta}{2\omega^2}.$

being

$$J_n = \int_0^{4K(m)} \mathrm{cn}^n(\psi, m)d\psi, \text{ for } n = 2, 4, 6, ...$$

The limit cycle is obtained from the algebraic equation

$$P(a) = 0.$$

6.4 Extended Averaging Method

This method was proposed in [5]. Let

$$\ddot{x} + \omega_0^2 \mathrm{sign}(x)\left|x\right|^\lambda = \varepsilon f(x, \dot{x}),\ x(0) = x_0 \text{ and } x'(0) = \dot{x}_0\ (\lambda \geq 0).$$

In the case when $\varepsilon = 0$ we have the unperturbed oscillator is expresses as

$$\ddot{x} + \omega_0^2 \mathrm{sign}(x)\left|x\right|^\lambda = 0,\ x(0) = x_0 \text{ and } x'(0) = \dot{x}_0.$$

Let $x_0 = A$ and $\dot{x}_0 = 0$. The energy-type first integral is rewritten as

$$\frac{1}{2}\dot{x}^2 + \frac{\omega_0^2}{\lambda + 1}\left|x\right|^{\lambda+1} = \frac{\omega_0^2}{\lambda + 1}\left|A\right|^{\lambda+1}.$$

The exact period of the oscillations equals

$$T_\lambda = \frac{4\sqrt{\pi}\sqrt{\frac{1}{2\lambda+2}}A^{\frac{1}{2}-\frac{\lambda}{2}}\Gamma\left(\frac{1}{\lambda+1}\right)}{\omega_0\Gamma\left(\frac{1}{2} + \frac{1}{\lambda+1}\right)}.$$

6.4.1 Trigonometric form

The approximate trigonometric solution to the i.v.p

$$\ddot{x} + \omega_0^2 \mathrm{sign}(x)\left|x\right|^\lambda = 0,\ x(0) = A \text{ and } x'(0) = 0$$

is given by

$$x(t) = A\cos\left(qA^{(\lambda-1)/2}\omega_0 t\right),\ q = \frac{\sqrt{\pi}\Gamma\left(\frac{1}{2} + \frac{1}{\lambda+1}\right)}{2\sqrt{\frac{1}{2\lambda+2}}\Gamma\left(\frac{1}{\lambda+1}\right)}.$$

The solution to the i.v.p.

$$\ddot{x} + \omega_0^2 \text{sign}(x)\,|x|^\lambda = \varepsilon f(x, \dot{x}),\; x(0) = x_0 \text{ and } x'(0) = \dot{x}_0$$

is expressed as

$$x(t) = a(t)\cos(\psi(t)),$$

where

$$\dot{a} = -\frac{\varepsilon}{2\pi\omega_0\omega(a)}\int_0^{2\pi} f\left(a\cos\theta, -a\omega_0\omega(a)\sin\theta\right)\,\sin\theta d\theta,\; a(0) = c_0.$$

$$\dot{\psi} = \frac{2\omega_0\Gamma\left(\frac{\lambda+4}{2}\right)}{\sqrt{\pi}(\lambda+2)\omega(a)\Gamma\left(\frac{\lambda+3}{2}\right)}\,|a|^{\lambda-1} + \tfrac{1}{2}\omega_0\omega(a)$$

$$-\frac{\varepsilon}{2\pi a\omega_0\omega(a)}\int_0^{2\pi} f\left(a\cos\theta, -a\omega_0\omega(a)\sin\theta\right)\cos\theta\, d\theta,\; \psi(0) = c_1.$$

$$\omega(a) := \frac{\Gamma\left(\frac{1}{2}+\frac{1}{\lambda+1}\right)|a|^{\frac{\lambda-1}{2}}}{\sqrt{2}\sqrt{\frac{1}{\pi\lambda+\pi}}\Gamma\left(\frac{1}{\lambda+1}\right)}.$$

The constants c_0 and c_1 are found from the initial conditions $x(0) = x_0$ and $x'(0) = \dot{x}_0$.

6.4.2 Elliptic form

The approximate elliptic solution to the i.v.p

$$\ddot{x} + \omega_0^2 \text{sign}(x)\,|x|^\lambda = 0,\; x(0) = A \text{ and } x'(0) = 0.$$

is given by

$$x(t) = A\text{cn}\left(\Omega t, m\right),$$

where

$$\Omega = \Omega(a) = \omega_0\Omega^* a^{\frac{\lambda-1}{2}} \text{and } \Omega^* = \frac{\sqrt{\frac{2(\lambda+1)}{\pi}}\Gamma\left(\frac{1}{2}+\frac{1}{\lambda+1}\right)}{\Gamma\left(\frac{1}{\lambda+1}\right)}K(m).$$

The modulus m is determined from the condition

$$\sqrt{1-m}K(m) = \frac{\sqrt{\pi}}{1+\lambda}\frac{\Gamma\left(\frac{1}{1+\lambda}\right)}{\Gamma\left(\frac{3+\lambda}{2+2\lambda}\right)}.$$

The value of m may be estimated as follows:

$$m = m(\lambda) = \begin{cases} 1.49471\lambda^5 - 5.29166\lambda^4 + 8.05441\lambda^3 - 7.37196\lambda^2 \\ +4.92236\lambda - 1.80785 & \text{if } 0 \leq \lambda \leq 1 \\ 4.73\text{e-}6\lambda^7 - 0.000205922\lambda^6 + 0.00374466\lambda^5 - 0.0369246\lambda^4 \\ +0.214972\lambda^3 - 0.755035\lambda^2 + 1.56917\lambda - 0.995731 & \text{if } 1 < \lambda \leq 10 \\ 0.806192642 + \frac{0.180426}{\lambda^2} - 0.0452437e^{-\lambda}\lambda + 0.00113865\log(\lambda) + \frac{0.0528631}{\log(\lambda+1)} & \text{if } \lambda > 10. \end{cases}$$

$$m = m(\lambda) \longrightarrow 0.806192642 \text{ as } \lambda \longrightarrow +\infty$$

See Figure 6.4.

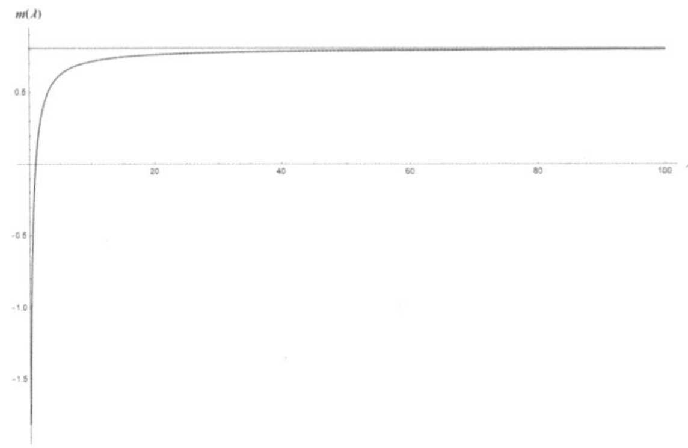

Figure 6.4: The value of m as a function of $\lambda \geq$

$$K(m) = \frac{\pi}{2} + 0.386997m + 0.200579m^2 + 0.212501m^3 + 0.355304m^4 - 0.00911757m^5$$

$$-0.665362m^6 - 0.109512m^7 + 0.792598m^8 + 0.461789m^9 \text{ for } -1 \leq m \leq 0.9$$

with max errorless than 0.21%.

The solution to the i.v.p

$$\ddot{x} + \omega_0^2 \text{sign}(x)\,|x|^\lambda = \varepsilon f(x, \dot{x}),\ x(0) = x_0 \text{ and } x'(0) = \dot{x}_0,$$

or

$$\ddot{x} + \omega_0^2\,|x|^{\lambda-1}\,x = \varepsilon f(x, \dot{x}),\ x(0) = x_0 \text{ and } x'(0) = \dot{x}_0,$$

is expressed as

$$x(t) = a(t)\text{cn}\left(\int_0^t \Omega(\tau)d\tau + \phi(t), m\right).$$

Let

$$\psi = \int_0^t \Omega(\tau)d\tau + \phi(t).$$

We have:

$$x'(t) = \dot{x} \approx \eta(t) := -a(t)\Omega(t)\text{sn}(\psi, m)\text{dn}(\psi, m) = -a\Omega\text{sndn} \text{ and } x''(t) = \ddot{x} \approx \eta'(t).$$

Using the above relations and assuming that $a(t) \geq 0$ yields

$$\begin{cases} \frac{1}{2}C\left((\lambda-3)m\mathrm{cn}^4+(\lambda-1-2(\lambda-1)m)\mathrm{cn}^2+(\lambda+1)(m-1)\right)a^{\lambda/2}\dot{a} = \\ \sqrt{a}\varepsilon f\left(a\mathrm{cn}, -Ca^{\frac{\lambda+1}{2}}\mathrm{sndn}\right)\mathrm{sndn} - a^{\lambda+\frac{1}{2}}\omega_0^2\left|\mathrm{cn}\right|^{\lambda-1}\mathrm{cnsndn} \\ +C^2a^{\lambda+\frac{1}{2}}\left(2\mathrm{cn}^2m-2m+1\right)\mathrm{cnsndn}. \\ \frac{1}{2}C\left((\lambda-3)m\mathrm{cn}^4+(\lambda-1-2(\lambda-1)m)\mathrm{cn}^2+(\lambda+1)(m-1)\right)a^{\frac{\lambda+1}{2}}\dot{\phi} = \\ \varepsilon f\left(a\mathrm{cn}, -Ca^{\frac{\lambda+1}{2}}\mathrm{sndn}\right)\mathrm{cn} - a^{\lambda}\omega_0^2\left|\mathrm{cn}\right|^{\lambda-1}\mathrm{cn}^2 + C^2a^{\lambda}\left(1-2m\mathrm{sn}^2\right)\mathrm{cn}^2. \\ C=\omega_0\Omega^*, \text{ where } \Omega^{st} = \frac{\sqrt{\frac{2(\lambda+1)}{\pi}}\Gamma\left(\frac{1}{2}+\frac{1}{\lambda+1}\right)}{\Gamma\left(\frac{1}{\lambda+1}\right)}K(m),\ \Omega^*=\Omega^*(\lambda). \end{cases}$$

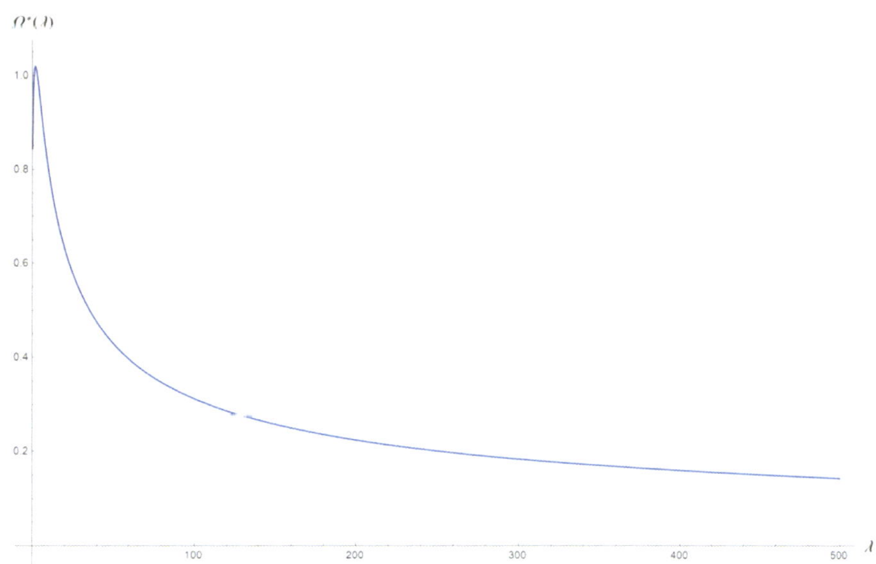

Figure 6.5: The number Ω^* as a function of λ

Figure. The number Ω^* as a function of λ.

The next step consists in integrating the above odes over $0 \leq \psi \leq 4K(m)$. The averaged odes are :

$a'^{\frac{1}{2}-\frac{\lambda}{2}} \int_0^{4K(m)} \mathrm{dn}(\psi|m)\mathrm{sn}(\psi|m) f\left(a(t)\mathrm{cn}(\psi|m), \omega_0\Omega^*\left(-a(t)^{\frac{\lambda+1}{2}}\right)\mathrm{dn}(\psi|m)\mathrm{sn}(\psi|m)\right)d\psi.$

$\theta'(t) = C(-\varepsilon a(t)^{-\frac{\lambda}{2}-\frac{1}{2}} J\left(\mathrm{cn}f\left(\mathrm{cn}a(t), -\mathrm{dnsn}\omega_0\Omega^* a(t)^{\frac{\lambda+1}{2}}\right)\right)+$

$\omega_0^2 a(t)^{\frac{\lambda}{2}-\frac{1}{2}} J\left(\mathrm{cn}^2\left|\mathrm{cn}\right|^{\lambda-1}\right) + D\omega_0^2\left(\Omega^*\right)^2 a(t)^{\frac{\lambda-1}{2}})$

$b = -Ca^{\frac{1+\lambda}{2}}.$

$D = \frac{1}{6m}C(\lambda+3)((1-2m)v_4 + 4(m-1)K(m)).$

$v_4 = E(\mathrm{am}(4K(m)|m)|m).$

$E = C^2 \int_0^{4K(m)} (1-2m\mathrm{sn}^2)\mathrm{cn}^2 d\psi =$
$2C^2((2m-1)E(\mathrm{am}(2K(m)|m)|m) - 2(m-1)K(m)).$
$I_\lambda = \int_0^{4K(m)} |\mathrm{cn}(\psi,m)|^{\lambda-1}\mathrm{cn}^2(\psi,m)d\psi.$

Finally, we solve the averaged odes for the initial conditions $a(0) = c_0$ and $\phi'(0) = c_1$. For the constant v_4 we have the following approximation:

$$v_4 := E(\mathrm{am}(4K(m)|m)|m) =$$

$$2\pi - 1.57117m - 0.293521m^2 - 0.113864m^3 - 0.0730956m^4 - 0.0980684m^5$$

$$-0.0292195m^6 + 0.11042m^7 + 0.029654m^8 - 0.123798m^9 - 0.0730665m^{10}$$

for $-1 \leq m \leq 0.9$ with max error$\mathcal{E} = 0.00022$.

The integral I_λ may be approximated as follows :

$$I_\lambda \approx \begin{cases} 358.701\lambda^{11} + 1383.94\lambda^{10} + 2017.38\lambda^9 + 1310.13\lambda^8 + 255.865\lambda^7 \\ -109.264\lambda^6 - 48.5823\lambda^5 - 0.0781657\lambda^4 + 3.1326\lambda^3 \\ -1.13226\lambda^2 + 0.385126\lambda + 3.2882 \;\text{ if } -1 \leq \lambda \leq 0.24. \\ -5.48\text{e} - 19\lambda^{11} + 3.35\text{e} - 16\lambda^{10} - 9\text{e} - 14\lambda^9 + 1.4\text{e}{-}11\lambda^8 - 1.4\text{e-}9\lambda^7 + 9.5\text{e} - 8\lambda^6 \\ -4.38\text{e} - 6\lambda^5 + 0.000139745\lambda^4 - 0.00304548\lambda^3 + 0.044629\lambda^2 \\ -0.436861\lambda + 3.43363 \text{ if } 0.24 \leq \lambda \leq 100. \end{cases}$$

See Figure 6.6. Assume that the function $f = f(x, \dot{x})$ is a polynomial in nthe variables x and $\dot{x}$.

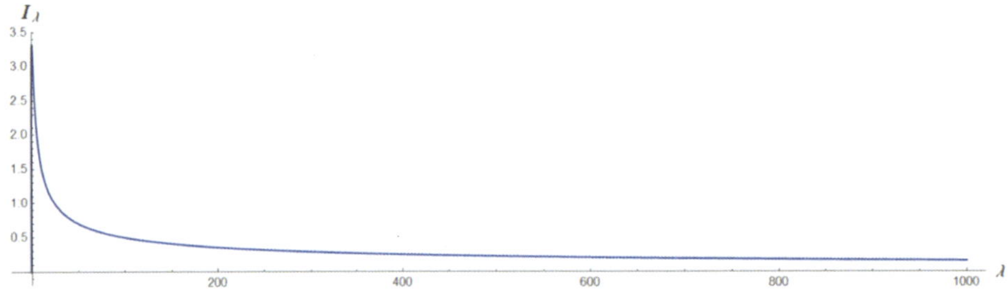

Figure 6.6: $I_\lambda = I_\lambda = \int_0^{4K(m)} |\mathrm{cn}(\psi,m)|^{\lambda-1}\mathrm{cn}^2(\psi,m)d\psi$ as a function of $\lambda \geq -1$

Then

$$f\,(a\mathrm{cn}, b\mathrm{sndn})\,\mathrm{sndn} = P(\mathrm{cn}^2) + \mathrm{cn}P_1(\mathrm{cn}^2) + \mathrm{sndn}P_2(\mathrm{cn}),$$

and

$$f\,(a\mathrm{cn}, b\mathrm{sndn})\,\mathrm{cn} = \hat{P}(\mathrm{cn}^2) + \mathrm{cn}\hat{P}_1(\mathrm{cn}^2) + \mathrm{sndn}\hat{P}_2(\mathrm{cn}),$$

being $P_1(\mathrm{cn}^2)$, $P_2(\mathrm{cn})$, $\hat{P}_1(\mathrm{cn}^2)$ and $\hat{P}_2(\mathrm{cn})$ polynomials. Then

$$\int_0^{4K(m)} f\,(a\mathrm{cn}, b\mathrm{sndn})\,\mathrm{sndn}d\psi = \int_0^{4K(m)} P(\mathrm{cn}^2)d\psi,$$

and

$$\int_0^{4K(m)} f\,(a\mathrm{cn}, b\mathrm{sndn})\,\mathrm{cn}d\psi = \int_0^{4K(m)} \hat{P}(\mathrm{cn}^2)d\psi.$$

Let

$$P(x) = \sum_{k=0}^{M} \eta_k x^k \text{ and } \hat{P}(x) = \sum_{l=0}^{N} \hat{\eta}_l x^l.$$

Then

$$\int_0^{4K(m)} f\,(a\mathrm{cn}, b\mathrm{sndn})\,\mathrm{sndn}d\psi = \sum_{k=0}^{M} \eta_k \int_0^{4K(m)} \mathrm{cn}^{2k}(\psi, m)d\psi,$$

and

$$\int_0^{4K(m)} f\,(a\mathrm{cn}, b\mathrm{sndn})\,\mathrm{cn}d\psi = \sum_{l=0}^{N} \hat{\eta}_k \int_0^{4K(m)} \mathrm{cn}^{2l}(\psi, m)d\psi.$$

Thus, we only need to evaluate the integrals

$$\int_0^{4K(m)} \mathrm{cn}^{2j}(\psi, m)d\psi \text{ for any } j. \tag{6.3}$$

The integrals (6.3) depend on λ since the modulus m depends on λ. We will make use of the following Chebyshev approximation :

$\int_0^{4K(m)} |\mathrm{cn}(\psi,m)|^{2\rho}\, d\psi \approx 4.77422 + 1.8196\lambda - 0.415422\lambda^2 + 0.0352511\lambda^3 - 3.83173\rho$
$-2.51402\lambda\rho + 0.525365\lambda^2\rho - 0.0426378\lambda^3\rho + 2.41205\rho^2 + 1.72846\lambda\rho^2 - 0.349497\lambda^2\rho^2$
$+0.0278796\lambda^3\rho^2 - 0.863984\rho^3 - 0.635335\lambda\rho^3 + 0.126711\lambda^2\rho^3 - 0.0100351\lambda^3\rho^3 + 0.176117\rho^4$
$+0.130888\lambda\rho^4 - 0.0259289\lambda^2\rho^4 + 0.00204624\lambda^3\rho^4 - 0.0202716\rho^5 - 0.0151462\lambda\rho^5 + 0.0029892\lambda^2\rho^5$
$-0.00023544\lambda^3\rho^5 + 0.00122609\rho^6 + 0.000918911\lambda\rho^6 - 0.000180936\lambda^2\rho^6 + 0.0000142341\lambda^3\rho^6$
$-0.0000302711\rho^7 - 0.0000227312\lambda\rho^7 + 4.47\text{e-}6\lambda^2\rho^7 - 3.51\text{e-}7\lambda^3\rho^7$ for $0 \le \lambda \le 5$ and $0 \le \rho \le 10$.

Alternativeñy, we may use the following quadrature formula :

$$\int_a^b f(x)dx \approx \tfrac{1}{480}(b-a)(24f(\tfrac{3a}{5}+\tfrac{2b}{5}) + 24f(\tfrac{2a}{5}+\tfrac{3b}{5}) + 24f(\tfrac{a+b}{2}) + 24f(\tfrac{1}{4}(3a+b)) + 24f(\tfrac{1}{5}(4a+b))$$

$$+20f(\tfrac{1}{10}(9a+b)) + 31f(\tfrac{1}{20}(19a+b)) + 24f(\tfrac{1}{4}(a+3b)) + 24f(\tfrac{1}{10}(7a+3b))$$

$$+25f(\tfrac{1}{20}(17a+3b)) + 24f(\tfrac{1}{5}(a+4b)) + 24f(\tfrac{1}{10}(3a+7b)) + 24f(\tfrac{1}{20}(13a+7b))$$

$$+20f(\tfrac{1}{10}(a+9b)) + 24f(\tfrac{1}{20}(11a+9b)) + 24f(\tfrac{1}{20}(9a+11b)) + 24f(\tfrac{1}{20}(7a+13b))$$

$$+25f(\tfrac{1}{20}(3a+17b)) + 31f(\tfrac{1}{20}(a+19b)) + 8f(a) + 8f(b)).$$

Example 25. Let

$$\ddot{x} + \omega_0^2 \text{sign}(x) |x|^{\lambda} = -\varepsilon \dot{x},\ x(0) = A \text{ and } x'(0) = -\varepsilon A.$$

The averaged differential equations are

$$\dot{a} = \frac{(\varepsilon (4K[m](-1+m) + (1-2m)\upsilon_4))}{3mr} a,\ \dot{\theta} = 0,$$

where

$$\upsilon_4 = E(\text{am}(4K(m)|m)|m)$$

Letting

$$\varepsilon = 0.1,,\ \lambda = 1.2,\ A = 2,\ \omega_0^2 = 2 \text{ and } 0 \le t \le 100$$

we get

$$m = 0.107573,\ \upsilon_4 = 6.11064,\ r = 6.32495.$$

The solution is expresed as

$$x(t) = 2.2.71828^{-0.047619t} \text{cn}\left(321.258\ - 321.2582.71828^{-0.0047619t} \middle| 0.107573\right)$$

In the case when $\lambda > 1$ we may proceed as follows. First, we use the chebyshevapproximation

$$\text{sign}(x) |x|^{\lambda} \approx px + qx^3 + rx^5 \text{ for } |x| \le M,$$

where

$$\begin{aligned} p &= \frac{2^{-\frac{3\lambda}{2}-\frac{1}{2}}}{(3+\sqrt{3})M} \left(32\left(\sqrt{3}-1\right)^{\lambda} M^{\lambda} + 2^{\lambda}\left(\left(9-5\sqrt{3}\right)2^{\frac{\lambda+1}{2}} + 4 - 4\sqrt{3}\right) M^{\lambda}\right) \\ q &= \frac{2}{(3+\sqrt{3})M^3}^{1-\frac{3\lambda}{2}} \left(2^{\lambda}\left(\left(7\sqrt{3}-15\right)2^{\lambda/2} + \sqrt{2}\left(3+5\sqrt{3}\right)\right) M^{\lambda} - 24\sqrt{2}\left(\sqrt{3}-1\right)^{\lambda} M^{\lambda}\right) \\ r &= \frac{2^{3-\frac{3\lambda}{2}}}{(3+\sqrt{3})M^5} \left(4\sqrt{2}\left(\sqrt{3}-1\right)^{\lambda} M^{\lambda} - 2^{\lambda}\left(\left(\sqrt{3}-3\right)2^{\lambda/2} + \sqrt{2} + \sqrt{6}\right) M^{\lambda}\right) \end{aligned}$$

For example, let

$$M = 1 \text{ and } \lambda = \sqrt{2}.$$

See Figure 6.7.

Next, we solve the i.v.p.

$$\ddot{x} + \omega_0^2 (px + qx^3 + rx^5) = \varepsilon f(x, \dot{x}),\ x(0) = x_0 \text{ and } x'(0) = \dot{x}_0 \tag{6.4}$$

In the case when $\varepsilon = 0$ we have the followintg approximate solution :

$$x(t) = c_0 \text{cn}\left(wt + c_1, m\right),$$

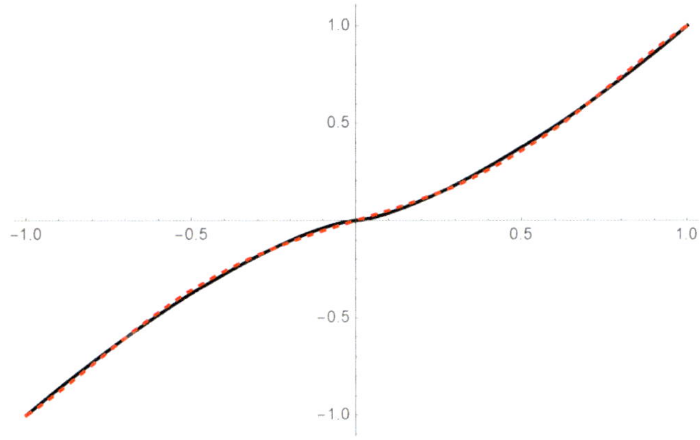

Figure 6.7: The polynomial approximation of $\operatorname{sign}(x)\,|x|^{\lambda}$ on $-M \leq x \leq M$

where

$$m = \frac{8c_0^2 q + 10c_0^4 r}{16c_0^2 q + 15c_0^4 r + 16p} \text{ and } w = \omega_0 \sqrt{c_0^2 q + \frac{15c_0^4 r}{16} + p}.$$

The solution to the i.v.p. (6.4) is assumed to be

$$x(t) = a(t)\operatorname{cn}(wt + \phi(t), m)$$

We apply now the elliptic averaging method using the formulas for (6.3).

6.5 Cubication-Quintification Method

Let

$$\ddot{x} + f(x) = 0,\ x(0) = x_0 \text{ and } x'(0) = \dot{x}_0.$$

Assume that $-M \leq f(x) \leq M$ and that $x = 0$ is the only equilibrium point. If the function $F(x)$ is odd, we approximate it as follows :

$$f(x) \approx Px + Qx^3 + Rx^5,$$

where

$$P = \tfrac{1}{3\sqrt{6}(1+\sqrt{3})M}[\left(3+\sqrt{3}\right) f\left(-\tfrac{M}{\sqrt{2}}\right) - \left(3+\sqrt{3}\right) f\left(\tfrac{M}{\sqrt{2}}\right) - 2\left(12+7\sqrt{3}\right) f\left(-\tfrac{1}{2}\sqrt{2-\sqrt{3}M}\right).$$
$$+2\left(12+7\sqrt{3}\right) f\left(\tfrac{1}{2}\sqrt{2-\sqrt{3}M}\right) + \left(6-4\sqrt{3}\right)\left(f\left(-\tfrac{1}{2}\sqrt{2+\sqrt{3}M}\right) - f\left(\tfrac{1}{2}\sqrt{2+\sqrt{3}M}\right)\right)].$$

$$Q = \tfrac{\sqrt{2}}{3M^3}[\left(\begin{array}{l} -8f\left(-\tfrac{M}{\sqrt{2}}\right) + 8f\left(\tfrac{M}{\sqrt{2}}\right) + \left(7+5\sqrt{3}\right) f\left(-\tfrac{1}{2}\sqrt{2-\sqrt{3}M}\right) - \left(7+5\sqrt{3}\right) f\left(\tfrac{1}{2}\sqrt{2-\sqrt{3}M}\right) \\ +\left(5\sqrt{3}-7\right)\left(f\left(-\tfrac{1}{2}\sqrt{2+\sqrt{3}M}\right) - f\left(\tfrac{1}{2}\sqrt{2+\sqrt{3}M}\right)\right) \end{array}\right.$$

$$R = \frac{4\sqrt{2}(\sqrt{3}-1)}{3M^5}[\left(\begin{array}{l} \left(1+\sqrt{3}\right) f\left(-\tfrac{M}{\sqrt{2}}\right) - \left(1+\sqrt{3}\right) f\left(\tfrac{M}{\sqrt{2}}\right) - \left(2+\sqrt{3}\right) f\left(-\tfrac{1}{2}\sqrt{2-\sqrt{3}M}\right) \\ +\left(2+\sqrt{3}\right) f\left(\tfrac{1}{2}\sqrt{2-\sqrt{3}M}\right) - f\left(-\tfrac{1}{2}\sqrt{2+\sqrt{3}M}\right) + f\left(\tfrac{1}{2}\sqrt{2+\sqrt{3}M}\right) \end{array}\right)]$$

Then, we solve the i.v.p.

$$\ddot{x} + Px + Qx^3 + Rx^5 = 0,\ x(0) = x_0 \text{ and } x'(0) = \dot{x}_0.$$

If $f(x)$ is not odd, $m \leq f(x) \leq M$ and

$$|f(x) - P_3(x)| \leq \frac{1}{192}\left(\frac{M-m}{2}\right)^4 \max_{\xi\in[m,M]}\left|f^{(4)}(\xi)\right| < 0.25,$$

where $P_3(x)$ is the Lagrange interpolation polynomial evaluated at nodes

$$(\eta_1, f(\eta_1)),\ (\eta_2, f(\eta_2)),\ (\eta_3, f(\eta_3)) \text{ and } (\eta_4, f(\eta_4)),$$

being

$$\eta_k = \frac{M+m}{2} + \frac{M-m}{2}\cos\left(\frac{2k+1}{8}\right),\ k = 0, 1, 2, 3,$$

we use the approximation

$$f(x) \approx n + px + qx^2 + rx^3 := P_3(x).$$

We then solve the i.v.p.

$$\ddot{x} + n + px + qx^2 + rx^3 = 0,\ x(0) = x_0 \text{ and } x'(0) = \dot{x}_0. \tag{6.5}$$

As we already seen, the exact solution to (6.5) may be written as

$$x(t) = c_0 + \frac{Q}{1 + R\wp(t \pm C; g_2, g_3)},$$

where

$$
\begin{aligned}
Q &= -\tfrac{6\left(c_0p+c_0^2q+c_0^3r+n\right)}{2c_0q+3c_0^2r+p},\ R = \tfrac{12}{2c_0q+3c_0^2r+p}. \\
g_2 &= \tfrac{1}{12}\left(-12c_0nr-6c_0^2pr-4c_0^3qr-3c_0^4r^2-4nq+p^2\right). \\
g_3 &= \tfrac{1}{216}\left(\begin{array}{c} 36c_0npr-12c_0nq^2+18c_0^2p^2r-6c_0^2pq^2+12c_0^3pqr \\ +9c_0^4pr^2-4c_0^3q^3-3c_0^4q^2r+27n^2r-6npq+p^3 \end{array}\right). \\
C &= \wp^{-1}(c_1;g_2,g_3),\ c_1 = \tfrac{5c_0p+2c_0qx_0+4c_0^2q+3c_0^2rx_0+3c_0^3r+6n+px_0}{12(c_0-x_0)}.
\end{aligned}
$$

The number c_0 is a solution to thre quartic

$$c_0^4+\frac{4q}{3r}c_0^3+\frac{2p}{r}c_0^2+\frac{4n}{r}c_0-\frac{12nx_0+6px_0^2+4qx_0^3+3rx_0^4+6\dot{x}_0^2}{3r}=0.$$

On the orther hand, we have the identity

$$\wp\left(t;g_2,g_3\right)=-\frac{1}{12}(4m+1)\omega+\frac{\omega}{2\left(1-\operatorname{cn}\left(\sqrt{\omega}t\right|m)\right)},$$

where

$$m=\frac{1}{4}\left(2\pm\frac{1}{\omega}\sqrt{3}\sqrt{4g_2+\omega^2}\right) \text{ and } \omega^6-12g_2\omega^4+256\left(g_2^3-27g_3^2\right)=0.$$

It is clear that

$$x(t)=c_0+\frac{Q}{1+R\wp(t\pm C;g_2,g_3)}=\bar{P}+\frac{\bar{Q}}{1+\bar{R}\operatorname{cn}\left(\sqrt{\omega}(t\pm C)\right|m)},$$

where

$$
\begin{aligned}
\bar{P} &= c_0-\tfrac{6(c_0(c_0(c_0r+q)+p)+n)}{2c_0q+3c_0^2r-(4m+1)\omega+p}. \\
\bar{Q} &= \tfrac{36\omega(c_0(c_0(c_0r+q)+p)+n)}{\left(2c_0q+3c_0^2r+(5-4m)\omega+p\right)\left(2c_0q+3c_0^2r-(4m+1)\omega+p\right)}. \\
\bar{R} &= \tfrac{6\omega}{2c_0q+3c_0^2r-4m\omega+p+5\omega}-1.
\end{aligned}
$$

Also,

$$x(t)=A+\frac{B}{1+u(t)},$$

where

$$
\begin{aligned}
&\ddot{u}+\alpha u+\beta u^3=0,\ u(0)=u_0,\ u'(0)=\dot{u}_0. \\
&\alpha = \tfrac{12Cp+24C\mathcal{P}q+36C\mathcal{P}^2r+9n^2+6np\mathcal{P}-6n\mathcal{P}^2q-18n\mathcal{P}^3r+3p^2\mathcal{P}^2+2p\mathcal{P}^3q-3p\mathcal{P}^4r+\mathcal{P}^4q^2}{12C-12n\mathcal{P}-6p\mathcal{P}^2-4\mathcal{P}^3q-3\mathcal{P}^4r}. \\
&\beta = -\tfrac{3(n+p\mathcal{P}+\mathcal{P}^2q+\mathcal{P}^3r)^2}{12C-12n\mathcal{P}-6p\mathcal{P}^2-4\mathcal{P}^3q-3\mathcal{P}^4r}. \\
&u_0 = \tfrac{x_0-\mathcal{P}-\mathcal{Q}}{\mathcal{P}-x_0},\ \dot{u}_0 = -\tfrac{\mathcal{Q}\dot{x}_0}{(\mathcal{P}-x_0)^2}. \\
&\mathcal{Q} = \tfrac{12C-12n\mathcal{P}-6p\mathcal{P}^2-4\mathcal{P}^3q-3\mathcal{P}^4r}{3(n+p\mathcal{P}+\mathcal{P}^2q+\mathcal{P}^3r)}. \\
&C = nx_0+\tfrac{px_0^2}{2}+\tfrac{qx_0^3}{3}+\tfrac{rx_0^4}{4}+\tfrac{\dot{x}_0^2}{2}.
\end{aligned}
$$

The number $\mathcal{P}$ obeys the following sextic :

$$
\begin{aligned}
&-\left(27nr^2-9pqr+2q^3\right)\mathcal{P}^6+3\left(36Cr^2-6nqr+9p^2r-2pq^2\right)\mathcal{P}^5-15\left(-12Cqr-9npr+2nq^2\right)\mathcal{P}^4 \\
&+30\left(4Cq^2+9n^2r\right)\mathcal{P}^3+90\left(-6Cnr+2Cpq+n^2q\right)\mathcal{P}^2+18\left(24C^2r-4Cnq+6Cp^2+3n^2p\right)\mathcal{P} \\
&+144C^2q+108Cnp+54n^3=0.
\end{aligned}
$$

Example 26. Let

$$\ddot{x} + a_0 + a_1 x + a_2 x^2 + a_3 x^3 + a_4 x^4 = 0, x(0) = x_0, x'(0) = \dot{x}_0 \text{ with } a_4 \neq 0.$$

We demand that

$$\frac{|a_4|}{128}(M-m)^4 < 1/3.$$

Choice :

$$a_0 = -\frac{1}{20}, a_1 = -\frac{3}{20}, a_2 = -\frac{3}{40}, a_3 = \frac{1}{20}, a_4 = \frac{1}{16}, \; x_0 = 1 \text{ and } \dot{x}_0 = 1/2$$

We have

$$\dot{x}^2 = -\frac{1}{40}(x-2)x(x+1)\left(x^2+2x+2\right)$$

and so

$$m = 0, \; M = 1 \text{ and } \frac{|a_4|}{128}(M-m)^4 = 0.008.$$

The approximation reads

$$f(x) \approx 0.3x^3 - 0.387499x^2 - 0.0250013x - 0.0578122.$$

See Figure 6.8.

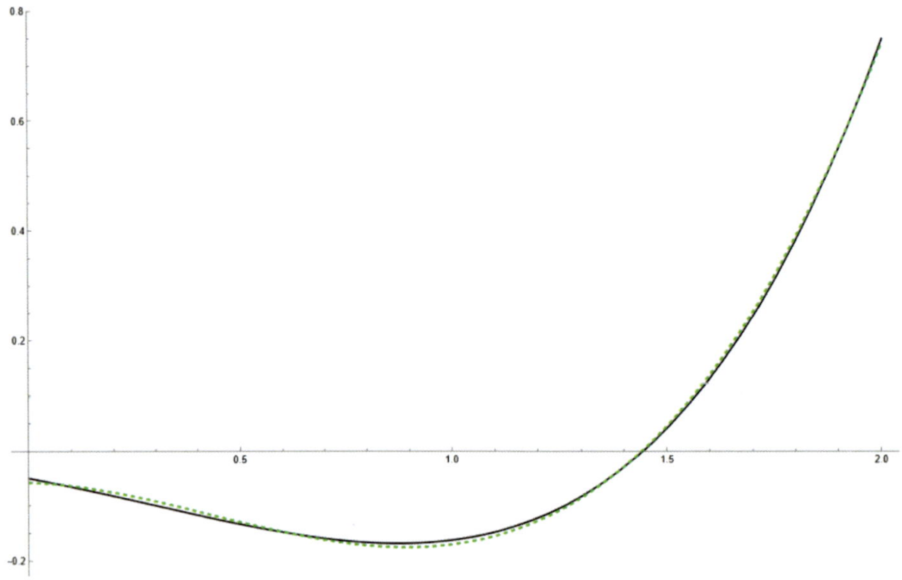

Figure 6.8

We solve the i.v.p.

$$\ddot{x} + 0.3x^3 - 0.387499x^2 - 0.0250013x - 0.0578122 = 0, \; x(0) = 1 \text{ and } x'(0) = 1/2.$$

See Figure 6.9.

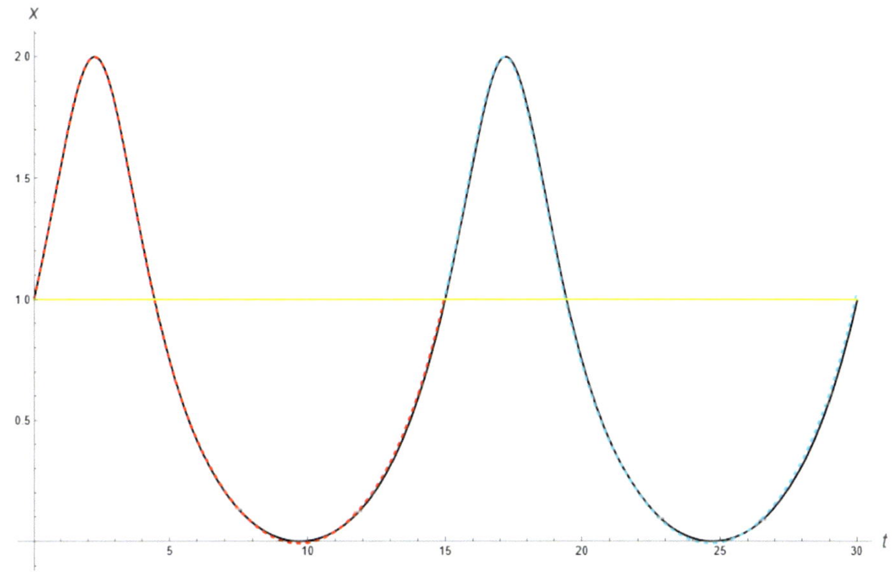

Figure 6.9: Period = 14.9583

The approximate analytical solutions to the i.v.p.

$$\ddot{x} - \frac{1}{20} - \frac{3}{20}x - \frac{3}{40}x^2 + \frac{1}{20}x^3 + \frac{1}{16}x^4 = 0, x(0) = 1, x'(0) = 1/2$$

are given as follows :

$$x_{\wp}(t) = 1.9993 - \frac{2.19701}{1+5.93174\wp(t+(5.26398+9.36417i);-0.00757155,0.000135824)}.$$
Error = 0.0199523

$$x_{\text{Jacobi}}(t) = -0.785763 + \frac{1.2147}{1-0.563851\text{cn}(1.00476-0.453585t|0.269037)}$$
Error = 0.0199523

$$x_{\text{Trigo}}(t) = -0.785763 - \frac{102.795}{\cos(2.79141-1.26014t)+46.7161\cos(0.930471-0.420048t)-84.6253}.$$
Error = 0.0171511

Example 27. Let

$$\ddot{x} + x + 2x^3 - x^5 + x^7 + x^9 = 0, \ x(0) = 1 \text{ and } x'(0) = 0.$$

We use the following approximation with $M = 1$:

$$x + 2x^3 - x^5 + x^7 + x^9 \approx 69/64x + 5/8x^3 + 35/16x^5.$$

Then, we solve the i.v.p.

$$\ddot{x} + 69/64x + 5/8x^3 + 35/16x^5 = 0, \quad x(0) = 1 \text{ and } x'(0) = 0. \tag{6.6}$$

The exact solution to the i.v.p. (6.6) is given by

$$m = \tfrac{33781-169\sqrt{33781}}{67562},\ \omega = \tfrac{\sqrt{33781}}{64}.$$
$$\lambda = \tfrac{10}{783}\left(169-\sqrt{33781}\right),\ \mu = \tfrac{169\sqrt{33781}-31171}{2610}.$$

$$x(t) = \frac{\sqrt{1+\lambda+\mu}\operatorname{cn}(\sqrt{\omega}t,m)}{\sqrt{1+\lambda \operatorname{cn}^2(\sqrt{\omega}t,m)+\mu \operatorname{cn}^4(\sqrt{\omega}t,m)}}.$$

The period of oscillations equals

$$T = \frac{32}{\sqrt[4]{33781}} K\left(\frac{1}{2} - \frac{169}{2\sqrt{33781}}\right) = 3.74585...$$

See Figure 6.10.

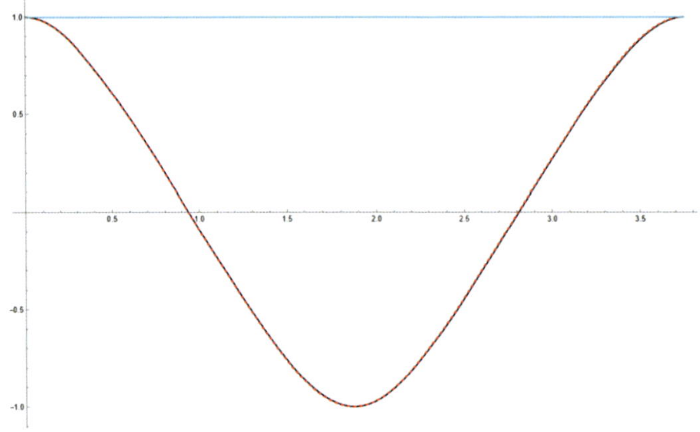

Figure 6.10: error= 0.00535963

Example 28. Let

$$\ddot{x} + x + 1/3\sqrt[3]{x} + 1/5\sqrt[5]{x} = 0,\ x(0) = 1 \text{ and } x'(0) = 0$$

In this case,

$$x + 1/3\sqrt[3]{x} + 1/5\sqrt[5]{x} \approx +2.57551x - 2.56922x^3 + 1.5719x^5.$$

The approximate solution equals

$$x(t) = \frac{\sqrt{1+\lambda+\mu}\operatorname{cn}(\sqrt{\omega}t, m)}{\sqrt{\sqrt{1+\lambda \operatorname{cn}^2(\sqrt{\omega}t, m)+\mu \operatorname{cn}^4(\sqrt{\omega}t, m)}}} \text{ with period } T = 4.92594,$$

where

$$\omega = 1.69239, \mu = 0.0761791, \lambda = 0.243255, \mu = -0.0824609.$$

See Figure 6.11.

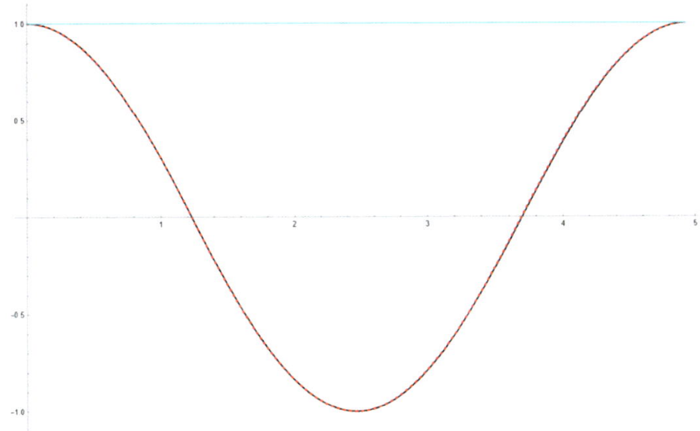

Figure 6.11: error= 0.00705968

Example 29. Evaluation of hyperelliptic integrals. Let us consider the integral

$$J=\int_\alpha^\beta \frac{dx}{\sqrt{(x-\alpha)(x-\beta)(x-\gamma)(x-\delta)(x-\varepsilon)}}\ \text{for}\ 0<\alpha<\beta<\gamma<\delta<\varepsilon.$$

This integral is difficult to evaluate in closed form. To get its approximate value we use the folowing approximation:

$$(x-\alpha)(x-\beta)(x-\gamma)(x-\delta)(x-\varepsilon)=P(x):=c_0+c_1x+c_2x^2+c_3x^3+c_4x^4,$$

where

$$\begin{aligned}
&c_0=\tfrac{1}{512}(\alpha+\beta)\left(\alpha^4+44\alpha^3\beta+166\alpha^2\beta^2+44\alpha\beta^3+\beta^4\right)-\alpha\beta\gamma\delta\varepsilon\\
&c_1=\tfrac{1}{256}\left(-25\alpha^4+4\beta\left(64\alpha\gamma\delta-75\alpha^3\right)-630\alpha^2\beta^2-300\alpha\beta^3-25\beta^4\right)+\\
&\varepsilon(\alpha\delta(\beta+\gamma)+\alpha\beta\gamma+\beta\gamma\delta)\\
&c_2=\tfrac{1}{32}\left(25\alpha^3+135\alpha^2\beta-32\delta(\alpha(\beta+\gamma)+\beta\gamma)+\alpha\beta(135\beta-32\gamma)+25\beta^3\right)-\\
&\varepsilon(\delta(\alpha+\beta+\gamma)+\alpha(\beta+\gamma)+\beta\gamma)\\
&c_3=-\tfrac{35\alpha^2}{16}+\varepsilon(\alpha+\beta+\gamma+\delta)+\delta(\alpha+\beta+\gamma)-\tfrac{37\alpha\beta}{8}+\alpha\gamma-\tfrac{35\beta^2}{16}+\beta\gamma\\
&c_4=\tfrac{3(\alpha+\beta)}{2}-\gamma-\delta-\varepsilon
\end{aligned}$$

Next, we write

$$P(x)=\lambda(x-a)(x-b)(x-c)(x-d),\ \text{where}\ d<c<a<b$$

and then we evaluate the integral

$$I=\int_a^b \frac{dx}{\sqrt{\lambda}\sqrt{(x-a)(x-b)(x-c)(d-x)}}=\frac{2K\left(\frac{(a-b)(c-d)}{(a-c)(b-d)}\right)}{\sqrt{\lambda}\sqrt{(a-c)(b-d)}}$$

For exmple, let

$$J=\int_\alpha^\beta \frac{dx}{\sqrt{(x-1)(x-2)(x-3)(x-5)(x-15)}}$$

A suitable approximation is

$$(x-1)(x-2)(x-3)(x-5)(x-15) \approx 18.5(x-3.67179)(x-3.26402)(2-x)(x-1) \qquad (6.7)$$

See Figure 6.12. The approximate value for the integral is then $(a=1,\ b=2, c=3.26402, d=$

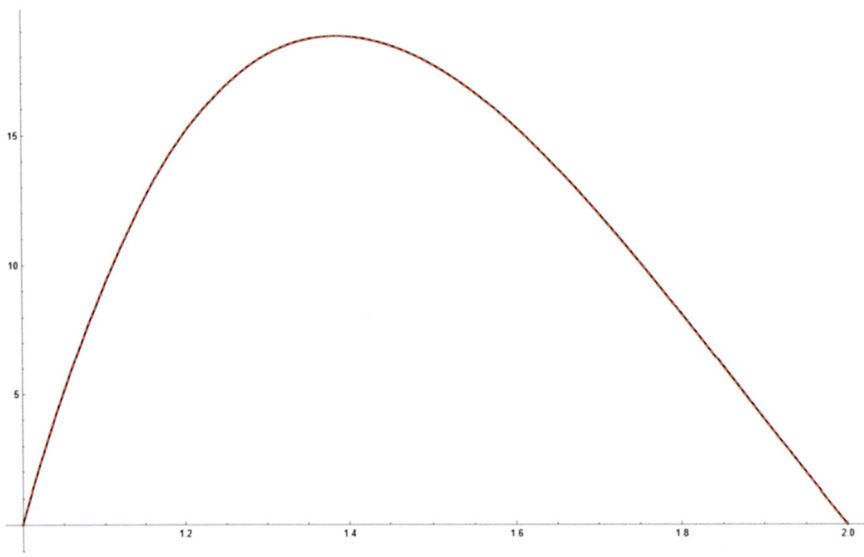

Figure 6.12: The approximation (6.7). Error= 0.00354632

$3.67179)$

$$J = \frac{2\sqrt{\frac{2}{37}}}{\sqrt{(a-c)(b-d)}} K\left(\frac{(a-b)(c-d)}{(a-c)(b-d)}\right) \approx 0.386208015675091.$$

The numerical integration gives the value

$$J = 0.3862855772969425,$$

which differs from the approximate value by 0.0000775616. We conclude that our approximation is excellent.

In general, some elliptic integrals have the form

$$J = \int_{\alpha}^{\beta} \frac{dx}{\sqrt{g(x)}}.$$

In order to get its approximate value, we consider the ode

$$\dot{x}^2 = g(x),\ x(\alpha) = 0.$$

This ode may bewritten in the form

$$\ddot{x} = \frac{1}{2} g'(x)^2,\ x(\alpha) = 0 \text{ and } x'(0) = \pm\sqrt{g(0)}.$$

Next, we approximate the function $f(x) = \frac{1}{2}g'(x)^2$ by means of a cubic polynomial. That way we will obtain the approximate value for the period of the solution and this will solve the problem.

6.6 Summary

In Chapter 6 we developed trigonometric and elliptic averaging methods tailored to the analysis of strongly nonlinear oscillators. The main conclusions are the following:

1. **Averaging as a unifying framework for nonlinear oscillations.** The chapter starts from the general perturbed oscillator

$$\ddot{x} + V'(x) = \varepsilon f(x, \dot{x}),$$

   and formulates the averaging method in terms of slowly varying amplitude and phase, $a(t)$ and $\varphi(t)$. This provides a unified description of the slow dynamics for a wide class of nonlinear oscillators, connecting the exact solution of the unperturbed system with the effective evolution under small perturbations.

2. **Trigonometric averaging for weakly to moderately nonlinear systems.** When the unperturbed motion is well approximated by a trigonometric waveform, the standard (Fourier-based) averaging procedure yields evolution equations for the amplitude and phase which are simple and explicit. This approach is effective for weakly nonlinear oscillators and provides clear formulas for frequency shifts, damping effects and steady-state responses, but its accuracy deteriorates as the nonlinearity becomes stronger.

3. **Elliptic averaging for strongly nonlinear oscillations.** To cope with strongly nonlinear regimes, the chapter develops elliptic averaging methods in which the carrier solution is expressed in terms of Jacobi elliptic functions (or, in some cases, Weierstrass elliptic functions). The amplitude and phase equations are then obtained by averaging over one period expressed as $4K(m)$, where $K(m)$ is the complete elliptic integral and m is the modulus. This construction naturally incorporates amplitude–frequency dependence and double–well effects into the averaged dynamics.

4. **Applications to Duffing and related oscillators.** The elliptic averaging framework is applied to several prototypes, including the Duffing oscillator and its cubic–quintic generalizations. In each case, explicit averaged systems for $(a(t), \varphi(t))$ are derived and used to approximate the full dynamics. Numerical comparisons show that elliptic averaging remains accurate for large amplitudes and deep nonlinear potentials, substantially extending the validity of classical trigonometric averaging.

5. **Cubication–quintification as a bridge to elliptic structures.** The chapter introduces the cubication–quintification method to approximate general nonlinearities by an equivalent cubic–quintic potential. Once this approximation is performed, the resulting oscillator fits naturally into the elliptic averaging framework. This provides a practical route to estimate

periods, frequency–amplitude relations and other dynamical quantities for nonlinear systems whose exact solutions are not available.

6. **Improved accuracy and extended domain of validity.** Systematic numerical tests demonstrate that elliptic averaging achieves significantly smaller errors than purely trigonometric averaging in the strongly nonlinear regime, while retaining analytical transparency. The method captures subtle features such as nonlinear frequency shifts and waveform asymmetries that are inaccessible to simpler approximations.

7. **Foundations for subsequent analytical and numerical studies.** Overall, Chapter 6 shows that averaging methods based on elliptic functions form a powerful tool for the analysis of strongly nonlinear oscillators. They bridge the gap between fully exact elliptic solutions and classical perturbative techniques, and they provide the basis for the more detailed numerical benchmarks and applications discussed in the following chapters.

Chapter 7

Numerical Methods

Second-order nonlinear ordinary differential equations (ODEs) appear frequently in mechanics, physics, engineering, and applied sciences. Analytical solutions are only available in very restricted cases (often involving special functions such as elliptic or hypergeometric functions). For general nonlinear problems, *numerical methods* provide the most practical and reliable approach.

The numerical treatment of a nonlinear second-order ODE of the form

$$\ddot{x}(t) = f(t, x(t), \dot{x}(t)), \tag{7.1}$$

involves discretizing the time domain and approximating derivatives by finite differences or through iterative schemes. Several classes of numerical methods are widely used:

Runge–Kutta Methods

Higher-order Runge–Kutta schemes are among the most popular for nonlinear ODEs due to their balance of accuracy and stability. A second-order ODE is typically reformulated as a system of two first-order equations, and then solved using explicit or implicit RK integrators. For stiff nonlinear systems, implicit Runge–Kutta methods (e.g., Radau or Gauss methods) offer enhanced stability.

7.1 Sixth-Order Runge–Kutta Method

The Runge–Kutta family of methods is among the most widely used approaches for the numerical integration of ordinary differential equations (ODEs). The sixth-order Runge–Kutta method (RK6) offers higher accuracy per step than the classical fourth-order RK4 scheme, while maintaining a single-step character.

We consider the initial value problem:

$$\frac{dy}{dt} = f(t, y), \qquad y(t_0) = y_0. \tag{7.2}$$

For a step size h, the RK6 update takes the form

$$y_{n+1} = y_n + h \sum_{i=1}^{s} b_i k_i, \tag{7.3}$$

where each stage k_i is an evaluation of f at intermediate points, depending on coefficients a_{ij} and c_i. The coefficients are typically organized in a *Butcher tableau.*

A widely used explicit sixth-order Runge–Kutta scheme with seven stages is given by:

$$\begin{aligned}
k_1 &= f(t_n,\, y_n)\,, \\
k_2 &= f\Big(t_n + \tfrac{h}{3},\, y_n + \tfrac{h}{3}k_1\Big)\,, \\
k_3 &= f\Big(t_n + \tfrac{h}{3},\, y_n + \tfrac{h}{6}k_1 + \tfrac{h}{6}k_2\Big)\,, \\
k_4 &= f\Big(t_n + \tfrac{h}{2},\, y_n + \tfrac{h}{8}k_1 + \tfrac{3h}{8}k_3\Big)\,, \\
k_5 &= f\Big(t_n + \tfrac{2h}{3},\, y_n + \tfrac{h}{2}k_1 - \tfrac{3h}{2}k_3 + 2hk_4\Big)\,, \\
k_6 &= f\Big(t_n + \tfrac{h}{3},\, y_n - \tfrac{3h}{2}k_1 + \tfrac{9h}{2}k_3 - 6hk_4 + \tfrac{4h}{3}k_5\Big)\,, \\
k_7 &= f\Big(t_n + h,\, y_n + \tfrac{h}{6}k_1 + 2hk_4 + \tfrac{h}{2}k_5 + \tfrac{h}{6}k_6\Big)\,.
\end{aligned} \tag{7.4}$$

The final update is

$$y_{n+1} = y_n + \frac{h}{90}\left(7k_1 + 32k_3 + 12k_4 + 32k_5 + 7k_7\right). \tag{7.5}$$

This scheme achieves sixth-order accuracy, with local truncation error $\mathcal{O}(h^7)$ and global error $\mathcal{O}(h^6)$. It is particularly valuable when high accuracy is required without excessively reducing the step size, making it suitable for nonlinear oscillators, stiff problems (with implicit variants), and long-time simulations. A more general scheme is described below. Consider the secondâ“order nonlinear ODE

$$\ddot{y}(t) = f(t, y(t), \dot{y}(t)), \qquad y(t_0) = y_0, \qquad \dot{y}(t_0) = v_0. \tag{7.6}$$

Introduce the firstâ“order system

$$\begin{aligned}
y_1(t) &= y(t), \quad \dot{y}_1(t) = y_2(t), \\
y_2(t) &= \dot{y}(t), \quad \dot{y}_2(t) = f(t, y_1(t), y_2(t)),
\end{aligned} \tag{7.7}$$

or, in vector form,

$$\dot{\mathbf{Y}}(t) = \mathbf{F}(t, \mathbf{Y}(t)), \qquad \mathbf{Y} = \begin{bmatrix} y_1 \\ y_2 \end{bmatrix}, \qquad \mathbf{F}(t, \mathbf{Y}) = \begin{bmatrix} y_2 \\ f(t, y_1, y_2) \end{bmatrix}. \tag{7.8}$$

We apply a *7â“stage explicit sixthâ“order Rungeâ“Kutta (RK6)* method with step size h. The method is defined by the following *Butcher tableau* (one widely used RK6 scheme):

$$
\begin{array}{c|ccccccc}
0 & & & & & & & \\
\frac{1}{3} & \frac{1}{3} & & & & & & \\
\frac{1}{3} & \frac{1}{6} & \frac{1}{6} & & & & & \\
\frac{1}{2} & \frac{1}{8} & 0 & \frac{3}{8} & & & & \\
\frac{2}{3} & \frac{1}{2} & 0 & -\frac{3}{2} & 2 & & & \\
\frac{1}{3} & -\frac{3}{2} & 0 & \frac{9}{2} & -6 & \frac{4}{3} & & \\
1 & \frac{1}{6} & 0 & 0 & 2 & \frac{1}{2} & \frac{1}{6} & \\
\hline
 & \frac{1}{90} & 0 & \frac{32}{90} & \frac{12}{90} & \frac{32}{90} & 0 & \frac{7}{90}
\end{array}
$$

Vector stage equations

Let t_n denote the current time, $\mathbf{Y}_n = \begin{bmatrix} y_{1,n} \\ y_{2,n} \end{bmatrix}$, and set

$$\mathbf{Y}_{n+1} = \mathbf{Y}_n + h \sum_{i=1}^{7} b_i \, \mathbf{k}_i, \qquad \mathbf{k}_i = \mathbf{F}\left(t_n + c_i h, \ \mathbf{Y}_n + h \sum_{j=1}^{i-1} a_{ij} \mathbf{k}_j \right). \tag{7.9}$$

With the tableau above, the stages read explicitly:

$$
\begin{aligned}
\mathbf{k}_1 &= \mathbf{F}(t_n, \ \mathbf{Y}_n), && (7.10)\\
\mathbf{k}_2 &= \mathbf{F}\Big(t_n + \tfrac{h}{3}, \ \mathbf{Y}_n + h\big(\tfrac{1}{3}\mathbf{k}_1\big)\Big), && (7.11)\\
\mathbf{k}_3 &= \mathbf{F}\Big(t_n + \tfrac{h}{3}, \ \mathbf{Y}_n + h\big(\tfrac{1}{6}\mathbf{k}_1 + \tfrac{1}{6}\mathbf{k}_2\big)\Big), && (7.12)\\
\mathbf{k}_4 &= \mathbf{F}\Big(t_n + \tfrac{h}{2}, \ \mathbf{Y}_n + h\big(\tfrac{1}{8}\mathbf{k}_1 + \tfrac{3}{8}\mathbf{k}_3\big)\Big), && (7.13)\\
\mathbf{k}_5 &= \mathbf{F}\Big(t_n + \tfrac{2h}{3}, \ \mathbf{Y}_n + h\big(\tfrac{1}{2}\mathbf{k}_1 - \tfrac{3}{2}\mathbf{k}_3 + 2\mathbf{k}_4\big)\Big), && (7.14)\\
\mathbf{k}_6 &= \mathbf{F}\Big(t_n + \tfrac{h}{3}, \ \mathbf{Y}_n + h\big(-\tfrac{3}{2}\mathbf{k}_1 + \tfrac{9}{2}\mathbf{k}_3 - 6\mathbf{k}_4 + \tfrac{4}{3}\mathbf{k}_5\big)\Big), && (7.15)\\
\mathbf{k}_7 &= \mathbf{F}\Big(t_n + h, \ \mathbf{Y}_n + h\big(\tfrac{1}{6}\mathbf{k}_1 + 2\mathbf{k}_4 + \tfrac{1}{2}\mathbf{k}_5 + \tfrac{1}{6}\mathbf{k}_6\big)\Big). && (7.16)
\end{aligned}
$$

The update is

$$\mathbf{Y}_{n+1} = \mathbf{Y}_n + \frac{h}{90}\Big(1\,\mathbf{k}_1 + 32\,\mathbf{k}_3 + 12\,\mathbf{k}_4 + 32\,\mathbf{k}_5 + 7\,\mathbf{k}_7\Big). \tag{7.17}$$

Componentwise (position–velocity) form

Write each stage as $\mathbf{k}_i = \begin{bmatrix} k_i^{(1)} \\ k_i^{(2)} \end{bmatrix}$. For this problem,

$$k_i^{(1)} = (y_2)^{(i)}, \qquad k_i^{(2)} = f\Big(t^{(i)}, \ (y_1)^{(i)}, \ (y_2)^{(i)}\Big), \tag{7.18}$$

where the stage abscissa and trial values are

$$t^{(i)} = t_n + c_i h, \qquad \begin{aligned} (y_1)^{(i)} &= y_{1,n} + h \sum_{j=1}^{i-1} a_{ij}\, k_j^{(1)}, \\ (y_2)^{(i)} &= y_{2,n} + h \sum_{j=1}^{i-1} a_{ij}\, k_j^{(2)}. \end{aligned} \tag{7.19}$$

Initialize $y_{1,n} = y_n$, $y_{2,n} = v_n = \dot{y}_n$. Then advance both position and velocity by

$$\begin{bmatrix} y_{1,n+1} \\ y_{2,n+1} \end{bmatrix} = \begin{bmatrix} y_{1,n} \\ y_{2,n} \end{bmatrix} + \frac{h}{90}\left(\begin{bmatrix} k_1^{(1)} \\ k_1^{(2)} \end{bmatrix} + 32 \begin{bmatrix} k_3^{(1)} \\ k_3^{(2)} \end{bmatrix} + 12 \begin{bmatrix} k_4^{(1)} \\ k_4^{(2)} \end{bmatrix} + 32 \begin{bmatrix} k_5^{(1)} \\ k_5^{(2)} \end{bmatrix} + 7 \begin{bmatrix} k_7^{(1)} \\ k_7^{(2)} \end{bmatrix} \right). \tag{7.20}$$

Accuracy and remarks. This explicit scheme is of order six: the local truncation error is $\mathcal{O}(h^7)$ and the global error is $\mathcal{O}(h^6)$. Because f depends on y *and* $\dot{y}$, specialized Runge–Kutta–Nyström formulas (which assume $f = f(t, y)$ only) do not apply; the firstâ“order vector RK6 above is the standard high-order choice. For stiff problems, an implicit (e.g., Gauss/Radau) variant is recommended.

Linear Multistep Methods

Predictor–corrector algorithms, such as the Adams–Bashforth–Moulton family, are efficient for long-time integration. These methods use information from multiple previous steps to construct approximations and are particularly useful when solving oscillatory nonlinear systems such as the Duffing oscillator.

7.2 Predictor–Corrector Methods for $\ddot{y} = f(t, y, \dot{y})$ (Explicit Coefficients)

Consider the nonlinear secondâ“order ODE

$$\ddot{y}(t) = f(t,\, y(t),\, \dot{y}(t)), \qquad y(t_0) = y_0, \quad \dot{y}(t_0) = v_0. \tag{7.21}$$

Introduce the firstâ“order system

$$\begin{aligned} y_1(t) &= y(t), \quad \dot{y}_1(t) = y_2(t), \\ y_2(t) &= \dot{y}(t), \quad \dot{y}_2(t) = f(t, y_1(t), y_2(t)), \end{aligned} \qquad \mathbf{Y}(t) = \begin{bmatrix} y_1 \\ y_2 \end{bmatrix}, \qquad \dot{\mathbf{Y}} = \mathbf{F}(t, \mathbf{Y}) := \begin{bmatrix} y_2 \\ f(t, y_1, y_2) \end{bmatrix}. \tag{7.22}$$

We discretize $t_n = t_0 + nh$ with constant step $h > 0$ and denote $\mathbf{Y}_n \approx \mathbf{Y}(t_n)$, $\mathbf{F}_n := \mathbf{F}(t_n, \mathbf{Y}_n) = \begin{bmatrix} y_{2,n} \\ f_n \end{bmatrix}$, $f_n := f(t_n, y_{1,n}, y_{2,n})$. A *PECE* step (Predict–Evaluate–Correct–Evaluate) uses an explicit *Adams–Bashforth* (AB) predictor followed by an implicit *Adams–Moulton* (AM) corrector evaluated at the predicted state.

Order-p ABM(PECE): explicit coefficients

We list standard constant–step AB/AM pairs of orders $p = 2, 3, 4$. All formulas apply componentwise to the vector $\mathbf{Y}$ in (7.22).

AB2–AM2 (order 2).

$$\textbf{Predict(AB2)}: \quad \mathbf{Y}^{\mathrm{P}}_{n+1} = \mathbf{Y}_n + \frac{h}{2}\Big(3\,\mathbf{F}_n - \mathbf{F}_{n-1}\Big), \tag{7.23}$$

$$\textbf{Evaluate}: \quad \mathbf{F}^{\mathrm{P}}_{n+1} = \mathbf{F}(t_{n+1}, \mathbf{Y}^{\mathrm{P}}_{n+1}), \tag{7.24}$$

$$\textbf{Correct(AM2)}: \quad \mathbf{Y}_{n+1} = \mathbf{Y}_n + \frac{h}{2}\Big(\mathbf{F}^{\mathrm{P}}_{n+1} + \mathbf{F}_n\Big), \tag{7.25}$$

$$\textbf{Evaluate}: \quad \mathbf{F}_{n+1} = \mathbf{F}(t_{n+1}, \mathbf{Y}_{n+1}). \tag{7.26}$$

AB3–AM3 (order 3).

$$\textbf{Predict(AB3)}: \quad \mathbf{Y}^{\mathrm{P}}_{n+1} = \mathbf{Y}_n + \frac{h}{12}\Big(23\,\mathbf{F}_n - 16\,\mathbf{F}_{n-1} + 5\,\mathbf{F}_{n-2}\Big), \tag{7.27}$$

$$\textbf{Evaluate}: \quad \mathbf{F}^{\mathrm{P}}_{n+1} = \mathbf{F}(t_{n+1}, \mathbf{Y}^{\mathrm{P}}_{n+1}), \tag{7.28}$$

$$\textbf{Correct(AM3)}: \quad \mathbf{Y}_{n+1} = \mathbf{Y}_n + \frac{h}{12}\Big(5\,\mathbf{F}^{\mathrm{P}}_{n+1} + 8\,\mathbf{F}_n - \mathbf{F}_{n-1}\Big), \tag{7.29}$$

$$\textbf{Evaluate}: \quad \mathbf{F}_{n+1} = \mathbf{F}(t_{n+1}, \mathbf{Y}_{n+1}). \tag{7.30}$$

AB4–AM4 (order 4).

$$\textbf{Predict(AB4)}: \quad \mathbf{Y}^{\mathrm{P}}_{n+1} = \mathbf{Y}_n + \frac{h}{24}\Big(55\,\mathbf{F}_n - 59\,\mathbf{F}_{n-1} + 37\,\mathbf{F}_{n-2} - 9\,\mathbf{F}_{n-3}\Big), \tag{7.31}$$

$$\textbf{Evaluate}: \quad \mathbf{F}^{\mathrm{P}}_{n+1} = \mathbf{F}(t_{n+1}, \mathbf{Y}^{\mathrm{P}}_{n+1}), \tag{7.32}$$

$$\textbf{Correct(AM4)}: \quad \mathbf{Y}_{n+1} = \mathbf{Y}_n + \frac{h}{24}\Big(9\,\mathbf{F}^{\mathrm{P}}_{n+1} + 19\,\mathbf{F}_n - 5\,\mathbf{F}_{n-1} + \mathbf{F}_{n-2}\Big), \tag{7.33}$$

$$\textbf{Evaluate}: \quad \mathbf{F}_{n+1} = \mathbf{F}(t_{n+1}, \mathbf{Y}_{n+1}). \tag{7.34}$$

Remark 7.1. The corrector (7.25), (7.29), (7.33) is implicit in $\mathbf{F}_{n+1}$; in PECE we *replace* $\mathbf{F}_{n+1}$ by the evaluation at the AB predictor $\mathbf{F}^{\mathrm{P}}_{n+1}$, avoiding a nonlinear solve. One may perform one or more *post–corrections* by recomputing $\mathbf{F}$ at the corrected state and reapplying AM to improve accuracy.

Componentwise (position–velocity) form

Write $\mathbf{Y}_n = \begin{bmatrix} y_{1,n} \\ y_{2,n} \end{bmatrix}$ with $y_{1,n} \approx y(t_n)$ and $y_{2,n} \approx \dot{y}(t_n)$, and $\mathbf{F}_n = \begin{bmatrix} y_{2,n} \\ f_n \end{bmatrix}$. For AB4–AM4, the predictor stages are:

$$\text{Predict } y_{1,n+1}^{\mathrm{P}} = y_{1,n} + \frac{h}{24}\Big(55\, y_{2,n} - 59\, y_{2,n-1} + 37\, y_{2,n-2} - 9\, y_{2,n-3}\Big), \tag{7.35}$$

$$\text{Predict } y_{2,n+1}^{\mathrm{P}} = y_{2,n} + \frac{h}{24}\Big(55\, f_n - 59\, f_{n-1} + 37\, f_{n-2} - 9\, f_{n-3}\Big), \tag{7.36}$$

$$\text{Evaluate } f_{n+1}^{\mathrm{P}} = f\Big(t_{n+1},\, y_{1,n+1}^{\mathrm{P}},\, y_{2,n+1}^{\mathrm{P}}\Big). \tag{7.37}$$

Then the AM4 corrector gives:

$$y_{1,n+1} = y_{1,n} + \frac{h}{24}\Big(9\, y_{2,n+1}^{\mathrm{P}} + 19\, y_{2,n} - 5\, y_{2,n-1} + y_{2,n-2}\Big), \tag{7.38}$$

$$y_{2,n+1} = y_{2,n} + \frac{h}{24}\Big(9\, f_{n+1}^{\mathrm{P}} + 19\, f_n - 5\, f_{n-1} + f_{n-2}\Big), \tag{7.39}$$

$$f_{n+1} = f\Big(t_{n+1},\, y_{1,n+1},\, y_{2,n+1}\Big). \tag{7.40}$$

The AB2–AM2 and AB3–AM3 formulas follow analogously by replacing the coefficients with those in (7.23)–(7.29).

Start–up and implementation

Multistep methods require initial history values. For ABk–AMk, obtain $\{\mathbf{Y}_1, \ldots, \mathbf{Y}_{k-1}\}$ (and thus $\{\mathbf{F}_0, \ldots, \mathbf{F}_{k-1}\}$) with a one–step method (e.g., RK4 or RK6) applied to (7.22). A practical PECE step at order 4 is:

1. **Predict** $(y_{1,n+1}^{\mathrm{P}}, y_{2,n+1}^{\mathrm{P}})$ via (7.35)–(7.36).
2. **Evaluate** $f_{n+1}^{\mathrm{P}} = f(t_{n+1}, y_{1,n+1}^{\mathrm{P}}, y_{2,n+1}^{\mathrm{P}})$.
3. **Correct** $(y_{1,n+1}, y_{2,n+1})$ via (7.38)–(7.39).
4. **Evaluate** $f_{n+1} = f(t_{n+1}, y_{1,n+1}, y_{2,n+1})$ and proceed.

Optional: perform one extra *post–correction* by recomputing AM4 with $y_{2,n+1}$ (and f_{n+1}) in place of the predicted values.

Local error, stability, and notes

The ABk–AMk PECE pair has global order $p = k$, provided f is smooth and the step size is sufficiently small. For strongly nonlinear or mildly stiff problems, reduce h or switch to implicit correctors with fixed–point/Newton solve (i.e., use *fully implicit* AM with $\mathbf{F}_{n+1}$ rather than $\mathbf{F}_{n+1}^{\mathrm{P}}$). Componentwise application as in (7.35)–(7.39) is standard for second–order oscillators (e.g., Duffing), and preserves the correct coupling between position and velocity.

Finite Difference Methods

In boundary-value problems, the time or spatial domain is discretized into a grid and derivatives are approximated using finite difference stencils. This approach transforms the ODE into a nonlinear algebraic system, which is then solved iteratively (e.g., with Newton–Raphson).

7.2.0.1 High-Accuracy Finite Differences for Second-Order Nonlinear ODEs

Consider the general second-order ODE

$$y''(x) = F(x, y(x), y'(x)), \qquad x \in [a, b], \tag{7.41}$$

subject to boundary conditions (Dirichlet, Neumann, or Robin). We discretize $[a, b]$ with a uniform grid $x_i = a + i\,h$, $i = 0, 1, \ldots, N$, $h = (b - a)/N$, and denote $y_i \approx y(x_i)$.

*Central finite-difference stencils (uniform grid)

For interior nodes i away from the boundary, the following *central* approximations achieve high order on uniform grids.

Second derivative $y''(x_i)$.

$$\text{2nd order (3-pt):} \quad y''(x_i) \approx \frac{y_{i-1} - 2y_i + y_{i+1}}{h^2}, \tag{7.42}$$

$$\text{4th order (5-pt):} \quad y''(x_i) \approx \frac{-y_{i-2} + 16y_{i-1} - 30y_i + 16y_{i+1} - y_{i+2}}{12\,h^2}, \tag{7.43}$$

$$\text{6th order (7-pt):} \quad y''(x_i) \approx \frac{2\,y_{i-3} - 27\,y_{i-2} + 270\,y_{i-1} - 490\,y_i + 270\,y_{i+1} - 27\,y_{i+2} + 2\,y_{i+3}}{180\,h^2}. \tag{7.44}$$

First derivative $y'(x_i)$ (needed if F depends on y').

$$\text{2nd order (3-pt):} \quad y'(x_i) \approx \frac{y_{i+1} - y_{i-1}}{2h}, \tag{7.45}$$

$$\text{4th order (5-pt):} \quad y'(x_i) \approx \frac{y_{i-2} - 8y_{i-1} + 8y_{i+1} - y_{i+2}}{12h}, \tag{7.46}$$

$$\text{6th order (7-pt):} \quad y'(x_i) \approx \frac{-y_{i-3} + 9y_{i-2} - 45y_{i-1} + 45y_{i+1} - 9y_{i+2} + y_{i+3}}{60h}. \tag{7.47}$$

Remark. Using (7.44) together with (7.47) yields a fully 6th-order spatial discretization of (7.41) on interior nodes.

*Compact (Pade) 4th-order scheme for y''

Compact schemes achieve higher accuracy with narrower stencils by solving a small linear system for the derivative values. A classical 4th-order compact tridiagonal formula for y'' on a uniform grid is

$$\frac{1}{10}\,y''_{i-1} + y''_i + \frac{1}{10}\,y''_{i+1} = \frac{6}{5\,h^2}\,(y_{i-1} - 2y_i + y_{i+1}), \qquad i = 1, \ldots, N-1, \tag{7.48}$$

which, after assembling for all interior i, produces a tridiagonal system for $\{y_i''\}$ that can be solved efficiently and then used in (7.41) (or in postprocessing).

*Imposing boundary conditions with high-order accuracy

Let the boundaries be at $i = 0$ ($x = a$) and $i = N$ ($x = b$).

Dirichlet. If $y(a) = \alpha$ and/or $y(b) = \beta$, set $y_0 = \alpha$, $y_N = \beta$ directly and use interior formulas from the closest index that preserves the stencil (e.g. for 6th order, start at $i = 3$ and stop at $i = N - 3$). Near boundaries, switch to lower-order or one-sided high-order stencils to avoid ghost points.

Neumann. If $y'(a) = g_a$, enforce with a one-sided 4th-order forward difference:

$$\frac{-25\,y_0 + 48\,y_1 - 36\,y_2 + 16\,y_3 - 3\,y_4}{12\,h} = g_a, \tag{7.49}$$

and analogously at $x = b$ (backward 4th-order):

$$\frac{25\,y_N - 48\,y_{N-1} + 36\,y_{N-2} - 16\,y_{N-3} + 3\,y_{N-4}}{12\,h} = g_b. \tag{7.50}$$

For 6th-order boundary treatment, longer one-sided stencils can be used (similar construction via Taylor expansion), or ghost points can be eliminated using the BC.

Robin. If $\alpha_L\,y(a) + \beta_L\,y'(a) = \gamma_L$, replace $y'(a)$ by (7.49) to obtain a linear relation among $\{y_0, y_1, y_2, y_3, y_4\}$; similarly at $x = b$.

*Nonlinear assembly (residual and Jacobian)

Discretizing (7.41) at interior nodes with (say) the 6th-order stencils (7.44) and (7.47) yields the nonlinear algebraic system $\mathbf{R}(\mathbf{y}) = \mathbf{0}$, $\mathbf{y} = (y_0, \ldots, y_N)^\top$, with component

$$\begin{aligned} R_i(\mathbf{y}) :=\; & \underbrace{\frac{2y_{i-3} - 27y_{i-2} + 270y_{i-1} - 490y_i + 270y_{i+1} - 27y_{i+2} + 2y_{i+3}}{180\,h^2}}_{\text{6th-order } y''(x_i)} - \\ & F\Big(x_i,\, y_i,\, \underbrace{\frac{-y_{i-3} + 9y_{i-2} - 45y_{i-1} + 45y_{i+1} - 9y_{i+2} + y_{i+3}}{60h}}_{\text{6th-order } y'(x_i)}\Big), \end{aligned} \tag{7.51}$$

for $i = 3, \ldots, N - 3$, together with boundary rows encoding the BCs. Solve with Newton's method:

$$\mathbf{J}(\mathbf{y}^{(k)})\,\delta\mathbf{y}^{(k)} = -\mathbf{R}(\mathbf{y}^{(k)}), \qquad \mathbf{y}^{(k+1)} = \mathbf{y}^{(k)} + \delta\mathbf{y}^{(k)}, \tag{7.52}$$

where the Jacobian is sparse (banded). For (7.51), entries in row i couple at most the indices

$i \pm 3$:

$$\frac{\partial R_i}{\partial y_j} = (\text{linear stencil coeffs from } y'') - \frac{\partial F}{\partial y}(x_i, y_i, y_i')\,\delta_{ij} - \frac{\partial F}{\partial y'}(x_i, y_i, y_i')\,\frac{\partial y_i'}{\partial y_j}, \tag{7.53}$$

$$\frac{\partial y_i'}{\partial y_{i\pm1}} = \mp\frac{45}{60h}, \quad \frac{\partial y_i'}{\partial y_{i\pm2}} = \pm\frac{9}{60h}, \quad \frac{\partial y_i'}{\partial y_{i\pm3}} = \mp\frac{1}{60h}. \tag{7.54}$$

This yields a septadiagonal structure for the 6th-order scheme.

Model problems (assembly patterns)

Linear variable-coefficient:

$$y'' + p(x)\,y' + q(x)\,y = r(x). \tag{7.55}$$

Discretize y'' and y' with (7.43)/(7.44), (7.46)/(7.47) and assemble a (banded) linear system.

Nonlinear: Duffing-type BVP.

$$y''^3 = 0, \qquad y(a) = \alpha_0, \quad y(b) = \beta_0. \tag{7.56}$$

Here $F(x, y, y'^3$ (no y' dependence), so (7.51) simplifies and the Jacobian has the same band but a simpler nonlinear term $-(\alpha + 3\beta y_i^2)\delta_{ij}$.

Accuracy, stability, and refinement

On smooth solutions, the interior truncation error of (7.43) is $\mathcal{O}(h^4)$ and of (7.44) is $\mathcal{O}(h^6)$. Global accuracy will be limited by (i) the boundary treatment order, (ii) nonlinearity (Newton tolerance), and (iii) conditioning of the assembled system. Use grid refinement studies and/or Richardson extrapolation to verify observed rates.

Notes on nonuniform grids

On graded meshes x_i, central coefficients change. Either (a) derive local coefficients via Taylor expansion about x_i using the specific $\{x_{i-k}, \ldots, x_{i+k}\}$, or (b) map to a uniform computational grid $\xi \in [0, 1]$ via $x = x(\xi)$ and discretize the transformed ODE, which includes metric factors x_ξ and $x_{\xi\xi}$.

The formulas above provide a practical, high-order toolkit for discretizing (7.41) with explicit coefficients that are easy to assemble into sparse linear/nonlinear systems and solve with banded direct or iterative methods.

Shooting and Collocation Methods

For boundary-value formulations, the shooting method converts the problem into an equivalent initial-value problem, solved iteratively until boundary conditions are satisfied. Collocation methods, on the other hand, approximate the solution by a basis (e.g., polynomials, splines) and enforce the ODE at collocation points.

Energy- and Structure-Preserving Methods

Many nonlinear second-order ODEs arise from conservative mechanical systems. In such cases, symplectic integrators are particularly valuable, as they preserve Hamiltonian structures and long-time stability of oscillatory solutions.

In practice, the choice of numerical method depends on the *nature of the nonlinearities*, the *required accuracy*, and whether the problem is posed as an *initial-value problem (IVP)* or a *boundary-value problem (BVP)*. Modern implementations combine adaptive step-size control, error estimation, and stability enhancements to provide efficient solvers for nonlinear dynamics.

When using a numerical method we need to estimate derivatives and integrals in discrete form.

7.3 Derivatives

Let

$$0 < h < 1 \text{ and } f(x) = f_x.$$

First Derivative Formulas

$$f'(a) = -\frac{f_a - f_{a+h}}{h}.$$
$$f'(a) = -\frac{f_{a-2h} - 4f_{a+h} + 3f_a}{2h}.$$
$$f'(a) = -\frac{-2f_{a-3h} + 9f_{a-2h} - 18f_{a+h} + 11f_a}{6h}.$$
$$f'(a) = -\frac{3f_{a-4h} - 16f_{a-3h} + 36f_{a-2h} - 48f_{a+h} + 25f_a}{12h}.$$
$$f'(a) = -\frac{-12f_{a-5h} + 75f_{a-4h} - 200f_{a-3h} + 300f_{a-2h} - 300f_{a+h} + 137f_a}{60h}.$$

$$f'(a) = -\frac{f_{a-h} - f_{a+h}}{2h}.$$
$$f'(a) = \frac{-f_{a-2h} - 8f_{a-h} + 8f_{a+h} + f_{a+2h}}{12h}.$$
$$f'(a) = -\frac{-f_{a-3h} + 9f_{a-2h} + 45f_{a-h} - 45f_{a+h} - 9f_{a+2h} + f_{a+3h}}{60h}.$$
$$f'(a) = \frac{-3f_{a-4h} + 32f_{a-3h} - 168f_{a-2h} - 672f_{a-h} + 672f_{a+h} + 168f_{a+2h} - 32f_{a+3h} + 3f_{a+4h}}{840h}.$$
$$f'(a) = -\frac{-2f_{a-5h} + 25f_{a-4h} - 150f_{a-3h} + 600f_{a-2h} + 2100f_{a-h} - 2100f_{a+h} - 600f_{a+2h} + 150f_{a+3h} - 25f_{a+4h} + 2f_{a+5h}}{2520h}.$$

$$f'(a) = \frac{f_a - f_{a-h}}{h}.$$
$$f'(a) = \frac{-4f_{a-h} + f_{a+2h} + 3f_a}{2h}.$$
$$f'(a) = \frac{-18f_{a-h} + 9f_{a+2h} - 2f_{a+3h} + 11f_a}{6h}.$$
$$f'(a) = \frac{-48f_{a-h} + 36f_{a+2h} - 16f_{a+3h} + 3f_{a+4h} + 25f_a}{12h}.$$
$$f'(a) = \frac{-300f_{a-h} + 300f_{a+2h} - 200f_{a+3h} + 75f_{a+4h} - 12f_{a+5h} + 137f_a}{60h}.$$

Second Derivative Formulas

$$f''(x)=\frac{-2f(h+x)+f(2h+x)+f(x)}{h^2}$$
$$f''(x)=\frac{-5f(h+x)+4f(2h+x)-f(3h+x)+2f(x)}{h^2}$$
$$f''(x)=\frac{-104f(h+x)+114f(2h+x)-56f(3h+x)+11f(4h+x)+35f(x)}{12h^2}$$
$$f''(x)=\frac{-154f(h+x)+214f(2h+x)-156f(3h+x)+61f(4h+x)-10f(5h+x)+45f(x)}{12h^2}$$

$$f''(x)=-\frac{f(x-2h)-16f(x-h)-16f(h+x)+f(2h+x)+30f(x)}{12h^2}$$
$$f''(x)=-\frac{-2f(x-3h)+27f(x-2h)-270f(x-h)-270f(h+x)+27f(2h+x)-2f(3h+x)+490f(x)}{180h^2}$$
$$f''(x)=-\frac{1}{5040h^2}[9f(x-4h)-128f(x-3h)+1008f(x-2h)-8064f(x-h)-8064f(h+x)$$
$$+1008f(2h+x)-128f(3h+x)+9f(4h+x)+14350f(x)]$$
$$f''(x)=-\frac{1}{25200h^2}[-8f(x-5h)+125f(x-4h)-1000f(x-3h)+6000f(x-2h)-42000f(x-h)$$
$$-42000f(h+x)+6000f(2h+x)-1000f(3h+x)+125f(4h+x)-8f(5h+x)+73766f(x)]$$

$$f''(x)=\frac{f(x-2h)-2f(x-h)+f(x)}{h^2}$$
$$f''(x)=\frac{-f(x-3h)+4f(x-2h)-5f(x-h)+2f(x)}{h^2}$$
$$f''(x)=\frac{11f(x-4h)-56f(x-3h)+114f(x-2h)-104f(x-h)+35f(x)}{12h^2}$$

Third Derivative Formulas

$$f'''(x)=-\frac{-3f(h+x)+3f(2h+x)-f(3h+x)+f(x)}{h^3}$$
$$f'''(x)=-\frac{-18f(h+x)+24f(2h+x)-14f(3h+x)+3f(4h+x)+5f(x)}{2h^3}$$
$$f'''(x)=-\frac{-71f(h+x)+118f(2h+x)-98f(3h+x)+41f(4h+x)-7f(5h+x)+17f(x)}{4h^3}$$

$$f'''(x)=-\frac{f(x-2h)-2f(x-h)+2f(h+x)-f(2h+x)}{2h^3}$$
$$f'''(x)=\frac{f(x-3h)-8f(x-2h)+13f(x-h)-13f(h+x)+8f(2h+x)-f(3h+x)}{8h^3}$$
$$f'''(x)=-\frac{7f(x-4h)-72f(x-3h)+338f(x-2h)-488f(x-h)+488f(h+x)-338f(2h+x)+72f(3h+x)-7f(4h+x)}{240h^3}$$
$$f'''(x)=\frac{1}{30240h^3}[205f(x-5h)-2522f(x-4h)+14607f(x-3h)-52428f(x-2h)+70098f(x-h)$$
$$-70098f(h+x)+52428f(2h+x)-14607f(3h+x)+2522f(4h+x)-205f(5h+x)]$$

$$f'''(x)=\frac{-f(x-3h)+3f(x-2h)-3f(x-h)+f(x)}{h^3}$$
$$f'''(x)=\frac{3f(x-4h)-14f(x-3h)+24f(x-2h)-18f(x-h)+5f(x)}{2h^3}$$
$$f'''(x)=\frac{-7f(x-5h)+41f(x-4h)-98f(x-3h)+118f(x-2h)-71f(x-h)+17f(x)}{4h^3}$$

Fourth Derivative Formulas

$$f^{(iv)}(x)=\frac{-4f(h+x)+6f(2h+x)-4f(3h+x)+f(4h+x)+f(x)}{h^4}$$
$$f^{(iv)}(x)=\frac{-14f(h+x)+26f(2h+x)-24f(3h+x)+11f(4h+x)-2f(5h+x)+3f(x)}{h^4}$$
$$f^{(iv)}(x)=\frac{-186f(h+x)+411f(2h+x)-484f(3h+x)+321f(4h+x)-114f(5h+x)+17f(6h+x)+35f(x)}{6h^4}$$

$$f^{(iv)}(x)=\frac{f(x-2h)-4f(x-h)-4f(h+x)+f(2h+x)+6f(x)}{h^4}$$
$$f^{(iv)}(x)=\frac{-f(x-3h)+12f(x-2h)-39f(x-h)-39f(h+x)+12f(2h+x)-f(3h+x)+56f(x)}{6h^4}$$
$$f^{(iv)}(x)=\frac{7f(x-4h)-96f(x-3h)+676f(x-2h)-1952f(x-h)-1952f(h+x)+676f(2h+x)-96f(3h+x)+7f(4h+x)+2730f(x)}{240h^4}$$
$$f^{(iv)}(x)=\frac{1}{15120h^4}[-82f(x-5h)+1261f(x-4h)-9738f(x-3h)+52428f(x-2h)-140196f(x-h)$$
$$-140196f(h+x)+52428f(2h+x)-9738f(3h+x)+1261f(4h+x)-82f(5h+x)+192654f(x)]$$

$$f^{(iv)}(x) = \frac{f(x-4h)-4f(x-3h)+6f(x-2h)-4f(x-h)+f(x)}{h^4}$$
$$f^{(iv)}(x) = \frac{-2f(x-5h)+11f(x-4h)-24f(x-3h)+26f(x-2h)-14f(x-h)+3f(x)}{h^4}$$
$$f^{(iv)}(x) = \frac{17f(x-6h)-114f(x-5h)+321f(x-4h)-484f(x-3h)+411f(x-2h)-186f(x-h)+35f(x)}{6h^4}$$

Fifth Derivative Formulas

$$f^{(v)}(x) = -\frac{-5f(h+x)+10f(2h+x)-10f(3h+x)+5f(4h+x)-f(5h+x)+f(x)}{h^5}$$
$$f^{(v)}(x) = -\frac{-40f(h+x)+95f(2h+x)-120f(3h+x)+85f(4h+x)-32f(5h+x)+5f(6h+x)+7f(x)}{2h^5}$$
$$f^{(v)}(x) = -\frac{-295f(h+x)+810f(2h+x)-1235f(3h+x)+1130f(4h+x)-621f(5h+x)+190f(6h+x)-25f(7h+x)+46f(x)}{6h^5}$$

$$f^{(v)}(x) = -\frac{f(x-3h)-4f(x-2h)+5f(x-h)-5f(h+x)+4f(2h+x)-f(3h+x)}{2h^5}$$
$$f^{(v)}(x) = \frac{f(x-4h)-9f(x-3h)+26f(x-2h)-29f(x-h)+29f(h+x)-26f(2h+x)+9f(3h+x)-f(4h+x)}{6h^5}$$
$$f^{(v)}(x) = -\frac{1}{288h^5}[13f(x-5h)-152f(x-4h)+783f(x-3h)-1872f(x-2h)+1938f(x-h)$$
$$-1938f(h+x)+1872f(2h+x)-783f(3h+x)+152f(4h+x)-13f(5h+x)]$$
$$f^{(v)}(x) = \frac{1}{12096h^5}[139f(x-6h)-1936f(x-5h)+12500f(x-4h)-48176f(x-3h)+101559f(x-$$
$$-99744f(x-h)+99744f(h+x)-101559f(2h+x)+48176f(3h+x)$$
$$-12500f(4h+x)+1936f(5h+x)-139f(6h+x)]$$

$$f^{(v)}(x) = \frac{-f(x-5h)+5f(x-4h)-10f(x-3h)+10f(x-2h)-5f(x-h)+f(x)}{h^5}$$
$$f^{(v)}(x) = \frac{5f(x-6h)-32f(x-5h)+85f(x-4h)-120f(x-3h)+95f(x-2h)-40f(x-h)+7f(x)}{2h^5}$$
$$f^{(v)}(x) = \frac{-25f(x-7h)+190f(x-6h)-621f(x-5h)+1130f(x-4h)-1235f(x-3h)+810f(x-2h)-295f(x-h)+46f(x)}{6h^5}.$$

chebyshevFormulas

$$f'(a) = \frac{1}{h}[40.3204f_{a+0.0495156h} - 10.7509f_{a+0.188255h} + 5.18334f_{a+0.38874h} - 3.29969f_{a+0.61126h}$$
$$+2.48832f_{a+0.811745h} - 2.11697f_{a+0.950484h} + f_{a+h} - 32.8245f_a]$$
$$f'(a) = \frac{1}{h}[1.16338f_{a-\frac{h}{2}} - 2.42249f_{a-0.46194h} + 2.96382f_{a-0.353553h}$$
$$-5.30632f_{a-0.191342h} + 5.16712f_{a+0.191342h}$$
$$-2.79235f_{a+0.353553h} + 2.13753f_{a+0.46194h} - 0.98753f_{a+\frac{h}{2}} + 0.0768394f_a]$$
$$f'(a) = \frac{1}{h}[40.3204f_{a+0.0495156h} - 10.7509f_{a+0.188255h} + 5.18334f_{a+0.38874h} - 3.29969f_{a+0.61126h}$$
$$+2.48832f_{a+0.811745h} - 2.11697f_{a+0.950484h} + f_{a+h} - 32.8245f_a]$$
$$f_x := f(x)$$

$f''(a) = \frac{1}{h^2}[-1033.13f_{a+0.0495156h} + 591.2f_{a+0.188255h} - 313.635f_{a+0.38874h} + 205.827f_{a+0.61126h}$
$-157.224f_{a+0.811745h} + 134.521f_{a+0.950484h} - 63.6491f_{a+h} + 636.092f_a]$
$f''(a) = \frac{1}{h^2}[-3.10802f_{a-\frac{h}{2}} + 7.68273f_{a-0.46194h} - 14.5656f_{a-0.353553h} + 53.5679f_{a-0.191342h}$
$+54.3145f_{a+0.191342h} - 15.846f_{a+0.353553h} + 9.27138f_{a+0.46194h} - 3.95605f_{a+\frac{h}{2}} - 87.3608f_a]$
$f''(a) = \frac{1}{h^2}[-1033.13f_{a+0.0495156h} + 591.2f_{a+0.188255h} - 313.635f_{a+0.38874h} + 205.827f_{a+0.61126h}$
$-157.224f_{a+0.811745h} + 134.521f_{a+0.950484h} - 63.6491f_{a+h} + 636.092f_a]$
$f_x := f(x)$

$f^{(3)}(a) = \frac{1}{h^3}[14868.2f_{a+0.0495156h} - 11064.8f_{a+0.188255h} + 7472.8f_{a+0.38874h} - 5286.73f_{a+0.61126h}$
$+4167.29f_{a+0.811745h} - 3615.06f_{a+0.950484h} + 1717.33f_{a+h} - 8258.99f_a]$
$f^{(3)}(a) = \frac{1}{h^3}[-230.562f_{a-\frac{h}{2}} + 493.63f_{a-0.46194h} - 598.519f_{a-0.353553h} + 515.858f_{a-0.191342h}$
$-525.262f_{a+0.191342h} + 612.417f_{a+0.353553h} - 511.992f_{a+0.46194h} + 240.662f_{a+\frac{h}{2}} + 3.76904f_a]$
$f^{(3)}(a) = \frac{1}{h^3}[14868.2f_{a+0.0495156h} - 11064.8f_{a+0.188255h} + 7472.8f_{a+0.38874h} - 5286.73f_{a+0.61126h}$
$+4167.29f_{a+0.811745h} - 3615.06f_{a+0.950484h} + 1717.33f_{a+h} - 8258.99f_a]$
$f_x := f(x)$

$f^{(4)}(a) = \frac{1}{h^4}[-140425.f_{a+0.0495156h} + 119392.f_{a+0.188255h} - 94409.f_{a+0.38874h} + 74418.5f_{a+0.61126h}$
$-61667.1f_{a+0.811745h} + 54718.4f_{a+0.950484h} - 26166.6f_{a+h} + 74138.8f_a]$
$f^{(4)}(a) = \frac{1}{h^4}[1982.93f_{a-\frac{h}{2}} - 4538.22f_{a-0.46194h} + 7004.7f_{a-0.353553h} - 11004.6f_{a-0.191342h}$
$-10971.6f_{a+0.191342h} + 6938.79f_{a+0.353553h} - 4445.96f_{a+0.46194h} + 1931.67f_{a+\frac{h}{2}} + 13102.3f_a]$
$f^{(4)}(a) = \frac{1}{h^4}[-140425.f_{a+0.0495156h} + 119392.f_{a+0.188255h} - 94409.f_{a+0.38874h} + 74418.5f_{a+0.61126h}$
$-61667.1f_{a+0.811745h} + 54718.4f_{a+0.950484h} - 26166.6f_{a+h} + 74138.8f_a]$
$f_x := f(x)$

$f^{(5)}(a) = \frac{1}{h^5}[876027.f_{a+0.0495156h} - 806107.f_{a+0.188255h} + 708238.f_{a+0.38874h} - 614525.f_{a+0.61126h}$
$+542526.f_{a+0.811745h} - 496826.f_{a+0.950484h} + 239861.f_{a+h} - 449194.f_a]$
$f^{(5)}(a) = \frac{1}{h^5}[46429.3f_{a-\frac{h}{2}} - 94094.1f_{a-0.46194h} + 87371.8f_{a-0.353553h} - 55648.5f_{a-0.191342h}$
$+55243.8f_{a+0.191342h} - 86857.5f_{a+0.353553h} + 93470.7f_{a+0.46194h} - 46095.1f_{a+\frac{h}{2}} + 179.601f_a]$
$f^{(5)}(a) = \frac{1}{h^5}[876027.f_{a+0.0495156h} - 806107.f_{a+0.188255h} + 708238.f_{a+0.38874h} - 614525.f_{a+0.61126h}$
$+542526.f_{a+0.811745h} - 496826.f_{a+0.950484h} + 239861.f_{a+h} - 449194.f_a]$
$f_x := f(x)$

7.3.1 General Formulas

Let

$$0 = \xi_0 < \xi_1 < \cdots < \xi_n = T \text{ and } f_j := f(\xi_j).$$

First Derivative Formulas

$f'(\xi_0) = \frac{f_3\xi_1^2(\xi_1-\xi_2)\xi_2^2+\xi_3^2(f_2(\xi_3-\xi_1)\xi_1^2+f_1\xi_2^2(\xi_2-\xi_3))}{\xi_1(\xi_1-\xi_2)\xi_2(\xi_1-\xi_3)(\xi_2-\xi_3)\xi_3},$

$f'(\xi_1) = \frac{f_1}{\xi_1} + \frac{f_1-f_2}{\xi_1-\xi_2} - \frac{f_2}{\xi_2} + \frac{f_1-f_3}{\xi_1-\xi_3} + \frac{f_3(\xi_1-\xi_2)}{\xi_2\xi_3} + \frac{(f_2-f_3)\xi_1}{\xi_2(\xi_3-\xi_2)},$

$f'(\xi_2) = \frac{f_1-f_2}{\xi_1-\xi_2} + \frac{f_3-f_1}{\xi_1-\xi_3} + \frac{f_2-f_3}{\xi_2-\xi_3} + \frac{(f_1-f_4)(\xi_3-\xi_2)}{(\xi_1-\xi_3)(\xi_1-\xi_4)} + \frac{f_2-f_4}{\xi_2-\xi_4} + \frac{(f_3-f_4)(\xi_1-\xi_2)}{(\xi_1-\xi_3)(\xi_4-\xi_3)}$,
$f'(\xi_j) = \frac{f_{j-1}-f_j}{\xi_{j-1}-\xi_j} + \frac{f_{j+1}-f_{j-1}}{\xi_{j-1}-\xi_{j+1}} + \frac{f_j-f_{j+1}}{\xi_j-\xi_{j+1}} + \frac{(f_{j-1}-f_{j+2})(\xi_{j+1}-\xi_j)}{(\xi_{j-1}-\xi_{j+1})(\xi_{j-1}-\xi_{j+2})}$
$+\frac{f_j-f_{j+2}}{\xi_j-\xi_{j+2}} + \frac{(f_{j+1}-f_{j+2})(\xi_{j-1}-\xi_j)}{(\xi_{j-1}-\xi_{j+1})(\xi_{j+2}-\xi_{j+1})}$.

Second Derivative Formulas

$f''(\xi_0) = \frac{2(\xi_3^3(f_1\xi_2-f_2\xi_1)+\xi_3(f_2\xi_1^3-f_1\xi_2^3)-f_3\xi_1\xi_2(\xi_1^2-\xi_2^2))}{\xi_1(\xi_1-\xi_2)\xi_2(\xi_1-\xi_3)(\xi_2-\xi_3)\xi_3}$.
$f''(\xi_1) = 2\frac{\frac{f_2}{\xi_2}-\frac{f_1}{\xi_1}}{\xi_2-\xi_1} + \frac{2}{\xi_3-\xi_1}(2\xi_1-\xi_2)(\frac{\frac{f_2-f_3}{\xi_2-\xi_3}-\frac{f_2}{\xi_2}}{\xi_3} - \frac{\frac{f_2}{\xi_2}-\frac{f_1}{\xi_1}}{\xi_2-\xi_1})$.
$f''(\xi_j) = \frac{2f_{j+2}(\xi_{j-1}-2\xi_j+\xi_{j+1})}{(\xi_{j-1}-\xi_{j+2})(\xi_{j+2}-\xi_j)(\xi_{j+2}-\xi_{j+1})} + \frac{2f_{j-1}(2\xi_j-\xi_{j+1}-\xi_{j+2})}{(\xi_{j-1}-\xi_j)(\xi_{j-1}-\xi_{j+1})(\xi_{j-1}-\xi_{j+2})}+$
$\frac{2f_{j+1}(\xi_{j-1}-2\xi_j+\xi_{j+2})}{(\xi_{j-1}-\xi_{j+1})(\xi_{j+1}-\xi_j)(\xi_{j+1}-\xi_{j+2})} + \frac{2f_j(\xi_{j-1}-3\xi_j+\xi_{j+1}+\xi_{j+2})}{(\xi_{j-1}-\xi_j)(\xi_j-\xi_{j+1})(\xi_j-\xi_{j+2})}$.

7.4 Integrals

For the numerical evaluation of integrals we use quadrature formulas. Some of them are :

$\int_a^b f(x)dx = -\frac{1}{2}(a-b)(f_a+f_b)$, $\mathcal{E} = -\frac{1}{12}(a-b)^3 f''$,
$\int_a^b f(x)dx = -\frac{1}{6}(a-b)\left(f_a+4f_{\frac{a}{2}+\frac{b}{2}}+f_b\right)$, $\mathcal{E} = -\frac{(a-b)^5 f^{(4)}}{2880}$.
$\int_a^b f(x)dx = -\frac{1}{8}(a-b)\left(f_a+3\left(f_{\frac{2a}{3}+\frac{b}{3}}+f_{\frac{a}{3}+\frac{2b}{3}}\right)+f_b\right)$, $\mathcal{E} = -\frac{(a-b)^5 f^{(4)}}{6480}$,
$\int_a^b f(x)dx = -\frac{1}{90}(a-b)\left(7f_a+4\left(8f_{\frac{3a}{4}+\frac{b}{4}}+3f_{\frac{a}{2}+\frac{b}{2}}+8f_{\frac{a}{4}+\frac{3b}{4}}\right)+7f_b\right)$, $\mathcal{E} = -\frac{(a-b)^7 f^{(6)}}{1935360}$,
$\int_a^b f(x)dx = -\frac{1}{288}(a-b)\left(19f_a+75f_{\frac{4a}{5}+\frac{b}{5}}+50f_{\frac{3a}{5}+\frac{2b}{5}}+50f_{\frac{2a}{5}+\frac{3b}{5}}+75f_{\frac{a}{5}+\frac{4b}{5}}+19f_b\right)$, $\mathcal{E} =$
$-\frac{11(a-b)^7 f^{(6)}}{37800000}$,
$\int_a^b f(x)dx = -\frac{1}{840}(a-b)\left(41f_a+216f_{\frac{5a}{6}+\frac{b}{6}}+27f_{\frac{2a}{3}+\frac{b}{3}}+272f_{\frac{a}{2}+\frac{b}{2}}+27f_{\frac{a}{3}+\frac{2b}{3}}+216f_{\frac{a}{6}+\frac{5b}{6}}+41f_b\right)$, $\mathcal{E}$
$-\frac{(a-b)^9 f^{(8)}}{1567641600}$,
where $f_x := f(x)$.

7.4.1 Chebyshev Formulas

$\int_a^b f(x)dx = -\frac{1}{6}(a-b)\left(f_a+f_b+4f_{\frac{a+b}{2}}\right)$,
$\int_a^b f(x)dx = -\frac{1}{18}(a-b)\left(f_a+f_b+8\left(f_{\frac{1}{4}(3a+b)}+f_{\frac{1}{4}(a+3b)}\right)\right)$,
$\int_a^b f(x)dx = -\frac{1}{3}(a-b)\left(f_{\frac{a+b}{2}}+f_{\frac{1}{4}((2+\sqrt{2})a-(-2+\sqrt{2})b)}+f_{\frac{1}{4}((2+\sqrt{2})b-(-2+\sqrt{2})a)}\right)$,
$\int_a^b f(x)dx = \frac{1}{90}(a-b)f_a+\left(\frac{545b}{26}-\frac{545a}{26}\right)f_{\frac{19a}{21}+\frac{2b}{21}}-\frac{2831}{111}af_{\frac{17a}{26}+\frac{9b}{26}}+\frac{2627}{103}bf_{\frac{17a}{26}+\frac{9b}{26}}-\frac{2831}{111}af_{\frac{9a}{26}+\frac{17b}{26}}+$
$\frac{2627}{103}bf_{\frac{9a}{26}+\frac{17b}{26}}-$
$\frac{545}{26}af_{\frac{2a}{21}+\frac{19b}{21}}+\frac{545}{26}bf_{\frac{2a}{21}+\frac{19b}{21}}+af_b-bf_b$,
$\int_a^b f(x)dx = \frac{1}{80}[(a-b)f_a+\left(\frac{253b}{19}-\frac{253a}{19}\right)f_{\frac{14a}{15}+\frac{b}{15}}-\frac{253}{19}af_{\frac{a}{15}+\frac{14b}{15}}+\frac{253}{19}bf_{\frac{a}{15}+\frac{14b}{15}}+af_b-bf_b+$
$\left(\frac{800b}{39}-\frac{800a}{39}\right)f_{\frac{a+b}{2}}$
$+\left(\frac{158b}{9}-\frac{158a}{9}\right)f_{\frac{1}{4}(3a+b)}+\left(\frac{158b}{9}-\frac{158a}{9}\right)f_{\frac{1}{4}(a+3b)}]$,
$\int_a^b f(x)dx = \frac{1}{13}[-((a-b)f_a)-\left(\frac{b}{3}-\frac{a}{3}\right)f_{\frac{19a}{20}+\frac{b}{20}}-\frac{118}{35}af_{\frac{13a}{16}+\frac{3b}{16}}+\frac{118}{35}bf_{\frac{13a}{16}+\frac{3b}{16}}-\frac{137}{56}af_{\frac{11a}{18}+\frac{7b}{18}}+$
$\frac{181}{74}bf_{\frac{11a}{18}+\frac{7b}{18}}-\frac{137}{56}af_{\frac{7a}{18}+\frac{11b}{18}}$
$+\frac{181}{74}bf_{\frac{7a}{18}+\frac{11b}{18}}-\frac{118}{35}af_{\frac{3a}{16}+\frac{13b}{16}}+\frac{118}{35}bf_{\frac{3a}{16}+\frac{13b}{16}}+\frac{1}{3}af_{\frac{a}{20}+\frac{19b}{20}}-\frac{1}{3}bf_{\frac{a}{20}+\frac{19b}{20}}-af_b+bf_b]$,
$\int_a^b f(x)dx = -\frac{2}{29}[(a-b)f_a+\left(\frac{35b}{69}-\frac{36a}{71}\right)f_{\frac{25a}{26}+\frac{b}{26}}+\frac{77}{25}af_{\frac{29a}{34}+\frac{5b}{34}}-\frac{77}{25}bf_{\frac{29a}{34}+\frac{5b}{34}}+\frac{169}{75}af_{\frac{9a}{13}+\frac{4b}{13}}-$
$\frac{169}{75}bf_{\frac{9a}{13}+\frac{4b}{13}}+\frac{169}{75}af_{\frac{4a}{13}+\frac{9b}{13}}-$

$\frac{169}{75}bf_{\frac{4a}{13}+\frac{9b}{13}} + \frac{77}{25}af_{\frac{5a}{34}+\frac{29b}{34}} - \frac{77}{25}bf_{\frac{5a}{34}+\frac{29b}{34}} - \frac{36}{71}af_{\frac{a}{26}+\frac{25b}{26}} + \frac{35}{69}bf_{\frac{a}{26}+\frac{25b}{26}} + af_b - bf_b + \frac{113}{38}af_{\frac{a+b}{2}} - \frac{113}{38}bf_{\frac{a+b}{2}}],$

$\int_a^b f(x)dx = \frac{1}{17}[-((a-b)f_a) - \left(\frac{14b}{23} - \frac{14a}{23}\right) f_{\frac{32a}{33}+\frac{b}{33}} - \frac{113}{39}af_{\frac{15a}{17}+\frac{2b}{17}} + \frac{113}{39}bf_{\frac{15a}{17}+\frac{2b}{17}} - \frac{111}{38}af_{\frac{17a}{29}+\frac{12b}{29}} + \frac{111}{38}bf_{\frac{17a}{29}+\frac{12b}{29}} - \frac{111}{38}af_{\frac{12a}{29}+\frac{17b}{29}} +$
$\frac{111}{38}bf_{\frac{12a}{29}+\frac{17b}{29}} - \frac{113}{39}af_{\frac{2a}{17}+\frac{15b}{17}} + \frac{113}{39}bf_{\frac{2a}{17}+\frac{15b}{17}} + \frac{14}{23}af_{\frac{a}{33}+\frac{32b}{33}} - \frac{14}{23}bf_{\frac{a}{33}+\frac{32b}{33}} - af_b + bf_b - \left(\frac{63a}{29} - \frac{63b}{29}\right) f_{\frac{1}{4}(3a+b)}$
$- \left(\frac{63a}{29} - \frac{63b}{29}\right) f_{\frac{1}{4}(a+3b)}],$

$\int_a^b f(x)dx = \frac{1}{19}[-((a-b)f_a) - \left(\frac{27b}{40} - \frac{27a}{40}\right) f_{\frac{39a}{40}+\frac{b}{40}} - \frac{86}{31}af_{\frac{19a}{21}+\frac{2b}{21}} + \frac{86}{31}bf_{\frac{19a}{21}+\frac{2b}{21}} - \frac{55}{26}af_{\frac{23a}{29}+\frac{6b}{29}} + \frac{55}{26}bf_{\frac{23a}{29}+\frac{6b}{29}} -$
$\frac{111}{38}af_{\frac{17a}{26}+\frac{9b}{26}} + \frac{111}{38}bf_{\frac{17a}{26}+\frac{9b}{26}} - \frac{111}{38}af_{\frac{9a}{26}+\frac{17b}{26}} + \frac{111}{38}bf_{\frac{9a}{26}+\frac{17b}{26}} - \frac{55}{26}af_{\frac{6a}{29}+\frac{23b}{29}} + \frac{55}{26}bf_{\frac{6a}{29}+\frac{23b}{29}} - \frac{86}{31}af_{\frac{2a}{21}+\frac{19b}{21}} + \frac{86}{31}bf_{\frac{2a}{21}+\frac{19b}{21}} + \frac{27}{40}af_{\frac{a}{40}+\frac{39b}{40}}$
$- \frac{27}{40}bf_{\frac{a}{40}+\frac{39b}{40}} - af_b + bf_b - \frac{145}{48}af_{\frac{a+b}{2}} + \frac{145}{48}bf_{\frac{a+b}{2}}],$

where $f_x := f(x)$.

7.4.2 Singular Integrals

Let us consider the integral

$$G(t) = \int_0^t (t-s)^{m-\alpha-1} f^{(m)}(s)ds \ ,$$

where m is a positive integer satisfying $m-1 < \alpha \le m$ and $f^{(m)}(s)$ is defined for $0 \le t \le s$. This integral has singularity at $s = t$.

We are requested to evaluate this integral. Let, for example, $0 < \alpha \le 1$ so that

$$G(t) = \int_0^t \frac{f'(s)}{(t-s)^\alpha} ds$$

We may approximate the function $f'(s)$ by means of a Chebyshev polynomial, say

$$f'(s) \approx P(s) = \sum_{k=0}^{n} a_k s^k,$$

and then

$$G(t) \approx \sum_{k=0}^{n} a_k \int_0^t \frac{s^k}{(t-s)^\alpha} ds = \sum_{k=0}^{n} \frac{k! a_k}{\Gamma(k+2-\alpha)} t^{k+1-\alpha}.$$

Alternatively, integratingt by parts gives

$$G(t) = \frac{t^{1-\alpha}}{1-\alpha} f'(0) + \frac{1}{1-\alpha} \int_0^t (t-s)^{1-\alpha} f''(s),$$

and we get a nonsingular integral, which may be evaluated using quadrature formulas. In some cases it is desiderable to obtain some analytical expression for the last integral. We may proceed as follows. Let $Q(s)$ be the Lagrange polynomial of the function $f(s)$ at the nodes

$$(\xi_j, f(\xi_j), \text{ where } \xi_j = T \sin^2\left(\frac{\pi j}{2n}\right), \ j = 0, 1, 2, \ldots n$$

$$Q(s) = \sum_{j=0}^{n} \left[\prod_{k \neq j} \frac{s - \xi_k}{\xi_j - \xi_k} \right] y_j,\ y_j = f(\xi_j).$$

If

$$Q(s) = \sum_{j=0}^{n} c_j s^j,$$

then

$$G(t) = \frac{t^{1-\alpha}}{1-\alpha} f'(0) + \frac{1}{1-\alpha} \sum_{j=2}^{n} j(j-1)c_j \int_0^t (t-s)^{1-\alpha} s^{j-2} ds,$$

$$G(t) = \frac{t^{1-\alpha}}{1-\alpha} f'(0) + \frac{1}{1-\alpha} \sum_{j=2}^{n} \frac{j!\Gamma(2-\alpha)}{\Gamma(j+1-\alpha)} c_j t^{j-\alpha}.$$

Example 30. In fractional calculus, the fractional Caputo derivative of order α of a function $f(t)$ is defined as

$$D_t^\alpha f(t) = \frac{1}{\Gamma(m-\alpha)} \int_0^t \frac{f^{(m)}(s}{(t-s)^{\alpha+1-m}})ds, \text{ where } m-1 < \alpha \leq m.$$

When $0 < \alpha < 1$ we obtain

$$D_t^\alpha f(t) = G(t) = \frac{t^{1-\alpha}}{1-\alpha} f'(0) + \frac{1}{1-\alpha} \int_0^t (t-s)^{1-\alpha} f''(s).$$

Similar results are obtained for $\alpha > 1$.

Let now

$$H(t) = \int_a^t \frac{f'(s)}{\exp\left(\alpha \frac{t-s}{1-\alpha}\right)} ds,\ a < 0 \text{ and } 0 < \alpha \leq 1.$$

This is a singular integral. Integration by parts gives

$$H(t) = \frac{1-\alpha}{\alpha} \left(\exp\left(\frac{t-a}{1-\alpha}\right) f'(a) - f'(t) \right) + \int_a^t \frac{1-\alpha}{\alpha} \exp\left(\frac{\alpha(t-s)}{1-\alpha}\right) f''(s) ds,$$

and we get a nonsingular integral, which may be evaluated as already explained.

7.5 Finite Differences

Suppose we are given that

$$\Phi(\ddot{\varphi}, \dot{\varphi}, \varphi, t) = 0,\ \varphi(0) = \varphi_0 \text{ and } \varphi'(0) = \dot{\varphi}_0.$$

We use the following FDM scheme :

$$\dot{\varphi}_1 = \varphi_1' = \tfrac{\varphi_1 - \varphi_0}{\Delta t}$$
$$\dot{\varphi}_i = \varphi_i' = \tfrac{3\varphi_i - 4\varphi_{i-1} + \varphi_{i-2}}{2\Delta t} + \mathrm{O}\left(\Delta t^2\right),\ i \geq 2$$

$$\begin{array}{l} \ddot{\varphi}_2 = \varphi_2'' = \frac{\varphi_2 - 2\varphi_1 + \varphi_0}{\Delta t^2} \\ \ddot{\varphi}_i = \varphi_i'' = \frac{2\varphi_i - 5\varphi_{i-1} + 4\varphi_{i-2} - \varphi_{i-3}}{\Delta t^2} + \mathrm{O}\left(\Delta t^2\right),\ i \geq 3 \end{array}$$

$$t_i = i\Delta t,\ i = 0, 1, ..., N,\ \Delta t = \frac{T}{N}.$$

The initial data are

$$\varphi_0 = \varphi(0),\ \varphi_1 = \varphi_0 + \varphi'(0)\Delta t$$

Letting $i = 2$ gives

$$\Phi\left(\frac{\varphi_2 - 2\varphi_1 + \varphi_0}{\Delta t^2}, \frac{\varphi_2 - \varphi_1}{\Delta t}, \varphi_2, \frac{2T}{N}\right) = 0.$$

We solve the nonlinear equation (7.5) using a suitable method. For example,

$$\varphi_2 = \varphi_1 + \frac{2\Psi\left(\varphi_1\right)\dot{\Psi}\left(\varphi_1\right)}{\Psi\left(\varphi_1\right)\ddot{\Psi}\left(\varphi_1\right) - 2\dot{\Psi}\left(\varphi_1\right)^2},$$

where

$$\Psi(z) := \Phi\left(\frac{z - 2\varphi_1 + \varphi_0}{\Delta t^2}, \frac{z - \varphi_1}{\Delta t}, z, 2\Delta t\right).$$

Let $i \geq 3$. We find the numbers φ_i recursively as follows : Let $\varphi_i = \zeta$. Then

$$\Phi\left(\frac{2\zeta - 5\varphi_{i-1} + 4\varphi_{i-2} - \varphi_{i-3}}{\Delta t^2}, \frac{3\zeta - 4\varphi_{i-1} + \varphi_{i-2}}{2\Delta t}, \zeta, i\Delta t\right) = 0.$$

We solve the last equation for ζ assuming that we already found the numbers φ_k $(k = 0, 1, 2, ..., i-1)$. The last equation may be solved using the same formula that allowed us to find φ_2,that is (7.5)-(7.5):

$$\varphi_i = \zeta = \varphi_{i-1} + \frac{2\Psi\left(\varphi_{i-1}\right)\dot{\Psi}\left(\varphi_{i-1}\right)}{\Psi\left(\varphi_{i-1}\right)\ddot{\Psi}\left(\varphi_{i-1}\right) - 2\dot{\Psi}\left(\varphi_{i-1}\right)^2},\ i = 3, 4, ..., N.$$

Once we obtained the constants $\varphi_k(k = 0, 1, 2, ..., N)$, we construct an interpolating function that will represent the solution to our nonlinear ode.

Example 31. Let us consider the Kapitza pendulum

$$\begin{cases} \ddot{\varphi} + 2\delta\dot{\varphi} + \ (F_0 \cos{(\gamma t)}\ \ - \omega_0^2) \sin{(\varphi)} = f(t), \\ \varphi(0) = \varphi_0 \text{ and } \dot{\varphi}(0) = \dot{\varphi}_0, \end{cases}$$

$$\varphi_0 = \theta(0),\ \varphi_1 = \varphi_0 + \ \dot{\theta}(0)\Delta t.$$

Letting $j = 2$ gives

$$\frac{\varphi_0 - 2\varphi_1 + \varphi_2}{\Delta t^2} + 2\delta\frac{\varphi_2 - \varphi_1}{\Delta t} + \left(F_0\cos(2\gamma\Delta t) - \omega_0^2\right)\sin\varphi_2 - f(2\Delta t) = 0.$$

Approximate solution :

$$\varphi_2 \approx \frac{\Delta t\left(2\delta\varphi_1 + f(2\Delta t)\Delta t + \Delta t\left(\varphi_0\cos\left(\varphi_0\right) - \sin\left(\varphi_0\right)\right)\left(F_0\cos(2\gamma\Delta t) - 1\right)\right) - \varphi_0 + 2\varphi_1}{\Delta t^2\cos\left(\varphi_0\right)\left(F_0\cos(2\gamma\Delta t) - 1\right) + 2\delta\Delta t + 1}$$

Let now $j \geq 3$. The discretized ode reads

$$\frac{2\zeta - \varphi_{j-3} + 4\varphi_{j-2} - 5\varphi_{j-1}}{\Delta t^2} - f(j\Delta t) + \sin(\zeta)\left(F_0\cos(\gamma j\Delta t) - \omega_0^2\right) + \frac{\delta\left(3\zeta + \varphi_{j-2} - 4\varphi_{j-1}\right)}{\Delta t} = 0,\ \zeta = \varphi_j$$

For a numerical illustration, let

$$\ddot{\varphi} + 0.2\dot{\varphi} + (\cos(\pi t) - 1)\sin(\varphi) = \cos(2t),\ \varphi(0) = \pi \text{ and } \varphi'(0) = 0.$$

In the Figure 7.1 we plotted both Runge-Kutta (black) and finite differences (dashed) solutions. We set $\Delta t = 0.01$.

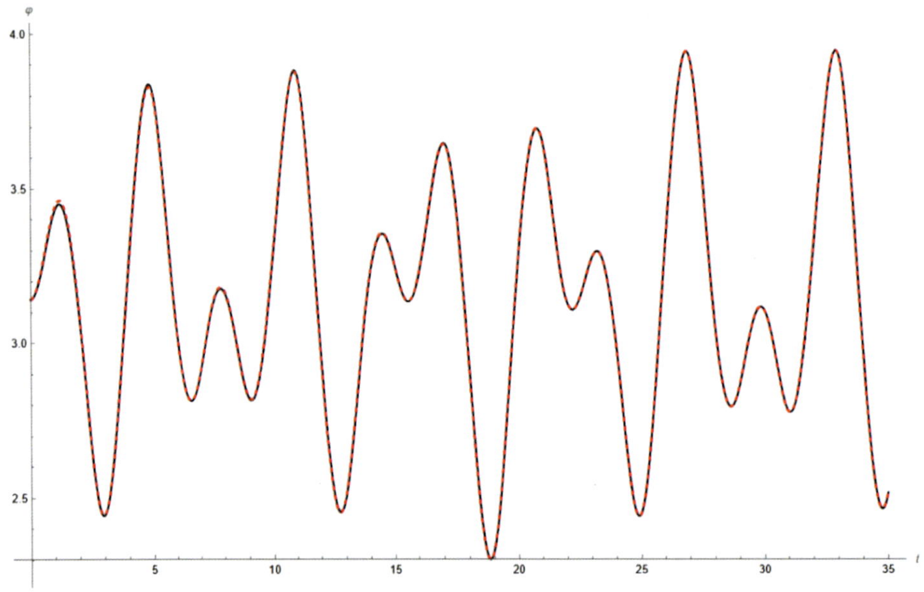

Figure 7.1: The distance between the two solutions equals 0.0115211

7.6 Fourier series Method

Let us consider the oscillator

$$\ddot{x} + f(x, \dot{x}, t) = 0,\ x(0) = x_0 \text{and } \acute{x}(0) = \dot{x}_0.$$

We will assume that this oscillator admits periodic solution with period $2T$.

First Step. We solve numerically the ode (7.6) and then we determine its period.

Second Step. The approximate solution has the form

$$x_{\text{app}}(t) = \frac{a_0}{2} + \sum_{n=1}^{N}\left[a_n \cos\left(\frac{n\pi}{T}t\right) + b_n \sin\left(\frac{n\pi}{T}t\right)\right],$$

where

$$\begin{aligned} a_0 &= \tfrac{1}{T}\textstyle\int_{-T}^{T} x(\tau)d\tau. \\ a_n &= \tfrac{1}{T}\textstyle\int_{-T}^{T} x(\tau)\cos\left(\frac{n\pi}{T}\tau\right)d\tau. \\ b_n &= \tfrac{1}{T}\textstyle\int_{-T}^{T} x(\tau)\sin\left(\frac{n\pi}{T}\tau\right)d\tau. \end{aligned}$$

The function $x(\tau)$ represents the numerical solution. We may use the following quadrature formula for evaluating the integrals (7.6) :

$\int_A^B \xi(\tau)d\tau = \frac{B-A}{1200}[31\xi\left(\frac{49A}{50}+\frac{B}{50}\right) + 20\xi\left(\frac{24A}{25}+\frac{B}{25}\right) + 25\xi\left(\frac{47A}{50}+\frac{3B}{50}\right) + 24\xi\left(\frac{23A}{25}+\frac{2B}{25}\right) + 24\xi\left(\frac{9A}{10}+\frac{B}{10}\right)$
$+24\xi\left(\frac{22A}{25}+\frac{3B}{25}\right) + 24\xi\left(\frac{43A}{50}+\frac{7B}{50}\right) + 24\xi\left(\frac{21A}{25}+\frac{4B}{25}\right) + 24\xi\left(\frac{41A}{50}+\frac{9B}{50}\right) + 24\xi\left(\frac{4A}{5}+\frac{B}{5}\right)$
$+24\xi\left(\frac{39A}{50}+\frac{11B}{50}\right) + 24\xi\left(\frac{19A}{25}+\frac{6B}{25}\right) + 24\xi\left(\frac{37A}{50}+\frac{13B}{50}\right) + 24\xi\left(\frac{18A}{25}+\frac{7B}{25}\right) + 24\xi\left(\frac{7A}{10}+\frac{3B}{10}\right)$
$+24\xi\left(\frac{17A}{25}+\frac{8B}{25}\right) + 24\xi\left(\frac{33A}{50}+\frac{17B}{50}\right) + 24\xi\left(\frac{16A}{25}+\frac{9B}{25}\right) + 24\xi\left(\frac{31A}{50}+\frac{19B}{50}\right) + 24\xi\left(\frac{3A}{5}+\frac{2B}{5}\right)$
$+24\xi\left(\frac{29A}{50}+\frac{21B}{50}\right) + 24\xi\left(\frac{14A}{25}+\frac{11B}{25}\right) + 24\xi\left(\frac{27A}{50}+\frac{23B}{50}\right) + 24\xi\left(\frac{13A}{25}+\frac{12B}{25}\right) + 24\xi\left(\frac{A}{2}+\frac{B}{2}\right)$
$+24\xi\left(\frac{12A}{25}+\frac{13B}{25}\right) + 24\xi\left(\frac{23A}{50}+\frac{27B}{50}\right) + 24\xi\left(\frac{11A}{25}+\frac{14B}{25}\right) + 24\xi\left(\frac{21A}{50}+\frac{29B}{50}\right) + 24\xi\left(\frac{2A}{5}+\frac{3B}{5}\right)$
$+24\xi\left(\frac{19A}{50}+\frac{31B}{50}\right) + 24\xi\left(\frac{9A}{25}+\frac{16B}{25}\right) + 24\xi\left(\frac{17A}{50}+\frac{33B}{50}\right) + 24\xi\left(\frac{8A}{25}+\frac{17B}{25}\right) + 24\xi\left(\frac{3A}{10}+\frac{7B}{10}\right)$
$+24\xi\left(\frac{7A}{25}+\frac{18B}{25}\right) + 24\xi\left(\frac{13A}{50}+\frac{37B}{50}\right) + 24\xi\left(\frac{6A}{25}+\frac{19B}{25}\right) + 24\xi\left(\frac{11A}{50}+\frac{39B}{50}\right) + 24\xi\left(\frac{A}{5}+\frac{4B}{5}\right)$
$+24\xi\left(\frac{9A}{50}+\frac{41B}{50}\right) + 24\xi\left(\frac{4A}{25}+\frac{21B}{25}\right) + 24\xi\left(\frac{7A}{50}+\frac{43B}{50}\right) + 24\xi\left(\frac{3A}{25}+\frac{22B}{25}\right) + 24\xi\left(\frac{A}{10}+\frac{9B}{10}\right)$
$+24\xi\left(\frac{2A}{25}+\frac{23B}{25}\right) + 25\xi\left(\frac{3A}{50}+\frac{47B}{50}\right) + 20\xi\left(\frac{A}{25}+\frac{24B}{25}\right) + 31\xi\left(\frac{A}{50}+\frac{49B}{50}\right) + 8\xi(A) + 8\xi(B)].$

For not too large integration interval, we may use the following formulas:

$\int_a^b f(x)dx = \frac{1}{8}(b-a)\left(3\left(f\left(\frac{1}{3}(2a+b)\right) + f\left(\frac{1}{3}(a+2b)\right)\right) + f(a) + f(b)\right).$

$\int_a^b f(x)dx = \frac{1}{12}(b-a)\left(2f\left(\frac{a+b}{2}\right) + 4f\left(\frac{1}{4}(3a+b)\right) + 4f\left(\frac{1}{4}(a+3b)\right) + f(a) + f(b)\right).$

$\int_a^b f(x)dx = \frac{1}{120}(b-a)[21f\left(\frac{3a}{5}+\frac{2b}{5}\right) + 21f\left(\frac{2a}{5}+\frac{3b}{5}\right) + 31f\left(\frac{1}{5}(4a+b)\right)$
$+31f\left(\frac{1}{5}(a+4b)\right) + 8f(a) + 8f(b)].$

$\int_a^b f(x)dx = \frac{1}{240}(b-a)[24f\left(\frac{3a}{5}+\frac{2b}{5}\right) + 24f\left(\frac{2a}{5}+\frac{3b}{5}\right) + 24f\left(\frac{a+b}{2}\right) + 20f\left(\frac{1}{5}(4a+b)\right) + 31f\left(\frac{1}{10}(9a+b)\right)$
$+25f\left(\frac{1}{10}(7a+3b)\right) + 20f\left(\frac{1}{5}(a+4b)\right) + 25f\left(\frac{1}{10}(3a+7b)\right)$
$+31f\left(\frac{1}{10}(a+9b)\right) + 8f(a) + 8f(b)].$

$\int_a^b f(x)dx = \frac{1}{360}(b-a)[24f\left(\frac{3a}{5}+\frac{2b}{5}\right) + 24f\left(\frac{2a}{5}+\frac{3b}{5}\right) + 24f\left(\frac{1}{3}(2a+b)\right)$
$+25f\left(\frac{1}{5}(4a+b)\right) + 31f\left(\frac{1}{15}(14a+b)\right) + 24f\left(\frac{1}{3}(a+2b)\right) + 20f\left(\frac{1}{15}(13a+2b)\right)$
$+25f\left(\frac{1}{5}(a+4b)\right) + 24f\left(\frac{1}{15}(11a+4b)\right) + 24f\left(\frac{1}{15}(8a+7b)\right) + 24f\left(\frac{1}{15}(7a+8b)\right)$
$+24f\left(\frac{1}{15}(4a+11b)\right) + 20f\left(\frac{1}{15}(2a+13b)\right) + 31f\left(\frac{1}{15}(a+14b)\right) + 8f(a) + 8f(b)].$

Using this last formula gives

$a_0 = \frac{1}{120}[24f(0) + 8f(-T) + 31f\left(-\frac{4T}{5}\right) + 20f\left(-\frac{3T}{5}\right) + 25f\left(-\frac{2T}{5}\right) + 24f\left(-\frac{T}{5}\right)$
$+24f\left(\frac{T}{5}\right) + 25f\left(\frac{2T}{5}\right) + 20f\left(\frac{3T}{5}\right) + 31f\left(\frac{4T}{5}\right) + 8f(T)].$

$a_n = \frac{1}{120}[24f(0) + 8(-1)^n(f(-T) + f(T)) + 24\left(f\left(-\frac{T}{5}\right) + f\left(\frac{T}{5}\right)\right)\cos\left(\frac{\pi n}{5}\right)$

$+25\left(f\left(-\frac{2T}{5}\right)+f\left(\frac{2T}{5}\right)\right)\cos\left(\frac{2\pi n}{5}\right)+20\left(f\left(-\frac{3T}{5}\right)+f\left(\frac{3T}{5}\right)\right)\cos\left(\frac{3\pi n}{5}\right)$
$+31\left(f\left(-\frac{4T}{5}\right)+f\left(\frac{4T}{5}\right)\right)\cos\left(\frac{4\pi n}{5}\right)\cos\left(\frac{\pi nt}{T}\right)]$
$b_n=\frac{1}{120}[24\left(f\left(\frac{T}{5}\right)-f\left(-\frac{T}{5}\right)\right)\sin\left(\frac{\pi n}{5}\right)+25\left(f\left(\frac{2T}{5}\right)-f\left(-\frac{2T}{5}\right)\right)\sin\left(\frac{2\pi n}{5}\right)$
$+20\left(f\left(\frac{3T}{5}\right)-f\left(-\frac{3T}{5}\right)\right)\sin\left(\frac{3\pi n}{5}\right)+31\left(f\left(\frac{4T}{5}\right)-f\left(-\frac{4T}{5}\right)\right)\sin\left(\frac{4\pi n}{5}\right)\sin\left(\frac{\pi nt}{T}\right)].$

Example 32. Let

$$\ddot{x}+4.44255+0.139892x-4.8427x^2+0.651693x^3=0,\ x(0)=1\text{and } x'(0)=0.4.$$

The exact solution reads

$$x_{\text{exact}}(t)=9.58844-\frac{9.31189}{1+0.137902\wp(t+(1.58281-1.11766i);5.12586,-2.35348)}$$

We have :

$$L=\int_{\rho}^{\infty}\frac{dx}{\sqrt{4x^3-g_2x-g_3}}=\int_{-1.31489}^{\infty}\frac{dx}{\sqrt{4x^3-5.12586x+2.35348}}=3.4523.$$

The approximate solution with $N=18$ equals

$x_{\text{app}}(t)=1.58589.+2.8328\sin\left(\frac{\pi t}{\mathrm{L}}\right)-0.236673\sin\left(\frac{2\pi t}{\mathrm{L}}\right)-1.38634\sin\left(\frac{3\pi t}{\mathrm{L}}\right)+0.309386\sin\left(\frac{4\pi t}{\mathrm{L}}\right)$
$+0.547392\sin\left(\frac{5\pi t}{\mathrm{L}}\right)-0.256245\sin\left(\frac{6\pi t}{\mathrm{L}}\right)-0.198749\sin\left(\frac{7\pi t}{\mathrm{L}}\right)+0.170027\sin\left(\frac{8\pi t}{\mathrm{L}}\right)$
$+0.0609109\sin\left(\frac{9\pi t}{\mathrm{L}}\right)-0.0994238\sin\left(\frac{10\pi t}{\mathrm{L}}\right)-0.0107169\sin\left(\frac{11\pi t}{\mathrm{L}}\right)+0.0531445\sin\left(\frac{12\pi t}{\mathrm{L}}\right)$
$-0.00367353\sin\left(\frac{13\pi t}{\mathrm{L}}\right)-0.0244369\sin\left(\frac{14\pi t}{\mathrm{L}}\right)+0.00877801\sin\left(\frac{15\pi t}{\mathrm{L}}\right)+0.0138582\sin\left(\frac{16\pi t}{\mathrm{L}}\right)$
$-0.00280764\sin\left(\frac{17\pi t}{\mathrm{L}}\right)-0.00182898\sin\left(\frac{18\pi t}{\mathrm{L}}\right)-0.372591\cos\left(\frac{\pi t}{\mathrm{L}}\right)-0.888107\cos\left(\frac{2\pi t}{\mathrm{L}}\right)$
$+0.571012\cos\left(\frac{3\pi t}{\mathrm{L}}\right)+0.53781\cos\left(\frac{4\pi t}{\mathrm{L}}\right)-0.419078\cos\left(\frac{5\pi t}{\mathrm{L}}\right)-0.25897\cos\left(\frac{6\pi t}{\mathrm{L}}\right)$
$+0.256002\cos\left(\frac{7\pi t}{\mathrm{L}}\right)+0.0985992\cos\left(\frac{8\pi t}{\mathrm{L}}\right)-0.146781\cos\left(\frac{9\pi t}{\mathrm{L}}\right)-0.0282417\cos\left(\frac{10\pi t}{\mathrm{L}}\right)$
$+0.0771193\cos\left(\frac{11\pi t}{\mathrm{L}}\right)-0.000920162\cos\left(\frac{12\pi t}{\mathrm{L}}\right)-0.0410587\cos\left(\frac{13\pi t}{\mathrm{L}}\right)+0.00682546\cos\left(\frac{14\pi t}{\mathrm{L}}\right)$
$+0.0197109\cos\left(\frac{15\pi t}{\mathrm{L}}\right)-0.010253\cos\left(\frac{16\pi t}{\mathrm{L}}\right)-0.0128607\cos\left(\frac{17\pi t}{\mathrm{L}}\right)+0.010271\cos\left(\frac{18\pi t}{\mathrm{L}}\right).$

Error$=0.0222824$.The two solutions are depicted in Figures 7.2 and 7.3.

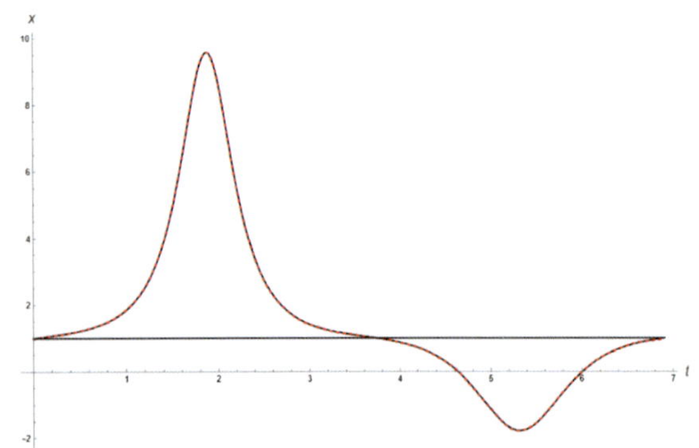

Figure 7.2: Comparision between $x_{exact}(t)$ and $x_{app}(t)$

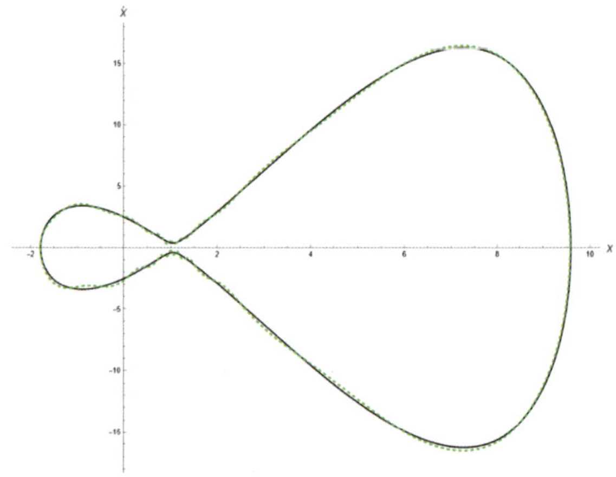

Figure 7.3: Comparision between the two parametric curves

It is clear that we may find the fourier series of any periodic function.

Example 33. Let

$$f(t) = \cos^{2/3}(t).$$

Letting $N = 15$ gives

$\cos(t)^{2/3} \approx 1.07119\cos(t) - 3.4\text{e\^{}-9}\cos(2t) - 0.0973789\cos(3t) - 3.4\text{e-8}\cos(4t) + 0.040095\cos(5t)$

$-1.52\text{e-7}\cos(6t) - 0.0226595\cos(7t) - 4.56\text{e-7}\cos(8t) + 0.0148441\cos(9t) - 1.\text{e-6}\cos(10t) - 0.0105981\cos(11t)$

$-2.2\text{e-6}\cos(12t) + 0.00801444\cos(13t) - 3.95\text{e-6}\cos(14t) - 0.00629884\cos(15t)$. See Figure 7.4.

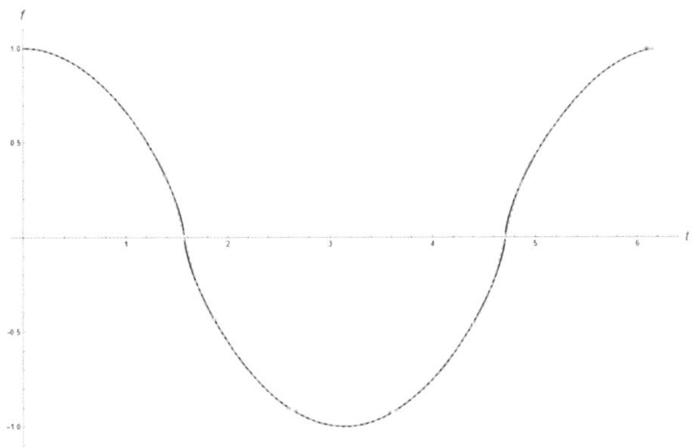

Figure 7.4

Example 34. Approximation of elliptic functions by means of trigonometric functions. Let us consider the Jacobi function

$$f(t) = \mathrm{cn}(t, m).$$

Our aim is to obtain the approximation

$$f(t) \approx A\cos\left(\frac{\pi}{2K(m)}t\right) + (1 - A)\cos\left(\frac{3\pi}{2K(m)}t\right).$$

To this end , we define

$$g(t) = A\cos\left(\frac{\pi}{2K(m)}t\right) + (1 - A)\cos\left(\frac{3\pi}{2K(m)}t\right) - \mathrm{cn}(t, m).$$

Next, we calculate the fourier series for $2T = 4K(m)$, that is, $T = 2K(m)$. Let

$$\theta = \frac{\pi}{2K(m)}t.$$

We have :

$g(t) = [RA + S]\cos\theta$+h.o.t,where

$$S = -0.0235114d_3 + 0.0235114d_7 + 0.0380423d_9 - 0.0323607d_8d_{10} - 0.01d_4\,(d_{10} + 2.23607)$$
$$+0.0161803d_{10}\,(d_{10} + 2.23607) - 0.01d_2\,(d_{20} + 2.23607) - 0.01d_6\,(2.23607d_{10} + d_{20})$$
$$+0.005\,(d_{20} + 2.23607)\,(2.23607d_{10} + d_{20}) - 0.0380423d_{d_{20}} - 0.0273819e_{13} - 0.025497e_{14}$$
$$-0.0214331e_{16} - 0.0192701e_{17} - 0.0170312e_{18} - 0.014725e_{19} - 0.0099476e_{21} - 0.00749525e_{22}$$
$$-0.00501333e_{23} - 0.00251162e_{24} + 0.00251162e_{26} + 0.00501333e_{27} + 0.00749525e_{28} + 0.0099476e_{29}$$
$$+0.014725e_{31} + 0.0170312e_{32} + 0.0192701e_{33} + 0.0214331e_{34} + 0.025497e_{36} + 0.0273819e_{37} - 0.02628$$

$$R = 0.0261803d_{10}^2 + 0.0025\,(d_{10} + 2.23607)^{\,2} + 0.0025\,(d_{20} + 2.23607)^{\,2} + 0.0025\,(2.23607d_{10} + d_{20})^{\,2}$$
$$+0.01\,(d_{10} + 2.23607)\,(d_{20} + 2.23607) + 0.272486.$$

$$d_j = \operatorname{cn}\left(\frac{K(m)}{5}j\middle|\, m\right),\; e_i = \operatorname{cn}\left(\frac{K(m)}{25}i\middle|\, m\right).$$

Each d_j and e_i may be approximated by means of a chebyshev polynomial. For example,

$$d_1 \approx -0.0574455m^3 + 0.00892759m^2 - 0.0259308m + 0.951057 \text{ for } 0 \le m \le 3/4.$$

We equate to zero the coefficient of $\cos\theta$ and solving the resulting equation for A gives

$$A = -0.275354(m - 1.37486)\left(m^2 - 1.30534m + 2.00244\right)\left(m^2 + 1.68483m + 1.31914\right)$$

A rational approximation :

$$A = \frac{1 - \frac{92m}{123}}{1 - \frac{17m}{26}}$$

We also may proceed using the following approach. Consider the integral

$$\mathcal{I}(m, A) = \int_0^{4K(m)} \left| A\cos\left(\frac{\pi}{2K(m)}t\right) + (1 - A)\cos\left(\frac{3\pi}{2K(m)}t\right) - \operatorname{cn}(t, m)\right| dt$$

For each $m \in [-1, 3/4]$ we minimize this integral for A, so that $A = A(m)$:

$$A(m) = A = \min_{A'} \mathcal{I}(m, A').$$

We will obtain a set of pairs $\{(m, A(m)\}$, $m \in [-1, 3/4]$. Using this pairs we construct an interpolation function. Then, we adjust it using a chebyshevpolynomial. The result is

m	A	m	A	m	A
$-1.$	1.04232	-0.4	1.02083	0.2	0.985989
-0.95	1.04082	-0.35	1.0186	0.25	0.981917
-0.9	1.03928	-0.3	1.01628	0.3	0.977558
-0.85	1.03769	-0.25	1.01386	0.35	0.97287
-0.8	1.03605	-0.2	1.01134	0.4	0.967802
-0.75	1.03436	-0.15	1.0087	0.45	0.962291
-0.7	1.03262	-0.1	1.00594	0.5	0.956256
-0.65	1.03082	-0.05	1.00305	0.55	0.949593
-0.6	1.02896	0.	1.	0.6	0.94216
-0.55	1.02703	0.05	0.99679	0.65	0.933766
-0.5	1.02504	0.1	0.993399	0.7	0.924132
-0.45	1.02297	0.15	0.989807	0.75	0.912839

Table . Optimal A values

See Figures 7.5 and 7.6

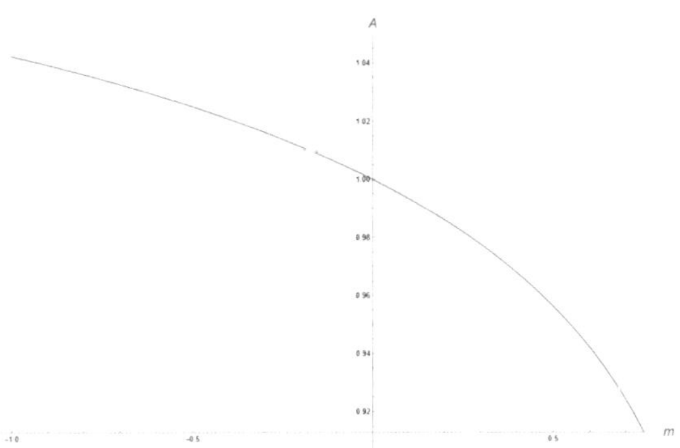

Figure 7.5: $A = A(m)$: optimal value

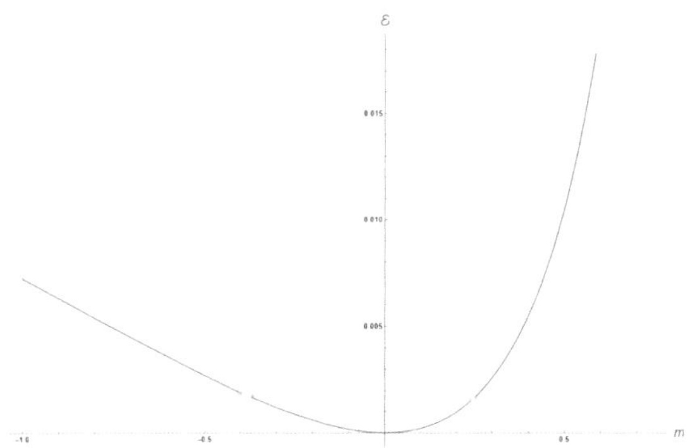

Figure 7.6: Error plot

Finally,

$$A = A(m) = \frac{1}{74}\left(\frac{11}{6} - m\right)\left(m^2 - \frac{21m}{29} + \frac{90}{17}\right)\left(m^2 + \frac{153m}{32} + \frac{277}{36}\right)$$

A rational approximation is

$$A = \frac{1 - \frac{23m}{37}}{1 - \frac{19m}{34}}.$$

For example, when $m = 0.75$ we get

$$\operatorname{cn}(t, m) \approx \frac{34}{37}\cos\left(\frac{\pi}{2K\left(\frac{3}{4}\right)}t\right) + \frac{3}{37}\cos\left(\frac{3\pi}{2K\left(\frac{3}{4}\right)}t\right) \text{ with error} \mathcal{E} = 0.018$$

See Figure. Other useful approximations are :

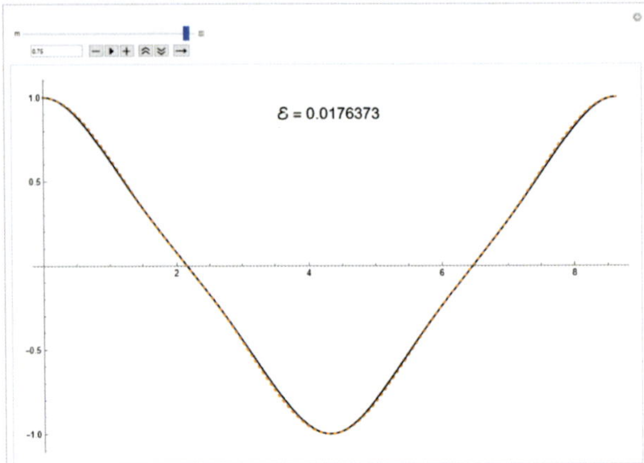

Figure 7.7

$$
\begin{aligned}
&\mathrm{cn}(t,m) \approx A\cos\left(\tfrac{\pi t}{2K(m)}\right) + B\cos\left(\tfrac{3\pi t}{2K(m)}\right) + (1-A-B)\cos\left(\tfrac{5\pi t}{2K(m)}\right),\\
&A = \frac{\frac{m^2}{76}-\frac{37m}{44}+1}{1-\frac{53m}{68}},\, B = \frac{\frac{4m}{65}-\frac{m^2}{68}}{1-\frac{23m}{31}}.\\
&\mathrm{cn}^2(t,m) \approx A + B\cos\left(\tfrac{\pi t}{K(m)}\right) + (1-A-B+1)\cos\left(\tfrac{2\pi t}{K(m)}\right),\\
&A = -\tfrac{m^8}{31} - \tfrac{m^7}{20} + \tfrac{m^6}{161} + \tfrac{m^5}{36} - \tfrac{m^4}{80} - \tfrac{m^3}{34} - \tfrac{m^2}{32} - \tfrac{m}{16} + \tfrac{1}{2},\\
&B = -\tfrac{m^8}{141} - \tfrac{m^7}{91} + \tfrac{m^6}{556} + \tfrac{m^5}{148} - \tfrac{m^4}{495} - \tfrac{m^3}{203} - \tfrac{m^2}{304} + \tfrac{1}{2}.\\
&\mathrm{cn}^3(t,m) \approx A\cos\left(\tfrac{\pi t}{2K(m)}\right) + B\cos\left(\tfrac{3\pi t}{2K(m)}\right) + (1-A-B)\cos\left(\tfrac{5\pi t}{2K(m)}\right),\\
&A = -\tfrac{m^6}{25} - \tfrac{2m^5}{33} - \tfrac{m^4}{88} - \tfrac{m^3}{307} - \tfrac{m^2}{23} - \tfrac{3m}{31} + \tfrac{3}{4},\\
&B = \tfrac{m^6}{99} + \tfrac{m^5}{63} + \tfrac{m^4}{254} + \tfrac{m^3}{334} + \tfrac{m^2}{59} + \tfrac{m}{21} + \tfrac{1}{4}.\\
&\mathrm{cn}^4(t,m) \approx A + B\cos\left(\tfrac{\pi t}{K(m)}\right) + (1-A-B+1)\cos\left(\tfrac{2\pi t}{K(m)}\right),\\
&A = -\tfrac{m^8}{135} - \tfrac{m^7}{82} + \tfrac{m^5}{268} - \tfrac{m^4}{150} - \tfrac{m^3}{72} - \tfrac{m^2}{44} - \tfrac{2m}{35} + \tfrac{3}{8},\\
&B = -\tfrac{m^8}{405} - \tfrac{m^7}{244} + \tfrac{m^5}{788} - \tfrac{m^4}{456} - \tfrac{m^3}{222} - \tfrac{m^2}{139} - \tfrac{m}{61} + \tfrac{1}{2}.\\
&\mathrm{cn}^5(t,m) \approx A\cos\left(\tfrac{\pi t}{2K(m)}\right) + B\cos\left(\tfrac{3\pi t}{2K(m)}\right) + (1-A-B)\cos\left(\tfrac{5\pi t}{2K(m)}\right),\\
&A = -\tfrac{m^6}{60} - \tfrac{m^5}{36} - \tfrac{m^4}{97} - \tfrac{m^3}{89} - \tfrac{m^2}{26} - \tfrac{2m}{21} + \tfrac{5}{8},\\
&B = \tfrac{m^6}{298} + \tfrac{m^5}{181} + \tfrac{m^4}{471} + \tfrac{m^3}{403} + \tfrac{m^2}{112} + \tfrac{m}{35} + \tfrac{5}{16}.\\
&|\mathrm{cn}| = (-A-B+1)\cos\left(\tfrac{2\pi t}{K(m)}\right) + A + B\cos\left(\tfrac{\pi t}{K(m)}\right),\\
&A = -\tfrac{2m^{10}}{33} - \tfrac{6m^9}{61} + \tfrac{m^8}{26} + \tfrac{3m^7}{28} - \tfrac{m^6}{82} - \tfrac{2m^5}{33} - \tfrac{m^4}{51} - \tfrac{m^3}{86} - \tfrac{m^2}{35} - \tfrac{m}{18} + \tfrac{29}{45},\\
&B = \tfrac{m^{10}}{132} + \tfrac{m^9}{80} - \tfrac{m^8}{252} - \tfrac{m^7}{80} + \tfrac{m^6}{507} + \tfrac{m^5}{119} + \tfrac{m^4}{251} + \tfrac{m^3}{241} + \tfrac{m^2}{111} + \tfrac{m}{43} + \tfrac{17}{41}.
\end{aligned}
$$

7.7 GalerkinMethod

This method is employed to solve boundary value problems. Let

$$\ddot{u} + F(t, u, \dot{u}) = 0,\ u(0) = u(T) = 0,\ 0 \le t \le T$$

The solution is assumed in the ansatz form

$$u(t) = \sum_{k=1}^{n-1} c_k \varphi_k(t),$$

where

$$0 = \xi_0 < \xi_1 < \cdots < \xi_{n-1} < \xi_n = T.$$

$$\varphi_k(t) = \frac{t - \xi_{k-1}}{\xi_k - \xi_{k-1}} \chi_{k-1}(t) + \frac{\xi_{k+1} - t}{\xi_{k+1} - \xi_k} \chi_k(t),$$

$$\chi_k(t) = 1 \text{ if } \xi_k \leq t \leq \chi_{k+1} \text{ and 0 otherwiswe.}$$

This methos converges in nthe case when $F(u, \dot{u})$ is a linear function : $F(u, \dot{u}) = au + b\dot{u}$. We will apply this approach for a nonlinear $F(u, \dot{u})$ having the form

$$F(u, \dot{u}) = \sum_{k,l} d_{k,l} u^k \dot{u}^l.$$

The first step is to integrate the residuals

$$R_j(t) = (u''(t) + F(u(t), \acute{u}(t)))\varphi_j(t),\ j = 1, 2, ..., n-1$$

over $0 \leq t \leq T$. We define

$$\xi_k = hk,\ h = \frac{T}{n}.$$

The following identities are useful for computing :

$$\varphi_j(t)\varphi_k(t) = 0 \text{ if } |j-k| \geq 2 \text{ and } 0 \leq t \leq T.$$

$$\int_0^T \varphi_j^p(t)dt = \frac{2T}{n(p+1)} \text{ for any } p \geq 1.$$

$$\int_0^T \varphi_j^r(t)\varphi_k^s(t)dt = \frac{Tr!s!}{n(r+s+1)!} \text{ for } |j-k| = 1,\ r+s \geq 1$$

$$\int_0^T u''(t)\varphi_j(t)dt = \frac{n}{T}\left(c_{j-1} - 2c_j + c_{j+1}\right),\ c_0 = c_n = 0,\ j = 1, 2, ..., n-1$$

$$\int_0^T u^{\mathcal{N}}(t)\varphi_j(t)dt = \frac{T}{n(1+\mathcal{N})(2+\mathcal{N})}\left[2(\mathcal{N}+1)c_j^{\mathcal{N}} + \sum_{k=0}^{\mathcal{N}-1}(k+1)c_j^k\left(c_{j-1}^{\mathcal{N}-k} + c_{j+1}^{\mathcal{N}-k}\right)\right],\ \mathcal{N} \geq 2.$$

$$\int_0^T u'(t)\varphi_j(t)dt = \frac{c_{j+1} - c_{j-1}}{2}$$

$$\int_0^T tu(t)\varphi_j(t)dt = \frac{T^2\left((2j-1)c_{j-1} + 8jc_j + (2j+1)c_{j+1}\right)}{12n^2}.$$

$$\int_0^T t^2u(t)\varphi_j(t)dt = \frac{T^3\left(\left(40j^2+4\right)c_j + (10(j-1)j+3)c_{j-1} + (10j(j+1)+3)c_{j+1}\right)}{60n^3}.$$

$$\int_0^T u(t)\acute{u}(t)\varphi_j(t)dt = \frac{1}{6}(c_{j+1} - c_{j-1})(c_{j-1} + c_j + c_{j+1})$$

For more details, see [1]. We also may use the flllowing approximation formula:
$\int_a^b f\,(x)dx = \frac{1}{240}(b-a)[24f\left(\frac{3a}{5} + \frac{2b}{5}\right) + 24f\left(\frac{2a}{5} + \frac{3b}{5}\right) + 24f\left(\frac{a+b}{2}\right) + 20f\left(\frac{1}{5}(4a+b)\right) + 31f\left(\frac{1}{10}(9a+b)\right)$

$+25f\left(\frac{1}{10}(7a+3b)\right)+20f\left(\frac{1}{5}(a+4b)\right)+25f\left(\frac{1}{10}(3a+7b)\right)$
$+31f\left(\frac{1}{10}(a+9b)\right)+8f(a)+8f(b)].$

Example 35. Let

$$\ddot{u}+q+pu+u^5=0, u(0)=u(T)=0,\ 0\leq t\leq T.$$

We have the following equations :

$$\frac{T}{42n}c_j^5+\frac{Tc_{j-1}}{21n}c_j^4+\frac{Tc_{j-1}^2}{14n}c_j^3+\frac{2Tc_{j-1}^3}{21n}c_j^2+\left(\frac{5Tc_{j-1}^4}{42n}+\frac{pT}{6n}+\frac{n}{T}\right)c_j$$
$$+[\frac{pTc_{j-2}}{6n}+\frac{2pTc_{j-1}}{3n}+\frac{Tc_{j-2}^5}{42n}+\frac{Tc_{j-1}c_{j-2}^4}{21n}+\frac{Tc_{j-1}^2c_{j-2}^3}{14n}+$$
$$\frac{2Tc_{j-1}^3c_{j-2}^2}{21n}+\frac{5Tc_{j-1}^4c_{j-2}}{42n}+\frac{nc_{j-2}}{T}+\frac{2Tc_{j-1}^5}{7n}-\frac{2nc_{j-1}}{T}+\frac{qT}{n}]=0,\ j=1,2,...,n-1.$$

$c_{-1}=c_0=c_n=0.$

Letting $j=1$ gives

$$\frac{qT}{n}+\left(\frac{n}{T}+\frac{pT}{6n}\right)c_1+\frac{T}{42n}c_1^5=0.$$

Solving this quintic we find c_1. Anothetr quintic is obtained for dtermining c_2 by putting $j=2$:

$$\frac{qT}{n}-\frac{2nc_1}{T}+\frac{2pTc_1}{3n}+\frac{2Tc_1^5}{7n}+\left(\frac{n}{T}+\frac{pT}{6n}+\frac{5Tc_1^4}{42n}\right)c_2+\frac{2Tc_1^3}{21n}c_2^2+\frac{Tc_1^2}{14n}c_2^3+\frac{Tc_1}{21n}c_2^4+\frac{T}{42n}c_2^5=0.$$

We choose the closest to c_1 root c_2 to the last quintic. We proceed in a similar way to find the other constants c_j. If $n=50$ and $T=3$ we obtain the following values :

$c_1=0.0434906, c_2=0.0793253, c_3=0.107118, c_4=0.126568, c_5=0.137467, c_6=0.139697,$

$c_7=0.133233, c_8=0.118146, c_9=0.0945982, c_{10}=0.0628433, c_{11}=0.0232239, c_{12}=-0.023833,$

$c_{13}=-0.07782, c_{14}=-0.138155, c_{15}=-0.204188, c_{16}=-0.275204, c_{17}=-0.350435, c_{18}=-0.42$

$c_{19}=-0.510179, c_{20}=-0.592861, c_{21}=-0.676067, c_{22}=-0.758651, c_{23}=-0.839326, c_{24}=-0.9$

$c_{25}=-0.988889, c_{26}=-1.05427, c_{27}=-1.1108, c_{28}=-1.1565, c_{29}=-1.18953, c_{30}=-1.20845,$

$c_{31}=-1.21235, c_{32}=-1.20106, c_{33}=-1.17512, c_{34}=-1.1357, c_{35}=-1.08449, c_{36}=-1.02339,$

$c_{37}=-0.954411, c_{38}=-0.879473, c_{39}=-0.800334, c_{40}=-0.718557, c_{41}=-0.635519,$

$c_{42}=-0.552433, c_{43}=-0.470386, c_{44}=-0.390368, c_{45}=-0.313294, c_{46}=-0.240018, c_{47}=-0.1$

$c_{48}=-0.107996, c_{49}=-0.0506777.$

See Figure 7.8 for a comprison with Runge-Kutta numerical solution.

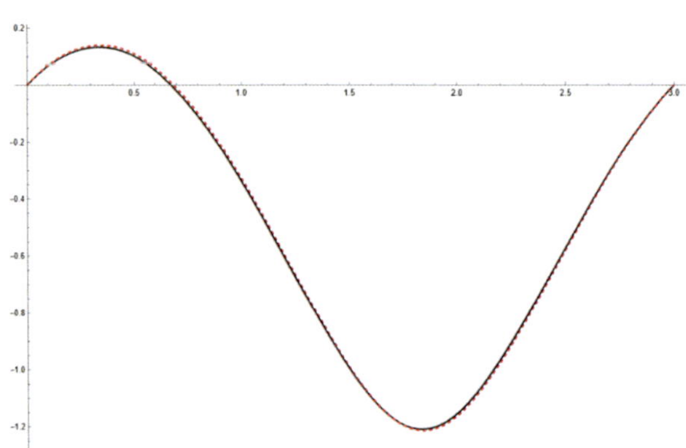

Figure 7.8: Comparision with Runge-Kutta numerical solution. error= 0.013

With $n=100$ and $T=3$ the erroris 0.0032. See Figure 7.9.

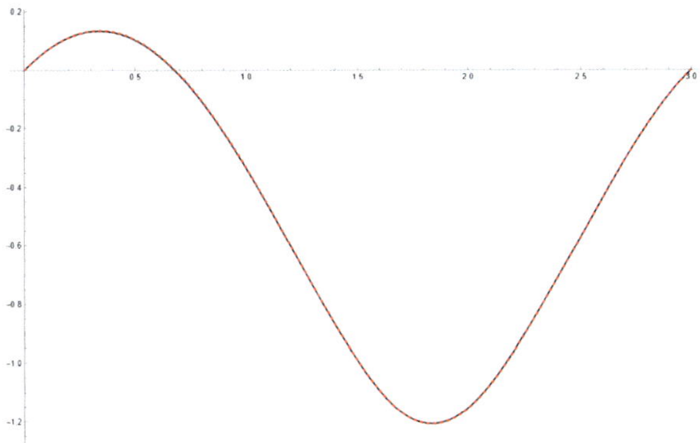

Figure 7.9: Comparision with Runge-Kutta numerical solution. error– 0.0032

Suppose now we are given that the oscillator

$$\ddot{x} + F(t, x, \dot{x}) = 0,\ x(0) = x_0 \text{ and } x'(0) = \dot{x}_0,$$

admits a periodic solution . We need to find its period. To this end, we consider the i.v.p

$$\ddot{u} + F(t, u, \dot{u}) = 0,\ u(0) = 0, u(T) = 0,\ 0 \leq t \leq T,$$

so that $u(0) = u(T) = 0$ and $u'(0) = u'(T) = 0$, that is

$$u'(0) = c_1 = 0 \text{ and } c_{n-1} = -c_1 = 0,$$

The period T is unknown. In this case, we will obtain a system of nonlinear equations whose unknowns will be c_2, c_3, ..., c_{n-2} and $T = c_n$.

Example 37. Let

$$\ddot{x} + x^5 + px + q = 0,\ x(0) = x_0 \text{and } x'(0) = \dot{x}_0.$$

We need to solve the following nonlinear equations:

$\frac{qT}{n} + \left[\frac{pT}{6n} + \frac{n}{T}\right] c_2 + \frac{T}{42n} c_2^5 = 0,$

$\frac{pTc_{j-2}}{6n} + \frac{2pTc_{j-1}}{3n} + c_j\left(\frac{5Tc_{j-1}^4}{42n} + \frac{pT}{6n} + \frac{n}{T}\right) + \frac{Tc_{j-2}^5}{42n} + \frac{Tc_{j-1}c_{j-2}^4}{21n} + \frac{Tc_{j-1}^2c_{j-2}^3}{14n} + \frac{2Tc_{j-1}^3c_{j-2}^2}{21n} + \frac{5Tc_{j-1}^4c_{j-2}}{42n} +$
$\frac{nc_{j-2}}{T} +$
$\frac{2Tc_{j-1}^5}{7n} + \frac{Tc_j^5}{42n} + \frac{Tc_{j-1}c_j^4}{21n} + \frac{Tc_{j-1}^2c_j^3}{14n} + \frac{2Tc_{j-1}^3c_j^2}{21n} - \frac{2nc_{j-1}}{T} + \frac{qT}{n} = 0,$

$j = 2, 3, ..., n-2.$

$\frac{pTc_{n-2}}{6n} + \frac{Tc_{n-2}^5}{42n} + \frac{nc_{n-2}}{T} + \frac{qT}{n} = 0.$

Let $n = 1000$, $p = 1$ and $q = 2$. We find that

$$T = 2.99949.$$

For the values $x_0 = 1$ and $\dot{x}_0 = 0.3$ the solutions u and x are depicted in Figure 44.

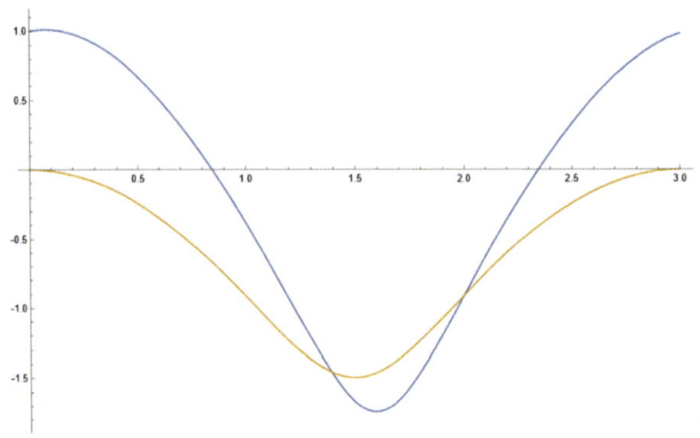

Figure 7.10: The period $T = 2.99946$

On the other hand, we have

$$x^5 + px + q = (x+1)\left(x^4 - x^3 + x^2 - x + 2\right)$$

and the real root equals $\rho = -1$. The period is very close to $3\,|\rho| = 3$. So, we may use the numerical solution $x = x(t)$ to find the exact period with initial guess $3\,|\rho|$:

$$T_0 = 3\,|\rho|\,,\, T_j = T_{j-1} - \frac{g(T_{j-1})}{g'(T_{j-1})},\, j = 1, 2, 3, ...$$

$$g(t) = (x(t) - x_0)^2 + (x'(t) - \dot{x}_0)^2.$$

This will give the value

$$T = 3.044,$$

which is a more accurate approximation.

Example 36. Solve a quintic. Let

$$P(x) := 12x^5 - 117x^4 + 24x^3 + 6x^2 - 25x + 19 = 0.$$

This equation may be written as

$$\frac{12x^5}{19} - \frac{117x^4}{19} + \frac{24x^3}{19} + \frac{6x^2}{19} - \frac{25x}{19} + 1 = 0$$

Consider the i.v.p.

$$\ddot{x} + \frac{12x^5}{19} - \frac{117x^4}{19} + \frac{24x^3}{19} + \frac{6x^2}{19} - \frac{25x}{19} + 1 = 0,\, x(0) = 0 \text{ and } x'(0) = 0.$$

Solving $x''(t) = 0$ gives

$$t_0 = 1.128448740979419.$$

See Figure 7.11

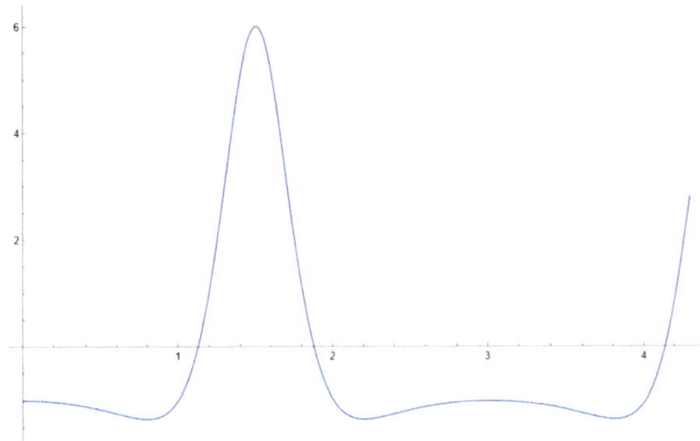

Figure 7.11: The function $y(t) = x''(t)$

Then, approximate solution to our quintic will be

$$\rho = x(t_0) = -0.7061243616183617.$$

Using Newton-Raphson we get

$$\rho_1 = \rho - \frac{P(\rho)}{P'(\rho)} = -0.7061265910688374 \text{ and } P(\rho_1) = -1.144 \times 10^{-10}.$$

Using this root we may find the period by solving $x(t) = 0$ near $t = 3\,|\rho_1|$:

$$T = 3.0097004112493217$$

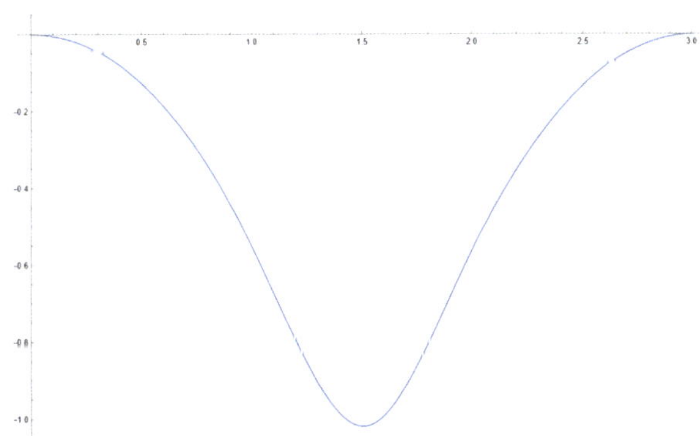

Figure 7.12: The periodic solution $x = x(t)$

$x(t) = 0.477661\cos(2.08764t) - 0.101192\cos(4.17529t) + 0.0268151\cos(6.26293t)$
$-0.00721765\cos(8.35058t) + 0.00166226\cos(10.4382t) - 0.397728.$

See Figure 7.13.

Let now

$$\ddot{x} + F(t, x, \dot{x}) = 0,\; x(0) = 0 \text{ and } x'(0) = \dot{x}_0,\; 0 \le t \le T,$$

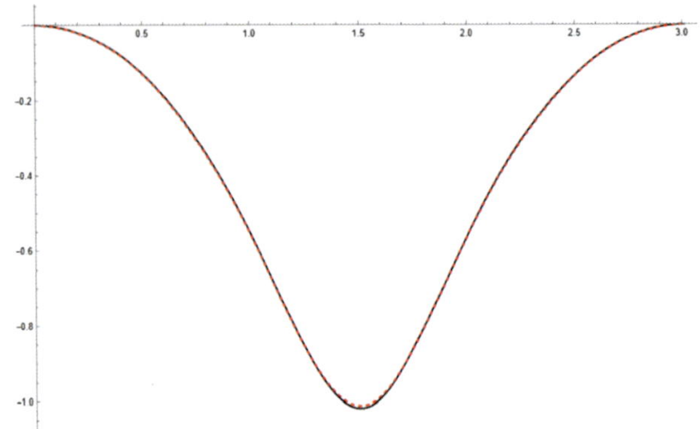

Figure 7.13: Runge-Kutta vs analytical solution (dashed)

We assume the solution as

$$x(t) = \sum_{k=1}^{n+1} c_k \varphi_k(t),\ c_1 = \frac{T}{n}\dot{x}_0$$

7.8 The Collocation Method

Let

$$\ddot{x} + F(t, x, \dot{x}) = 0,\ x(0) = x_0 \text{ and } x'(0) = \dot{x}_0,\ 0 \leq t \leq T.$$

7.8.1 First Variant

We solve the problem numerically using the Runge-Kutta algorithm. Then, we construct the lagrange interpolating polynomial $P(t)$ at nodes (ξ_j, η_j), where

$$\xi_j = T \sin^2\left(\frac{\pi j}{2n}\right),\ j = 0, 1, 2, \ldots n.$$

$$P(t) = \sum_{j=0}^{n} \left[\prod_{k \neq j} \frac{t - \xi_k}{\xi_j - \xi_k}\right] \eta_j$$

Let $x(t)$ be the numerical solution. Then

$$\eta_j = x(\xi_j) \text{ forf all } j.$$

The time interval should not be too large, since we will need a polynomil of high degree and that is difficult to handle. We may approximate the solution on intervals of length 1 or less. Let $r \leq t \leq r + 1$. In that case,

A cubic approximating polynomial on $r \leq t \leq r + 1$ is
$P_{3,r}(t) = \frac{1}{3}\left((r+1)\left(8\eta_1 r(4r+1) - (4r+3)\left(8\eta_2 r - \eta_3(4r+1)\right)\right) - \eta_0 r(4r+1)(4r+3)\right) +$
$\frac{1}{3}\left(\eta_0(16r(3r+2)+3) - 8\eta_1(2r(6r+5)+1) + 8\eta_2(2r(6r+7)+3) - \eta_3(16r(3r+4)+19)\right)t+$
$\frac{8}{3}\left(-2\eta_0(3r+1) + \eta_1(12r+5) - \eta_2(12r+7) + 2\eta_3(3r+2)\right)t^2 + \frac{16}{3}\left(\eta_0 - 2\eta_1 + 2\eta_2 - \eta_3\right)t^3,$
where
$\eta_j = x(\xi_j),\ \xi_j = \frac{1}{2}(1+2r) + \frac{1}{2}\cos\left(\frac{j\pi}{3}\right),\ j = 01, 2, 3.$

A quintic approximating polynomial on $r \le t \le r+1$is

$P_{5,r}(t) = \left(-\frac{256r^5}{5} - \frac{512r^4}{5} - \frac{336r^3}{5} - 16r^2 - r\right)\eta_0 + \left(\frac{512r^5}{5} + \frac{4077r^4}{19} + \frac{1306r^3}{9} + \frac{1687r^2}{48} + \frac{42r}{19}\right)\eta_1 +$
$\left(-\frac{512r^5}{5} - \frac{6725r^4}{28} - \frac{146329r^3}{804} - \frac{851r^2}{18} - \frac{55r}{18}\right)\eta_2 + \left(\frac{512r^5}{5} + \frac{7611r^4}{28} + \frac{11038r^3}{45} + \frac{2368r^2}{29} + \frac{110r}{19}\right)\eta_3 +$
$\left(-\frac{512r^5}{5} - \frac{5651r^4}{19} - \frac{38539r^3}{124} - \frac{2461r^2}{18} - \frac{377r}{18}\right)\eta_4 + \left(\frac{256r^5}{5} + \frac{768r^4}{5} + \frac{848r^3}{5} + \frac{416r^2}{5} + 17r + 1\right)\eta_5 +$
$[\left(256r^4 + \frac{2048r^3}{5} + \frac{1008r^2}{5} + 32r + 1\right)\eta_0 + \left(-512r^4 - \frac{13733r^3}{16} - \frac{37439r^2}{86} - \frac{2179r}{31} - \frac{42}{19}\right)\eta_1 +$
$\left(512r^4 + \frac{6725r^3}{7} + \frac{146329r^2}{268} + \frac{4066r}{43} + \frac{55}{18}\right)\eta_2 + \left(-512r^4 - \frac{7611r^3}{7} - \frac{16189r^2}{22} - \frac{4736r}{29} - \frac{110}{19}\right)\eta_3 +$
$\left(512r^4 + \frac{19035r^3}{16} + \frac{40093r^2}{43} + \frac{11758r}{43} + \frac{377}{18}\right)\eta_4 + \left(-256r^4 - \frac{3072r^3}{5} - \frac{2544r^2}{5} - \frac{832r}{5} - 17\right)\eta_5]t +$
$[\left(-512r^3 - \frac{3072r^2}{5} - \frac{1008r}{5} - 16\right)\eta_0 + \left(1024r^3 + \frac{21887r^2}{17} + \frac{37439r}{86} + \frac{1687}{48}\right)\eta_1$
$+ \left(-1024r^3 - \frac{61966r^2}{43} - \frac{146329r}{268} - \frac{851}{18}\right)\eta_2 +$
$(1024r^3 + \frac{70130r^2}{43} + \frac{10189r}{22} + \frac{2368}{29}\eta_3 + \left(-1024r^3 - \frac{30337r^2}{17} - \frac{40093r}{43} - \frac{2461}{18}\right)\eta_4$
$+ \left(512r^3 + \frac{4608r^2}{5} + \frac{2544r}{5} + \frac{416}{5}\right)\eta_5)]t^2 +$
$[\left(512r^2 + \frac{2048r}{5} + \frac{336}{5}\right)\eta_0 + \left(-1024r^2 - \frac{13733r}{16} - \frac{1306}{9}\right)\eta_1 + \left(1024r^2 + \frac{6725r}{7} + \frac{146329}{804}\right)\eta_2 +$
$\left(-1024r^2 - \frac{7611r}{7} - \frac{11038}{45}\right)\eta_3 + \left(1024r^2 + \frac{19035r}{16} + \frac{29215}{94}\right)\eta_4 + \left(-512r^2 - \frac{3072r}{5} - \frac{848}{5}\right)\eta_5]t^3 +$
$[\left(-256r - \frac{512}{5}\right)\eta_0 + \left(512r + \frac{4077}{19}\right)\eta_1 + \left(-512r - \frac{6725}{28}\right)\eta_2 + \left(512r + \frac{7611}{28}\right)\eta_3 + \left(-512r - \frac{5651}{19}\right)\eta_4 +$
$\left(256r + \frac{768}{5}\right)\eta_5]t^4 + \left(\frac{256\eta_0}{5} - \frac{512\eta_1}{5} + \frac{512\eta_2}{5} - \frac{512\eta_3}{5} + \frac{512\eta_4}{5} - \frac{256\eta_5}{5}\right)t^5,$

where

$\eta_j = x(\xi_j)$, $\xi_j = \frac{1}{2}(1+2r) + \frac{1}{2}\cos\left(\frac{j\pi}{5}\right)$, $j = 0, 1, 2, 3, 4, 5.$

Example 38. Let

$$\ddot{x} - 0.2\ (1 - x^2)\dot{x} + 3x + x^3 = 0.2\cos t,\ x(0) = x'(0) = 0,\ 0 \le t \le 13.$$

The aproximate polynomial solution reads

$P(t) = 1.01e{-}14t^{19} - 1.28e{-}12t^{18} + 7.3e{-}11t^{17} - 2.52e{-}9t^{16} + 5.84e{-}8t^{15} - 9.56e{-}7t^{14} +$
$0.0000114247t^{13} - 0.000101378t^{12} + 0.00067546t^{11} - 0.00340711t^{10} + 0.0130809t^9$
$-0.0380325t^8 + 0.0812674t^7 - 0.12122t^6 + 0.129727t^5 - 0.122733t^4 + 0.0416518t^3$
$+0.0935967t^2 + 0.000347954t.$

See Figure 7.14.

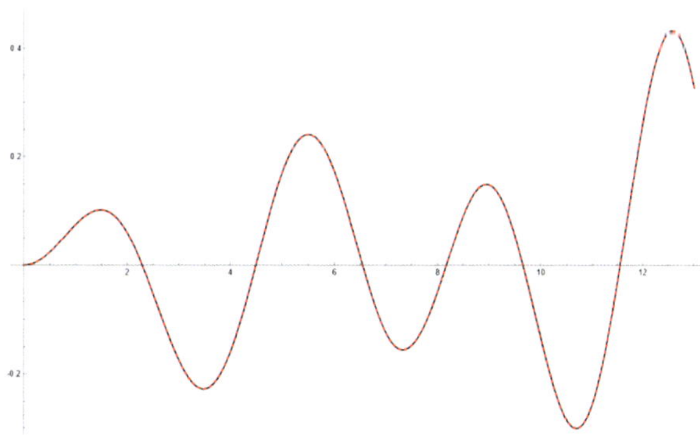

Figure 7.14: Comparion between the Runge-Kutta numerical

We also may approximate tghe solution by means of quintic splines (see Figure 7.15).

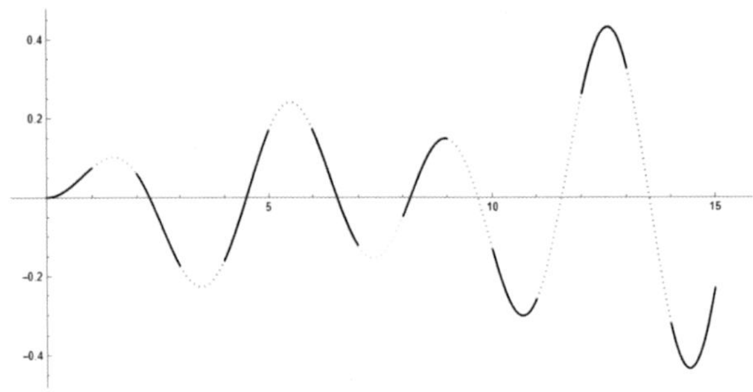

Figure 7.15: Aproximation by quintic slipenes

The splines are given parametrically $(0 \leq t \leq 1)$ as follows :

t	$t\,(0.00792635t^4 - 0.0439465t^3 + 0.0119204t^2 + 0.0989397t + 0.0000622675)$
$t+1$	$0.0086678t^5 + 0.0114513t^4 - 0.0847301t^3 - 0.0478964t^2 + 0.0975456t + 0.0749023$
$t+2$	$-0.0134687t^5 + 0.0530422t^4 + 0.0365217t^3 - 0.146857t^2 - 0.163426t + 0.0599404$
$t+3$	$-0.00558581t^5 - 0.0409483t^4 + 0.119428t^3 + 0.143289t^2 - 0.202682t - 0.174248$
$t+4$	$0.0179609t^5 - 0.0540011t^4 - 0.0864468t^3 + 0.201621t^2 + 0.250626t - 0.160747$
$t+5$	$0.000408494t^5 + 0.06788t^4 - 0.13372t^3 - 0.198614t^2 + 0.268193t + 0.169012$
$t+6$	$-0.0225t^5 + 0.0460736t^4 + 0.127659t^3 - 0.191089t^2 - 0.256809t + 0.173159$
$t+7$	$0.00851125t^5 - 0.0912382t^4 + 0.101594t^3 + 0.240667t^2 - 0.184048t - 0.123507$
$t+8$	$0.0263956t^5 - 0.0268559t^4 - 0.168592t^3 + 0.0858171t^2 + 0.279785t - 0.0480212$
$t+9$	$-0.0177993t^5 + 0.110955t^4 - 0.0298332t^3 - 0.316377t^2 - 0.0299972t + 0.148529$
$t+10$	$-0.0287604t^5 - 0.014761t^4 + 0.239068t^3 + 0.077322t^2 - 0.397366t - 0.134523$
$t+11$	$0.024548t^5 - 0.116098t^4 - 0.0808906t^3 + 0.422326t^2 + 0.27195t - 0.25902$
$t+12$	$0.0180992t^5 + 0.102016t^4 - 0.327201t^3 - 0.260885t^2 + 0.531887t + 0.262815$
$t+13$	$-0.0325485t^5 + 0.101851t^4 + 0.214624t^3 - 0.458491t^2 - 0.473547t + 0.326732$
$t+14$	$0.0159436t^5 - 0.214717t^4 + 0.334297t^3 + 0.455582t^2 - 0.501514t - 0.321379$

These splines are useful for analyzing the solution. For example, we may estimate the zeros, the period, the local extrema , etc.

7.8.2 Second Variant

We make use of cubic Hermite interpolation. Let

$$0 = \xi_0 < \xi_1 < \cdots < \xi_{n-1} < \xi_n = T.$$

If $f(t)$ is a smooth funtion on $[0, T]$, then

$f(t) = \frac{-1}{(\xi_j - \xi_j)^3} \times$

$[(t-\xi_j)\left((t-\xi_j)(\xi_j - 3\xi_{j+1} + 2t) f_{j+1} + (\xi_j - \xi_{j+1})(t-\xi_{j+1})\left((t-\xi_{j+1})\dot{f}_j + (t-\xi_j)\dot{f}_{j+1}\right)\right)$

$-(t-\xi_{j+1})^2 (2t - 3\xi_j + \xi_{j+1}) f_j],$

where

$$f_j = f(\xi_j) \text{ and } \dot{f} = f'(\xi_j) \text{ for all } j.$$

Given that

$$\ddot{x} + F(t, x, \dot{x}) = 0,\ x(0) = x_0 \text{ and } x'(0) = \dot{x}_0,\ 0 \le t \le T,$$

we assume the solution as

$$x(t) = \sum_{j=0}^{n+1} \varphi_j(t),$$

where

$\varphi_j(t) = \frac{1}{(\xi_{j+1}-\xi_j)^3}[(t-\xi_j)((t-\xi_j)(2+\xi_j-3\xi_{j+1}t)c_{j+1}+$
$(\xi_j - \xi_{j+1})(t-\xi_{j+1})((t-\xi_j)d_{j+1}+(t-\xi_{j+1})d_j) - (t-\xi_{j+1})^2(2t-3\xi_j+\xi_{j+1})c_j]\chi_j(t),$

and

$$\chi_j(t) = 1 \text{ if } \xi_j \le t < \xi_{j+1} \text{ and } 0 \text{ otherwise.}$$

The numbers c_j and d_j are found by solving the system

$$R(\xi_j) = R'(\xi_j) = 0,\ \ c_0 = x_0 \text{ and } d_0 = \dot{x}_0.$$

where

$$R(t) := x''(t) + F(t, x(t), x'(t)).$$

7.8.3 Third Variant

We make use of quintic Hermite interpolation. Let

$$0 = \xi_0 < \xi_1 < \cdots < \xi_{n-1} < \xi_n = T.$$

If $f(t)$ is a smooth funtion on $[0, T]$, then

$$f(t) = \sum_{j=0}^{n} \left[\varphi_{j,1}(t) f_j(\xi_j) + \varphi_{j,2}(t) f_j'(\xi_j) + \varphi_{j,3}(t) f''(\xi_j)\right],$$

being

$\varphi_{j,1}(t) = \frac{(t-\xi_{j+1})^3\left(10\xi_j^2+\xi_{j+1}^2+3t\xi_{j+1}-5\xi_j(\xi_{j+1}+3t)+6t^2\right)}{(\xi_j-\xi_{j+1})^5}\chi_j(t) - \frac{(t-\xi_{j-1})^3\left(\xi_{j-1}^2+\xi_{j-1}(3t-5\xi_j)+5\xi_j(2\xi_j-3t)+6t^2\right)}{(\xi_{j-1}-\xi_j)^5}\chi_{j-1}(t),$
$\varphi_{j,2}(t) = -\left(\frac{(\xi_{j-1}-4\xi_j+3t)(t-\xi_{j-1})^3}{(\xi_{j-1}-\xi_j)^4}\chi_{j-1}(t) + \frac{(t-\xi_{j+1})^3(-4\xi_j+\xi_{j+1}+3t)}{(\xi_j-\xi_{j+1})^4}\chi_j(t)\right)(t-\xi_j),$

and

$\varphi_{j,3}(t) = \frac{1}{2}\left(\frac{(t-\xi_{j+1})^3}{(\xi_j-\xi_{j+1})^3}\chi_j(t) - \frac{(t-\xi_{j-1})^3}{(\xi_{j-1}-\xi_j)^3}\chi_{j-1}(t)\right)(t-\xi_j)^2.$

$$\chi_j(t) = 1 \text{ if } \xi_j \le t < \xi_{j+1} \text{ and } 0 \text{ otherwise.}$$

Given that

$$\ddot{x} + F(t, x, \dot{x}, \ddot{x}) = 0,\ x(0) = x_0,\ \ x'(0) = \dot{x}_0 \text{ and } x''(0) = \ddot{x}_0,\ 0 \le t \le T,$$

we assume the solution as

$$x(t) = \sum_{j=0}^{n} \left[A_j \varphi_{j,1}(t) + B_j \varphi_{j,2}(t) + C_j \varphi_{j,3}(t)\right].$$

The numbers A_j, B_j and C_j are found from the system

$$R(\xi_j) = R'(\xi_j) = R''(\xi_j) = 0, \; A_0 = x_0, \, B_0 = \dot{x}_0 \text{ and } C_0 = \ddot{x}_0.$$

where

$$R(t) := x'''(t) + F(t, x(t), x'(t), x''(t)).$$

$\int_a^b f\,(x)dx = \frac{1}{240}(b-a)[24f\left(\frac{3a}{5}+\frac{2b}{5}\right)+24f\left(\frac{2a}{5}+\frac{3b}{5}\right)+24f\left(\frac{a+b}{2}\right)+20f\left(\frac{1}{5}(4a+b)\right)+31f\left(\frac{1}{10}(9a+b\right.$
$+25f\left(\frac{1}{10}(7a+3b)\right)+20f\left(\frac{1}{5}(a+4b)\right)+25f\left(\frac{1}{10}(3a+7b)\right)$
$+31f\left(\frac{1}{10}(a+9b)\right)+8f(a)+8f(b)].$

7.9 Summary

In Chapter 7 we developed and compared several numerical methods for Duffing-type oscillators and related nonlinear systems, with the aim of constructing a reliable benchmark for the analytical results of the thesis. The main conclusions are the following:

1. **A coherent numerical framework for Duffing-type equations.** The chapter established a coherent framework for the numerical integration of Duffing and generalized Duffing oscillators, formulating the problems as initial–value systems and specifying appropriate non-dimensionalizations, tolerances and error measures. This framework makes the numerical experiments reproducible and comparable across different parameter regimes.

2. **High-order time–stepping schemes as accurate reference solvers.** High–order explicit Runge–Kutta methods (in particular fourth- and sixth-order variants, and embedded adaptive schemes) were implemented and used as reference solvers. With suitably small time steps or tight tolerances, these methods produced solutions with very small global errors, effectively serving as "numerical ground truth" against which other methods and analytical approximations can be tested.

3. **Spectral, Galerkin and collocation methods for periodic responses.** For periodic or quasi-periodic motions, spectral approaches based on Fourier and Chebyshev expansions, as well as Galerkin and collocation formulations, were shown to achieve exponential or very high algebraic convergence when the solution is sufficiently smooth. These methods are particularly effective for computing steady periodic orbits and associated quantities such as periods, amplitudes and phase shifts.

4. **Quantitative comparison of accuracy, stability and efficiency.** The different numerical methods were compared systematically in terms of accuracy, long-time stability

and computational cost. High-order Runge–Kutta schemes provide a good compromise between simplicity and accuracy for general time-dependent simulations, whereas spectral and collocation methods are superior for accurately resolving periodic solutions. The study also highlighted the limitations of low-order schemes and the need for careful step-size control in strongly nonlinear regimes.

5. **Validation of analytical and approximate solutions.** The numerical benchmark was used to validate the exact elliptic solutions and the perturbative and averaging approximations developed in earlier chapters. In the regimes where those analytical approaches are expected to be valid, the agreement with the numerical reference solutions was typically better than 10^{-4} in relative terms, confirming the correctness and practical usefulness of the analytical formulas. In contrast, the numerical results clearly identify parameter regions where perturbative or trigonometric approximations lose accuracy, thus delimiting their domain of applicability.

6. **Guidelines for the numerical treatment of nonlinear oscillators.** Based on the comparative study, the chapter provides practical recommendations for selecting numerical schemes for Duffing-type problems: high-order explicit Runge–Kutta or symplectic integrators for general dynamical studies, spectral or collocation methods for precise periodic orbits, and adaptive strategies in the presence of strong nonlinearity or near bifurcation points. These guidelines are directly applicable to a wide class of nonlinear oscillators beyond the specific models treated here.

7. **Numerical foundations for subsequent fractional and applied studies.** Overall, Chapter 7 furnishes a robust numerical foundation for the rest of the thesis. The benchmark solutions and the comparative analysis of numerical schemes are essential for assessing the accuracy of the fractional models and applied case studies considered in later chapters, and for supporting the broader conclusions on the dynamics of Duffing-type systems.

Chapter 8

Geodesics and Planetary Motions

Geodesics

8.1 Motivation and overview

General relativity (GR) replaces the Newtonian concept of gravitation as a force by the geometric statement that free motion occurs along *geodesics* of a curved spacetime manifold [42, 43, 116, 131]. In spherically symmetric fields, the geodesic problem reduces to a two-dimensional effective dynamics in the plane of motion. A key point for this chapter is that, after passing to the inverse radius

$$u(\varphi) = \frac{1}{r(\varphi)},$$

the relativistic orbital equation takes the form of a nonlinear Binet-type equation whose nonlinearities encode curvature effects and (when present) electromagnetic charge contributions.

Historically, the most famous Solar System test of GR is the anomalous perihelion advance of Mercury [102, 136]. Mathematically, the perihelion advance appears because the radial variable does not repeat after a 2π change in the polar angle: the orbit is nearly elliptic, but not closed. Instead, the motion is periodic in φ with a period $T \neq 2\pi$. The perihelion shift per revolution is therefore the *period mismatch*

$$\delta\varphi = T - 2\pi. \tag{8.1}$$

This observation is the backbone of the exact method used in this chapter.

The chapter develops a unified route from the geodesic equations to an integrable nonlinear oscillator model of Duffing–Helmholtz type. It then shows how exact periodic solutions can be constructed in closed form using either Jacobi elliptic functions or the Weierstrass elliptic function, depending on the nature of the polynomial invariants that arise from the first integral.

8.2 From geodesics to a Duffing–Helmholtz Binet equation

8.2.1 Massive particles in Reissner–Nordström spacetime

For a charged spherically symmetric gravitational field (Reissner–Nordström), the orbital equation for a massive test particle can be written in the form [87]

$$\frac{d^2u}{d\varphi^2} + \left(1 + \frac{Q}{L^2}\right) u = \frac{3GM}{c^2}u^2 - 2Qu^3 + \frac{GM}{c^2L^2}, \tag{8.2}$$

where M is the central mass, G is Newton's constant, c is the speed of light, L is the (specific) angular momentum constant, and Q is the charge parameter (in the conventions of the chosen reduction).

The physical meaning of (8.2) is standard: the $-u$ term corresponds to the Newtonian inverse-square attraction written in Binet form; the u^2 term represents the leading GR correction responsible for perihelion precession, and the u^3 term carries the additional structure introduced by the charge parameter.

8.2.2 Light deflection

The corresponding equation for null geodesics (light rays) may be written as a similar nonlinear Binet equation, e.g. in a model form

$$\frac{d^2u}{d\varphi^2} + b^2u - 3Mu^2 - 2Q^2u^3, \tag{8.3}$$

where b is related to the impact parameter and the precise coefficients depend on the normalization used [87].

8.2.3 Unified Duffing–Helmholtz form

Both (8.2) and (8.3) are special cases of the general cubic Duffing–Helmholtz equation

$$\frac{d^2u}{d\varphi^2} + a_0 + a_1u + a_2u^2 + a_3u^3 = 0, \tag{8.4}$$

where the constants (a_0, a_1, a_2, a_3) are determined by the spacetime parameters and integrals of motion. The value of writing the geodesic problem as (8.4) is that the equation inherits a large integrability machinery from nonlinear oscillator theory: once a first integral is obtained, solutions can be expressed in terms of elliptic functions.

8.3 Aphelion–perihelion data and the first integral

Let P and A denote perihelion and aphelion radii, respectively. Then the inverse radius satisfies

$$u(0) = \frac{1}{P}, \qquad u'(0) = 0, \qquad u(\varphi_A) = \frac{1}{A}, \qquad u'(\varphi_A) = 0,$$

where φ_A is the angle at aphelion.

A standard way to proceed is to integrate (8.4) once and factor the resulting polynomial using the turning points $u = 1/P$ and $u = 1/A$. As in [113], one obtains a first integral of the form

$$\left(\frac{du}{d\varphi}\right)^2 = \frac{1}{2}a_3\left(u - \frac{1}{A}\right)\left(\frac{1}{P} - u\right)\left(b_0 + b_1 u + u^2\right), \tag{8.5}$$

with

$$b_0 = \frac{6a_1A^2P^2 + 4a_2A^2P + 3a_3A^2 + 4a_2AP^2 + 3a_3AP + 3a_3P^2}{3a_3A^2P^2}, \tag{8.6}$$

$$b_1 = \frac{4a_2AP + 3a_3A + 3a_3P}{3a_3AP}. \tag{8.7}$$

Equation (8.5) is the central object: it shows that the dynamics is governed by a quartic polynomial in u (after expansion), therefore the solution is elliptic in the generic case.

8.3.1 Geometric meaning of the polynomial structure

The factors $\left(u - \frac{1}{A}\right)$ and $\left(\frac{1}{P} - u\right)$ represent the two turning points (aphelion and perihelion). The remaining quadratic factor

$$u^2 + b_1 u + b_0$$

determines the additional structure of the orbit: depending on whether it has real roots or complex conjugate roots, the resulting integral for the period is represented more naturally by Jacobi or Weierstrass elliptic functions.

8.4 Illustrations of the orbital families

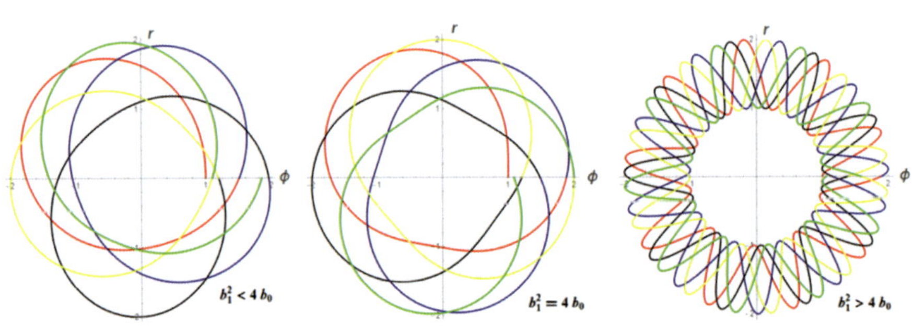

Figure 8.1: Orbits for $a_3 > 0$ (initial trajectory in red).

Figures 8.1–8.2 illustrate the qualitative change of the phase portrait induced by the sign of a_3. In the bounded regime the solution is periodic in φ and therefore the orbit precesses: it returns to perihelion after an angle $T \neq 2\pi$. In contrast, in unstable regimes one may obtain open trajectories or qualitatively different families, as expected from the change in the effective potential.

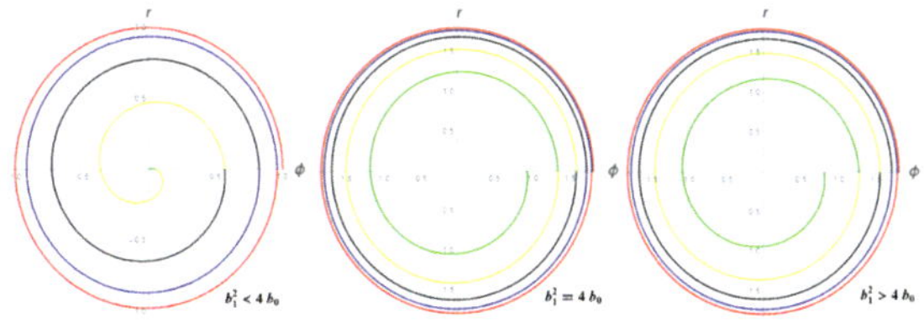

Figure 8.2: Trajectories for $a_3 < 0$ (initial trajectory in red).

8.5 Exact solution for null geodesics: discriminant splitting

Define the "Hamiltonian residual" (zero along exact solutions)

$$H(\varphi) = \left(\frac{du}{d\varphi}\right)^2 - \frac{1}{2}a_3\left(u - \frac{1}{A}\right)\left(\frac{1}{P} - u\right)\left(b_0 + b_1 u + u^2\right). \tag{8.8}$$

The key classification is controlled by the discriminant of the quadratic factor

$$\Delta := b_1^2 - 4b_0. \tag{8.9}$$

If $\Delta \geq 0$, the quadratic has real roots and Jacobi elliptic functions appear naturally. If $\Delta < 0$, the quadratic has complex conjugate roots and the standard representation uses the Weierstrass $\wp$-function [23, 132].

8.5.1 Case I: $\Delta = b_1^2 - 4b_0 \geq 0$ (Jacobi elliptic form)

Let the quadratic factor be written as

$$u^2 + b_1 u + b_0 = (u - r_1)(u - r_2). \tag{8.10}$$

Then (8.5) becomes

$$\left(\frac{du}{d\varphi}\right)^2 = \frac{1}{2}a_3\left(u - \frac{1}{A}\right)\left(\frac{1}{P} - u\right)(u - r_1)(u - r_2). \tag{8.11}$$

A constructive method is to seek $u(\varphi)$ in the rational ansatz

$$u(\varphi) = \frac{1}{A} + \frac{B}{1 + v(\varphi)^2}, \tag{8.12}$$

where v satisfies a (cubic) Duffing equation

$$v''(\varphi) + \alpha v(\varphi) + \beta v(\varphi)^3 = 0, \qquad v(0) = v_0, \quad v'(0) = 0, \tag{8.13}$$

and v_0 is fixed by $u(0) = 1/P$. Matching the coefficients yields [87]

$$\alpha = \frac{a_3(APr_2r_1 - 2Ar_1 + Ar_2 + Pr_1 - 2Pr_2 + 1)}{8AP}, \quad \beta = -\frac{a_3(A - P)(Ar_1 - 1)}{4A^2P}, \quad B = \frac{Ar_2 - 1}{A}. \tag{8.14}$$

The solution of (8.13) is expressed via the Jacobi elliptic cosine cn:

$$v(\varphi) = v_0 \,\mathrm{cn}\left(\sqrt{\alpha + \beta v_0^2}\,\varphi,\ \frac{\beta v_0^2}{2(\alpha + \beta v_0^2)}\right). \tag{8.15}$$

Consequently, $u(\varphi)$ is periodic, and its period is

$$T = \frac{2}{\sqrt{\alpha + \beta v_0^2}} K\left(\frac{\beta v_0^2}{2(\alpha + \beta v_0^2)}\right), \tag{8.16}$$

where $K(m)$ is the complete elliptic integral of the first kind. The perihelion shift per revolution is therefore the exact quantity

$$\delta\varphi = T - 2\pi = \frac{2}{\sqrt{\alpha + \beta v_0^2}} K\left(\frac{\beta v_0^2}{2(\alpha + \beta v_0^2)}\right) - 2\pi, \tag{8.17}$$

in agreement with the definition (8.1).

8.5.2 Case II: $\Delta = b_1^2 - 4b_0 < 0$ (Weierstrass form)

When the quadratic factor has no real roots, a convenient exact ansatz is

$$u(\varphi) = \frac{1}{P} + \frac{B}{1 + C\,\wp(\varphi; g_2, g_3)}, \tag{8.18}$$

where $\wp$ is the Weierstrass elliptic function, satisfying

$$\wp'(\varphi)^2 = 4\wp(\varphi)^3 - g_2\wp(\varphi) - g_3.$$

Substituting (8.18) into the differential equation and enforcing vanishing of the residual yields explicit expressions for B, C, g_2 and g_3 (see [87] for the full algebraic formulas). The periodicity of $\wp$ implies that u is periodic in φ, and the period is computed from the standard elliptic integral representation [23, 132].

8.6 Particular case: $a_3 = 0$ (quadratic nonlinearity)

When $a_3 = 0$, (8.4) becomes

$$\frac{d^2u}{d\varphi^2} + a_0 + a_1 u + a_2 u^2 = 0, \tag{8.19}$$

which still yields elliptic solutions (now associated with a cubic polynomial after one integration). This case is important because it includes the classical Schwarzschild Binet equation used in many textbook derivations of Mercury's perihelion advance.

Figures 8.3–8.4 show representative families for the quadratic model. Even in this simpler setting, the distinction between bounded and unbounded motion is controlled by the turning points and the sign/structure of the effective polynomial under the square root in the first

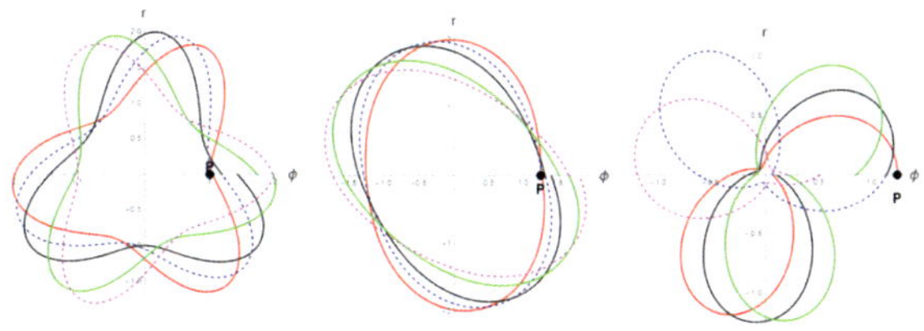

Figure 8.3: Orbits $r(\varphi) = 1/u(\varphi)$ of (8.19) for $a_2 > 0$.

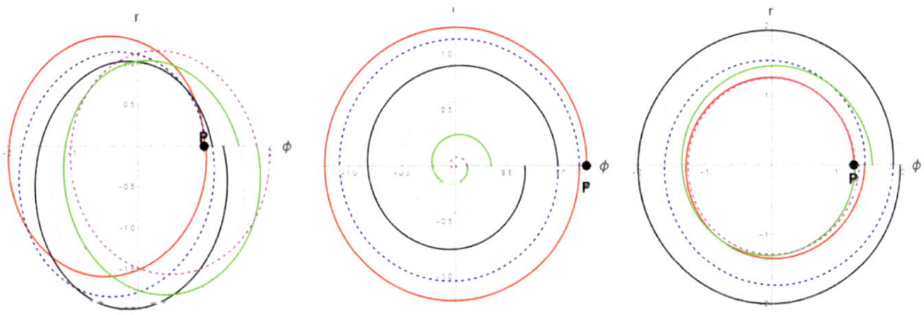

Figure 8.4: Orbits $r(\varphi) = 1/u(\varphi)$ of (8.19) for $a_2 < 0$.

integral.

8.7 Application: Einstein equation for planetary motion (Solar System)

For Schwarzschild planetary motion one obtains the relativistic Binet equation [117, 136]

$$u''(\varphi) = \frac{GM}{L^2} - u(\varphi) + \frac{3GM}{c^2} u(\varphi)^2, \tag{8.20}$$

which is a special case of (8.19). The key fact is that the exact period in φ is generically not 2π, so that a perihelion shift is produced. In weak fields one may expand the elliptic integral formulas to recover the classical approximation

$$\delta\varphi \approx \frac{6\pi GM}{c^2 a(1-e^2)},$$

but the elliptic approach provides an exact (non-perturbative) expression.

Figure 8.5 illustrates the geometric meaning of the shift: the orbit is nearly an ellipse, but its axis rotates slowly, producing the observed precession.

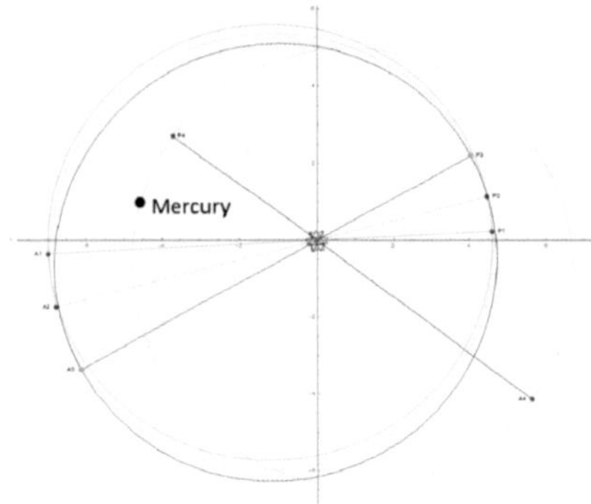

Figure 8.5: Mercury's perihelion advance over successive revolutions (illustration).

8.8 Motion in the Reissner–Nordström metric

Returning to (8.2), we identify the Duffing–Helmholtz parameters by rewriting the equation as (8.4). A convenient identification is

$$a_1 = 1 + \frac{Q}{L^2}, \qquad a_2 = -\frac{3GM}{c^2}, \qquad a_3 = 2Q,$$

with a_0 determined by the turning-point constraints (aphelion/perihelion). Once these constants are fixed, the exact solutions follow from the general construction described in the discriminant cases above.

8.9 Concluding remarks

The central message of this chapter is that relativistic orbital dynamics in spherically symmetric spacetimes can be reduced to an integrable nonlinear oscillator problem. The Duffing–Helmholtz structure produces a first integral with polynomial factorization governed by the orbital turning points. Exact closed-form solutions follow in terms of elliptic functions: Jacobi elliptic functions when the auxiliary quadratic has real roots ($\Delta \geq 0$), and the Weierstrass elliptic function when it has complex roots ($\Delta < 0$). The perihelion shift emerges transparently as a mismatch between the true period in φ and 2π, and can therefore be computed exactly from complete elliptic integrals.

Chapter 9

Applications of Duffing–Type Oscillators

The Duffing oscillator is a paradigmatic model in nonlinear dynamics, originally introduced to describe the motion of a mechanical oscillator with a nonlinear elastic restoring force. Despite its apparent simplicity, the Duffing equation captures a wide range of nonlinear phenomena, including amplitude–dependent frequencies, jump resonance, hysteresis, multistability, and chaos. For this reason, it has become a canonical testbed for analytical, numerical, and experimental studies in nonlinear science.

A further and conceptually deeper extension is obtained by introducing fractional derivatives into the Duffing equation. The resulting fractional Duffing oscillator incorporates memory effects and hereditary behavior, reflecting the fact that the present state of many real systems depends on their entire past evolution. This fractional generalization significantly broadens the range of applicability of Duffing–type models, especially in systems exhibiting viscoelasticity, anomalous dissipation, nonlocal interactions, or complex internal microstructure.

In this chapter we provide a unified overview of the main application areas where both classical and fractional Duffing–type oscillators play a fundamental role. Rather than organizing the discussion into separate sections, we summarize the applications through a set of descriptive items, each highlighting a major domain of relevance. Detailed analytical and numerical examples illustrating these applications will be developed in the subsequent sections of this chapter.

- **Mechanical and structural systems.** Duffing–type equations arise naturally in mechanical and structural engineering when linear elasticity assumptions are violated. Large deflections of beams, plates, and shells introduce nonlinear stiffness terms that are well approximated by cubic or cubic–quintic restoring forces. Such models are essential for describing post–buckling behavior, snap–through instabilities, and nonlinear vibration control. Fractional Duffing models further enhance realism by accounting for viscoelastic damping with memory effects, which is particularly important for polymers, composite materials, and structures subjected to long–term cyclic loading.

- **Nonlinear circuits and electronic devices.** In electrical engineering, the Duffing oscillator models nonlinear LC and RLC circuits containing nonlinear inductors, capacitors, or active elements. The nonlinear resonance curves predicted by the Duffing equation explain jump phenomena and hysteresis observed in experimental frequency responses. The introduction of fractional–order elements, often referred to as *fractances*, leads to fractional Duffing circuit models capable of capturing dielectric absorption, electrode polarization, and fractal electrical networks. These models are increasingly relevant in modern analog circuits, sensors, and resonant devices.

- **Nonlinear waves and soliton reductions.** Duffing–type oscillators frequently emerge as reduced models in nonlinear wave theory. Traveling–wave reductions of nonlinear partial differential equations, such as the Korteweg–de Vries and nonlinear Schrödinger equations, often lead to ordinary differential equations of Duffing type. These reductions provide insight into the existence and structure of periodic, quasi–periodic, and solitary waves, many of which are naturally expressed in terms of Jacobi elliptic functions. Fractional Duffing equations are closely connected with fractional soliton models, where anomalous dispersion and nonlocal effects play a dominant role.

- **Control systems, robotics, and energy harvesting.** The Duffing oscillator is widely used as a benchmark system in nonlinear control theory due to its multistability and sensitivity to parameter variations. These features make it an ideal test problem for feedback control, synchronization, and stabilization strategies. In robotics, fractional Duffing models describe systems with compliant joints and damping governed by fractional constitutive laws. Nonlinear energy harvesters based on piezoelectric or magnetoelastic mechanisms are also commonly modeled as Duffing–type oscillators, where nonlinear stiffness broadens the operational frequency bandwidth and fractional damping improves performance predictions.

- **Biological and medical systems.** Duffing oscillators have been employed to model biological rhythms such as cardiac oscillations, neural activity, and motor control. Nonlinear resonance and bifurcation phenomena observed in experimental data are naturally explained within the Duffing framework. Fractional Duffing models are particularly suitable for biological tissues, which exhibit viscoelastic behavior and long–memory effects. Applications include arterial wall dynamics, postural control, and signal propagation in complex neural networks.

- **Materials science and complex media.** In materials science, Duffing and fractional Duffing equations describe nonlinear vibrations in smart materials, metamaterials, and nanostructures. At micro– and nano–scales, dissipation mechanisms often deviate from classical viscous laws, making fractional models essential. In nanomechanical resonators, fractional Duffing equations explain experimentally observed anomalous frequency shifts and damping behavior. In metamaterials, Duffing–type nonlinearities enable tunable band gaps, wave localization, and amplitude–dependent dynamic responses.

The applications outlined above demonstrate the unifying role of Duffing–type oscillators across diverse scientific and engineering disciplines. The classical Duffing equation highlights the effects of nonlinear stiffness and multistability, while its fractional generalization captures memory and anomalous dissipation in complex systems. In the remainder of this chapter, we will demonstrate how the analytical and numerical results developed throughout this book can be applied to concrete problems drawn from these application areas.

9.1 Simple pendulum (nonlinear conservative oscillator)

The simple pendulum is the cleanest example of a real mechanical system whose exact dynamics is governed by a nonlinear autonomous second–order ODE of the form $u'' + F(u) = 0$.

Mechanical set-up. A point mass m is attached to a massless rod (or string) of length ℓ, moving in a vertical plane under gravity g. The configuration is described by the angular displacement $\theta(t)$ measured from the downward vertical.

Lagrangian derivation

The Cartesian position of the bob is

$$x(t) = \ell \sin\theta(t), \qquad y(t) = -\ell\cos\theta(t).$$

Differentiate to obtain the speed:

$$\dot{x} = \ell\dot{\theta}\cos\theta, \qquad \dot{y} = \ell\dot{\theta}\sin\theta \quad \Longrightarrow \quad v^2 = \dot{x}^2 + \dot{y}^2 = \ell^2\dot{\theta}^2.$$

Hence the kinetic energy is

$$T = \frac{1}{2}mv^2 = \frac{1}{2}m\ell^2\dot{\theta}^2.$$

Taking the gravitational potential to be zero at the lowest point, the height above the lowest point is $\ell(1 - \cos\theta)$, so the potential energy is

$$V = mg\ell(1 - \cos\theta).$$

Therefore the Lagrangian is

$$\mathcal{L}(\theta, \dot{\theta}) = T - V = \frac{1}{2}m\ell^2\dot{\theta}^2 - mg\ell(1 - \cos\theta). \tag{9.1}$$

Euler–Lagrange equation. Compute

$$\frac{\partial\mathcal{L}}{\partial\dot{\theta}} = m\ell^2\dot{\theta}, \qquad \frac{d}{dt}\Big(\frac{\partial\mathcal{L}}{\partial\dot{\theta}}\Big) = m\ell^2\ddot{\theta}, \qquad \frac{\partial\mathcal{L}}{\partial\theta} = -mg\ell\sin\theta.$$

The Euler–Lagrange equation $\frac{d}{dt}(\partial\mathcal{L}/\partial\dot{\theta}) - \partial\mathcal{L}/\partial\theta = 0$ gives the exact nonlinear pendulum ODE:

$$\ddot{\theta} + \frac{g}{\ell}\sin\theta = 0. \tag{9.2}$$

This is precisely of the conservative form $u'' + F(u) = 0$ with $u = \theta$ and $F(\theta) = (g/\ell)\sin\theta$.

Energy integral (why motion is periodic or not)

Multiplying (9.2) by $\dot{\theta}$ and integrating, or using Noether's theorem for time invariance, yields the conserved energy:

$$E = \frac{1}{2}m\ell^2\dot{\theta}^2 + mg\ell(1-\cos\theta), \qquad \frac{dE}{dt} = 0. \tag{9.3}$$

Interpretation for *dummies*:

- The first term is kinetic energy; the second is a potential well.
- If $E < 2mg\ell$ then $\theta(t)$ oscillates back and forth (closed phase curves).
- If $E = 2mg\ell$ the motion lies on the separatrix (critical trajectory).
- If $E > 2mg\ell$ the bob rotates continuously (open phase curves).

Connection to Duffing-type polynomial models (small angles)

For small oscillations, expand $\sin\theta$:

$$\sin\theta = \theta - \frac{\theta^3}{6} + \frac{\theta^5}{120} - \cdots$$

Substitute into (9.2) to obtain the polynomial hierarchy

$$\ddot{\theta} + \frac{g}{\ell}\theta - \frac{g}{6\ell}\theta^3 + \frac{g}{120\ell}\theta^5 - \cdots = 0. \tag{9.4}$$

Keeping only the cubic term gives a Duffing-type model; keeping cubic and quintic terms gives a cubic–quintic Duffing model. This is exactly why Duffing and Duffing–Helmholtz equations are *canonical*: they are the simplest polynomial normal forms capturing nonlinear stiffness while preserving conservative structure.

9.2 Spring pendulum with nonlinear stiffness

The spring pendulum is a classical mechanical system that naturally leads to nonlinear oscillators with mixed–parity restoring forces. It provides a direct physical realization of the Duffing–Helmholtz equation when geometric and elastic nonlinearities are retained.

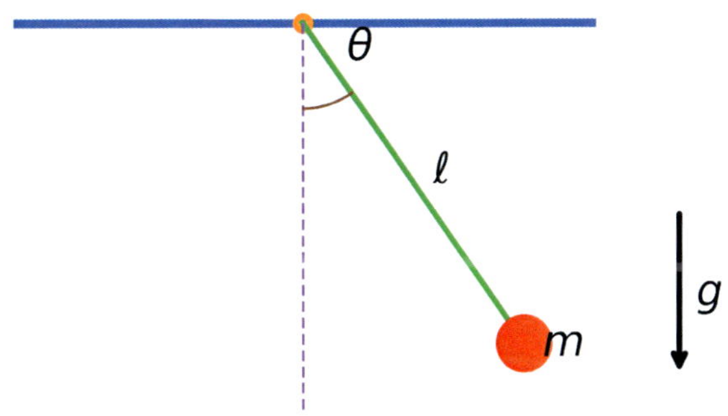

Figure 9.1: Simple pendulum model. The generalized conservative oscillator viewpoint starts from a Lagrangian (9.1), yields the nonlinear ODE (9.2), and provides the energy integral (9.3). Polynomial Duffing-type equations arise from the small-angle expansion (9.4).

9.2.1 Mechanical model and coordinates

We consider a point mass m attached to a vertical nonlinear spring of unstretched length ℓ. The mass is free to oscillate in a vertical plane. Let $r(t)$ denote the instantaneous elongation of the spring measured from its natural length, and let $\theta(t)$ denote the angular deviation from the vertical.

The position of the mass is described in polar coordinates by

$$x = (\ell + r)\sin\theta, \qquad y = -(\ell + r)\cos\theta.$$

The spring force is assumed to be nonlinear of the form

$$F_s(r) = -kr - \alpha r^2,$$

where $k > 0$ is the linear stiffness and α measures the quadratic nonlinearity of the spring.

9.2.2 Lagrangian formulation

The kinetic energy is

$$T = \frac{m}{2}\left(\dot{r}^2 + (\ell + r)^2\dot{\theta}^2\right).$$

The potential energy consists of gravitational and elastic contributions:

$$V = mg(\ell + r)\cos\theta + \frac{k}{2}r^2 + \frac{\alpha}{3}r^3.$$

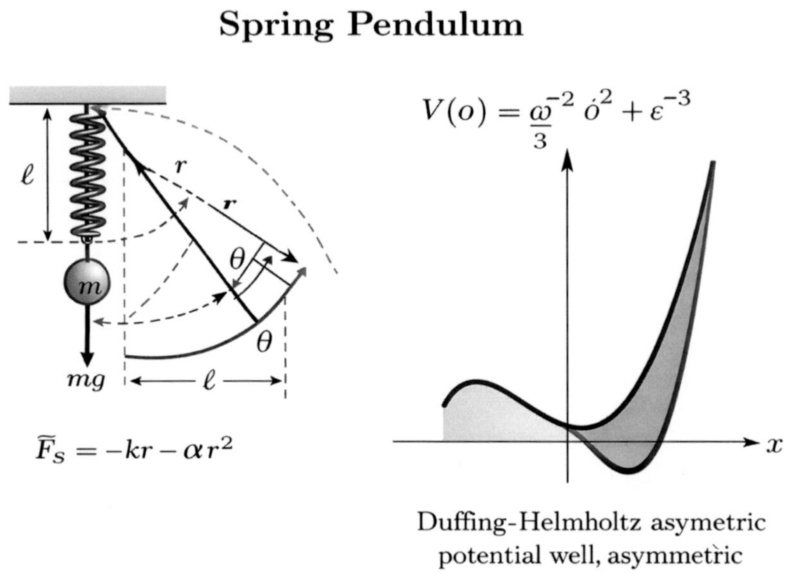

Figure 9.2: Spring pendulum with nonlinear stiffness. Geometric coupling between angular motion and radial extension generates mixed–parity nonlinearities. The associated effective potential exhibits an asymmetric Duffing–Helmholtz potential.

The Lagrangian of the system is therefore

$$\mathcal{L}(r,\theta,\dot{r},\dot{\theta}) = \frac{m}{2}\left(\dot{r}^2 + (\ell + r)^2\dot{\theta}^2\right) - mg(\ell + r)\cos\theta - \frac{k}{2}r^2 - \frac{\alpha}{3}r^3. \tag{9.5}$$

9.2.3 Reduction to a single Duffing–Helmholtz equation

We now restrict attention to small angular oscillations ($|\theta| \ll 1$) and weak radial motion. Expanding $\cos\theta \approx 1 - \theta^2/2$ and neglecting higher–order terms, the angular equation decouples at leading order, while the radial equation becomes

$$m\ddot{r} + kr + \alpha r^2 + \frac{mg}{\ell}r + \mathcal{O}(r^3) - 0. \tag{9.6}$$

After rescaling time and collecting coefficients, we obtain the generalized Duffing equation

$$\ddot{u} + a_0 + a_1 u + a_2 u^2 + a_3 u^3 = 0, \tag{9.7}$$

where $u(t)$ is a shifted radial displacement and the coefficients a_j depend explicitly on (m, g, ℓ, k, α).

9.2.4 Physical interpretation

This derivation shows that the spring pendulum is not merely a pedagogical example, but a concrete realization of a mixed–parity Duffing oscillator. The quadratic term arises from elastic asymmetry, while the cubic term originates from geometric coupling.

Why this matters.

- The model explains experimentally observed asymmetric oscillations.
- It provides a physical basis for the Duffing–Helmholtz model.
- It justifies the use of Duffing–type reductions in nonlinear mechanics.
- It connects directly with later analyses of separatrix splitting and chaos under damping and forcing.

9.3 The Lagrange top (heavy symmetric top)

The Lagrange top is a rigid symmetric top (with principal moments $I_1 = I_2 \neq I_3$) whose tip is fixed at a point O and whose center of mass C lies on the body symmetry axis at distance ℓ from O. The top is acted upon by uniform gravity.

Generalized coordinates (Euler angles)

We describe the orientation of the body by Euler angles (ϕ, θ, ψ):

- ϕ is the precession angle about the fixed vertical axis $\hat{\mathbf{Z}}$,
- θ is the inclination (nutation) angle between $\hat{\mathbf{e}}_3$ and $\hat{\mathbf{Z}}$,
- ψ is the spin angle about the symmetry axis $\hat{\mathbf{e}}_3$.

These are the natural coordinates for the rotational dynamics of a symmetric top.

Angular velocity and kinetic energy

For a symmetric top, the angular velocity has a standard form:

$$\omega_3 = \dot{\psi} + \dot{\phi}\cos\theta, \qquad \omega_1^2 + \omega_2^2 = \dot{\theta}^2 + \dot{\phi}^2 \sin^2\theta. \tag{9.8}$$

Hence the kinetic energy is

$$T = \frac{I_1}{2}\left(\dot{\theta}^2 + \dot{\phi}^2 \sin^2\theta\right) + \frac{I_3}{2}\left(\dot{\psi} + \dot{\phi}\cos\theta\right)^2. \tag{9.9}$$

Potential energy

The center of mass has height $z_C = \ell\cos\theta$ above the fixed point, so the potential energy is

$$V = mg\ell\cos\theta. \tag{9.10}$$

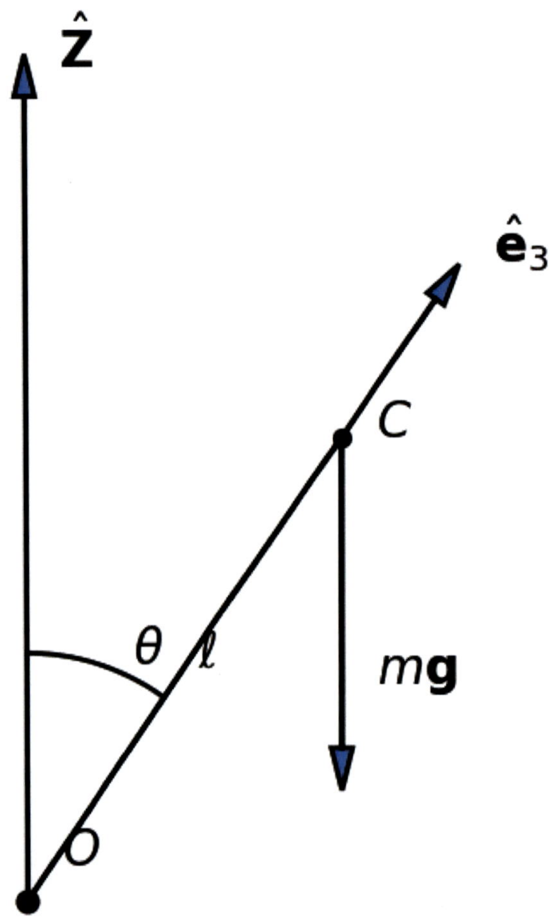

Figure 9.3: Lagrange top geometry: fixed point O, vertical axis $\hat{\mathbf{Z}}$, body symmetry axis $\hat{\mathbf{e}}_3$, tilt angle θ, and center of mass C at distance ℓ along $\hat{\mathbf{e}}_3$.

Lagrangian and cyclic coordinates

The Lagrangian is

$$\mathcal{L} = T - V = \frac{I_1}{2}\left(\dot{\theta}^2 + \dot{\phi}^2 \sin^2\theta\right) + \frac{I_3}{2}\left(\dot{\psi} + \dot{\phi}\cos\theta\right)^2 - mg\ell\cos\theta. \tag{9.11}$$

Observe that ϕ and ψ do not appear explicitly in $\mathcal{L}$, so they are cyclic coordinates and yield two conserved momenta:

$$p_\psi := \frac{\partial\mathcal{L}}{\partial\dot{\psi}} = I_3(\dot{\psi} + \dot{\phi}\cos\theta) \equiv \text{const}, \tag{9.12}$$

$$p_\phi := \frac{\partial\mathcal{L}}{\partial\dot{\phi}} = I_1\dot{\phi}\sin^2\theta + I_3(\dot{\psi} + \dot{\phi}\cos\theta)\cos\theta \equiv \text{const}. \tag{9.13}$$

Interpretation (very important): p_ψ is the conserved spin about $\hat{\mathbf{e}}_3$, and p_ϕ is the conserved component of total angular momentum along the vertical $\hat{\mathbf{Z}}$.

Reduction to a single second-order ODE (nutation equation)

From (9.12), we can solve

$$\dot{\psi} + \dot{\phi}\cos\theta = \frac{p_\psi}{I_3}.$$

Insert this into (9.13) to obtain $\dot{\phi}$ explicitly:

$$\dot{\phi} = \frac{p_\phi - p_\psi\cos\theta}{I_1\sin^2\theta}. \tag{9.14}$$

Now use the energy integral $E = T + V = \text{const}$. Substituting (9.14) into T gives

$$E = \frac{I_1}{2}\dot{\theta}^2 + \underbrace{\frac{(p_\phi - p_\psi\cos\theta)^2}{2I_1\sin^2\theta} + \frac{p_\psi^2}{2I_3} + mg\ell\cos\theta}_{=:U_{\text{eff}}(\theta)}. \tag{9.15}$$

Thus the dynamics reduces to the 1D conservative system

$$\frac{I_1}{2}\dot{\theta}^2 + U_{\text{eff}}(\theta) = E, \qquad \Longleftrightarrow \qquad I_1\ddot{\theta} + U_{\text{eff}}'(\theta) = 0, \tag{9.16}$$

where $U_{\text{eff}}(\theta)$ is the effective potential. This is the "master equation" for nutation.

For dummies: what this reduction means

For dummies. The full rigid-body motion looks complicated, but after using the two conserved momenta p_ψ and p_ϕ we reduce everything to *one particle* moving in the one-dimensional potential $U_{\text{eff}}(\theta)$.

- If U_{eff} has a well (a minimum), then $\theta(t)$ oscillates: this is stable nutation.

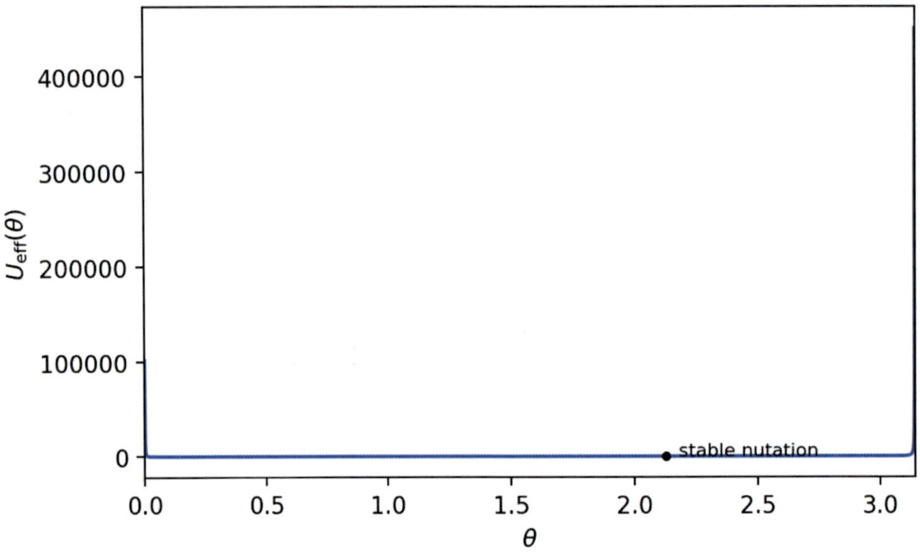

Figure 9.4: Example of an effective potential $U_{\mathrm{eff}}(\theta)$. A local minimum corresponds to a stable nutation regime (bounded oscillations of θ).

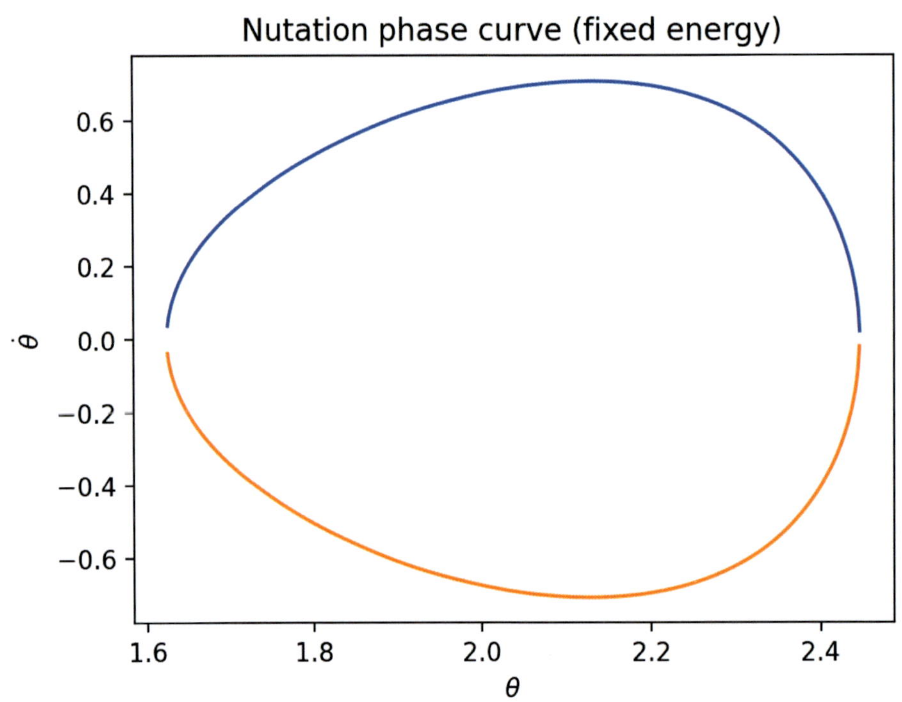

Figure 9.5: A phase curve in the $(\theta, \dot{\theta})$ plane at fixed energy. Turning points occur where $E = U_{\mathrm{eff}}(\theta)$ (i.e., $\dot{\theta} = 0$).

- If the energy level E lies above a barrier, θ can pass over it: this corresponds to large-angle nutation (sometimes "flips").

- Steady precession corresponds to an equilibrium of (9.16), i.e. a critical point $U'_{\text{eff}}(\theta_*) = 0$.

Connection to Duffing-type reductions (your book's philosophy)

On any invariant nutation interval $\theta \in [\theta_-, \theta_+]$ where the motion is bounded, one may approximate the smooth function $U'_{\text{eff}}(\theta)$ on that interval (using, e.g., Chebyshev approximation) by a low-degree polynomial in the shifted variable $\eta = \theta - \theta_0$. Then (9.16) becomes an approximate Duffing-type oscillator for $\eta(t)$ (quadratic/cubic/quintic terms depending on the truncation), which is precisely the analytical strategy developed in this monograph.

9.4 Kapitza pendulum (vertically vibrated pivot)

Kapitza pendulum parametric excitation high–frequency averaging effective potential inverted equilibrium stability multiple time scales nonlinear oscillator

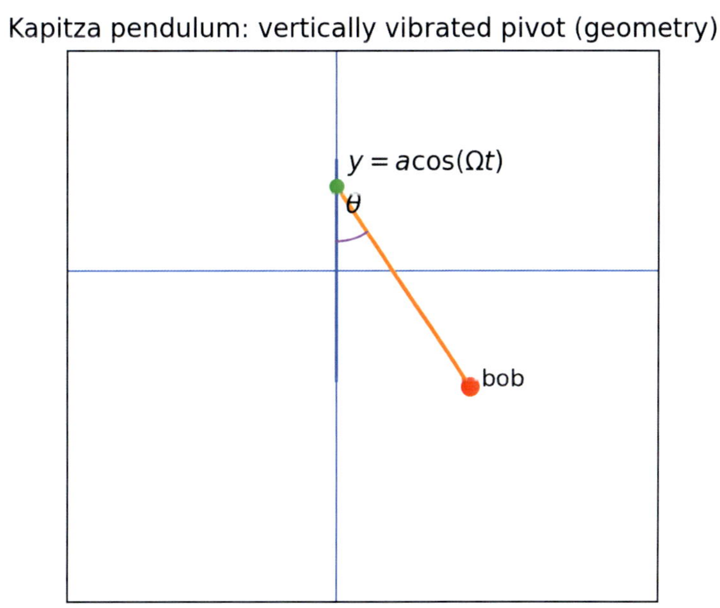

Figure 9.6: Kapitza pendulum: a simple pendulum whose pivot oscillates vertically as $y = a\cos(\Omega t)$.

Model and coordinates (for dummies). Take a point mass m attached to a rigid rod of length l. Let $\theta(t)$ be the angle from the downward vertical. The pivot moves vertically according to

$$y_p(t) = a\cos(\Omega t),$$

so the bob position is

$$x(t) = l\sin\theta(t), \qquad y(t) = y_p(t) - l\cos\theta(t).$$

This is a time–dependent Lagrangian system with a fast drive at frequency Ω.

Kinetic and potential energies. Differentiate:

$$\dot{x} = l\cos\theta\,\dot{\theta}, \qquad \dot{y} = \dot{y}_p + l\sin\theta\,\dot{\theta}, \qquad \dot{y}_p(t) = -a\Omega\sin(\Omega t).$$

Hence

$$T = \frac{m}{2}\left(\dot{x}^2 + \dot{y}^2\right) = \frac{m}{2}\left(l^2\dot{\theta}^2 + 2l\dot{y}_p\sin\theta\,\dot{\theta} + \dot{y}_p^2\right),$$

and the gravitational potential is

$$V = mg\,y = mg\big(y_p(t) - l\cos\theta\big).$$

Therefore the Lagrangian is

$$L(\theta,\dot{\theta},t) = T - V = \frac{m}{2}\left(l^2\dot{\theta}^2 + 2l\dot{y}_p\sin\theta\,\dot{\theta} + \dot{y}_p^2\right) - mg\big(y_p(t) - l\cos\theta\big). \tag{9.17}$$

Equation of motion (exact). Compute the Euler–Lagrange equation $\frac{d}{dt}(\partial L/\partial\dot{\theta}) - \partial L/\partial\theta = 0$. We have

$$\frac{\partial L}{\partial\dot{\theta}} = m(l^2\dot{\theta} + l\dot{y}_p\sin\theta), \qquad \frac{d}{dt}\frac{\partial L}{\partial\dot{\theta}} = m\Big(l^2\ddot{\theta} + l\ddot{y}_p\sin\theta + l\dot{y}_p\cos\theta\,\dot{\theta}\Big),$$

and

$$\frac{\partial L}{\partial\theta} = m(l\dot{y}_p\cos\theta\,\dot{\theta}) - mg(l\sin\theta).$$

The mixed terms cancel, giving the exact forced nonlinear oscillator

$$\ddot{\theta} + \frac{g}{l}\sin\theta = -\frac{\ddot{y}_p(t)}{l}\sin\theta \iff \ddot{\theta} + \left(\frac{g}{l} + \frac{a\Omega^2}{l}\cos(\Omega t)\right)\sin\theta = 0. \tag{9.18}$$

This is Mathieu-type in the small-angle limit, but it is fully nonlinear because of $\sin\theta$.

Small-angle linearization (quick intuition). For $|\theta| \ll 1$, $\sin\theta \approx \theta$ and (9.18) becomes

$$\ddot{\theta} + \left(\frac{g}{l} + \frac{a\Omega^2}{l}\cos(\Omega t)\right)\theta = 0,$$

a classical parametric resonance model (stability tongues in the (a,Ω) plane).

High-frequency stabilization mechanism (Kapitza averaging)

averaging fast–slow decomposition

The famous Kapitza effect is that the *inverted* equilibrium $\theta = \pi$ can become stable when the excitation is fast and strong enough.

Assume the high frequency regime

$$\Omega \gg \sqrt{\frac{g}{l}},$$

and decompose the motion into a slow part plus a small fast ripple:

$$\theta(t) = \Theta(t) + \xi(t), \qquad \xi \text{ is small and oscillates at frequency } \Omega.$$

A standard averaging calculation (expand $\sin(\Theta + \xi)$ to second order and choose ξ to cancel the leading fast forcing) yields an effective equation for the slow angle $\Theta(t)$ of the form

$$\ddot{\Theta} = -\frac{1}{ml^2}\frac{d}{d\Theta}U_{\text{eff}}(\Theta), \tag{9.19}$$

with effective potential

$$U_{\text{eff}}(\Theta) = mgl\cos\Theta + \frac{ma^2\Omega^2}{4}\sin^2\Theta \quad \text{(up to an irrelevant additive constant).} \tag{9.20}$$

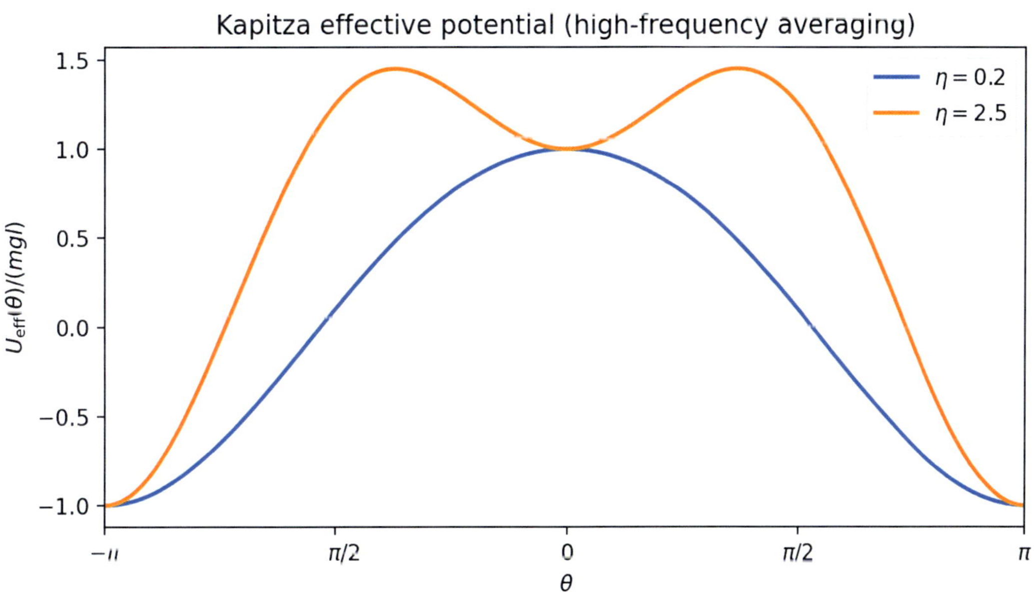

Figure 9.7: Effective potential $U_{\text{eff}}(\Theta)$: increasing $\eta = a^2\Omega^2/(gl)$ creates a local minimum near $\Theta = \pi$ (inverted stability).

Stability of down and up equilibria (explicit parameter conditions). Equilibria satisfy $U'_{\text{eff}}(\Theta) = 0$:

$$U'_{\text{eff}}(\Theta) = -mgl\sin\Theta + \frac{ma^2\Omega^2}{2}\sin\Theta\cos\Theta = \sin\Theta\left(-mgl + \frac{ma^2\Omega^2}{2}\cos\Theta\right).$$

Hence $\Theta = 0$ (down) and $\Theta = \pi$ (up) are always equilibria. Their stability is decided by U''_{eff}:

$$U''_{\text{eff}}(\Theta) = -mgl\cos\Theta + \frac{ma^2\Omega^2}{2}\left(\cos^2\Theta - \sin^2\Theta\right).$$

- **Downward equilibrium** $\Theta = 0$:

$$U''_{\text{eff}}(0) = -mgl + \frac{ma^2\Omega^2}{2}.$$

  In the usual gravity-dominated regime, $\Theta = 0$ remains stable; very strong excitation can modify the effective frequency.

- **Inverted equilibrium** $\Theta = \pi$:

$$U''_{\text{eff}}(\pi) = +mgl + \frac{ma^2\Omega^2}{2}.$$

  In the averaged model, the inverted point becomes a stable equilibrium once the effective minimum forms, which occurs when the curvature of U_{eff} near π is positive and the barrier exists. A convenient dimensionless parameter is

$$\eta := \frac{a^2\Omega^2}{g\,l}. \tag{9.21}$$

  The classical Kapitza threshold is

$$\eta > 2 \iff a^2\Omega^2 > 2gl, \tag{9.22}$$

  which ensures the inverted configuration is stabilized by the fast vibration (within the averaging assumptions).

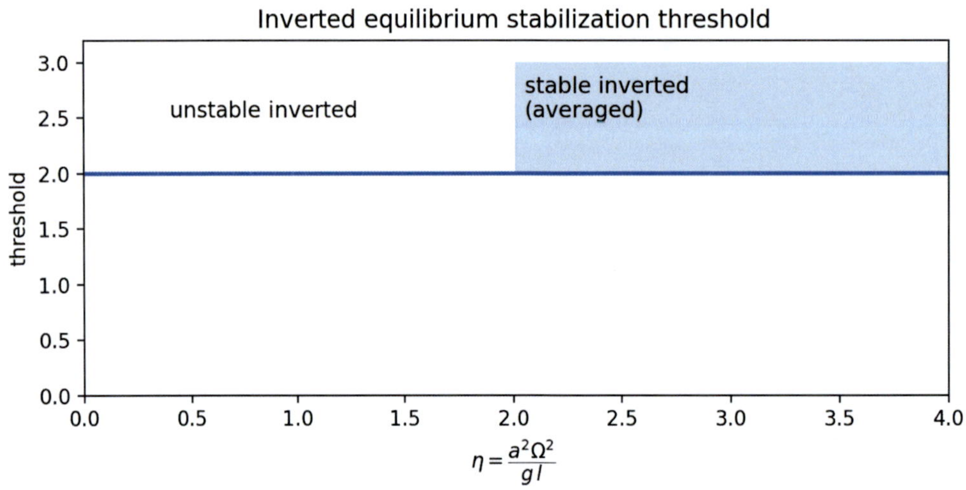

Figure 9.8: Averaged stabilization condition for the inverted equilibrium: $\eta = a^2\Omega^2/(gl) > 2$.

Why this belongs in a Duffing-type Equation (9.18) is a nonlinear parametric oscillator. Near an equilibrium (down or up) one can expand $\sin\theta$ and obtain polynomial models

$$\ddot{\theta} + \omega_0^2\theta + \alpha_3\theta^3 + \cdots + \Big(\text{parametric forcing}\Big)\theta = 0,$$

so Duffing-type truncations arise naturally. Moreover, the effective potential viewpoint (9.20) is exactly the same organizing principle used for conservative Duffing and Duffing–Helmholtz oscillators: equilibria, wells, barriers, and separatrices are read directly from U_{eff}.

Worked numeric example (plug numbers). Let $l = 1$ m and $g = 9.81$ m/s^2. If $a = 1$ cm $= 0.01$ m, then the threshold (9.22) gives

$$\Omega > \sqrt{\frac{2gl}{a^2}} \approx \sqrt{\frac{19.62}{10^{-4}}} \approx 443 \text{ rad/s} \quad \text{(about 70.5 Hz)}.$$

Thus, sufficiently fast vibrations with centimeter amplitude can stabilize the inverted pendulum in practice.

9.5 Rolling cylinder inside a fixed circular guide

rolling cylinder nonholonomic constraint Lagrangian mechanics conservative oscillator small oscillations Duffing-type equation

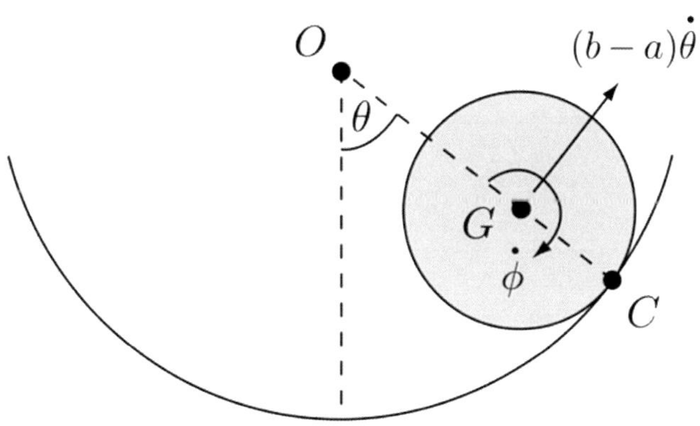

Figure 9.9: A small solid cylinder of radius a and mass m rolls without slipping on the inside surface of a fixed circular cylinder of radius $b > a$. The generalized coordinate is the angular displacement θ of the centre G from the vertical equilibrium position.

Geometry and kinematics. Let O be the centre of the fixed cylinder and G the centre of the rolling cylinder. The distance $OG = b - a$ is constant. We choose as generalized coordinate the angle θ measured from the downward vertical.

The centre of mass G moves along a circle of radius $b - a$, so its speed is

$$v_G = (b - a)\dot{\theta}.$$

Let ϕ be the rotation angle of the rolling cylinder about its own centre. The rolling constraint (no slipping) gives

$$(b-a)\dot{\theta} = a\dot{\phi}. \tag{9.23}$$

Kinetic energy. The kinetic energy has translational and rotational parts:

$$T = \frac{1}{2}mv_G^2 + \frac{1}{2}I\dot{\phi}^2,$$

where for a solid cylinder $I = \frac{1}{2}ma^2$. Using (9.23), we obtain

$$T = \frac{1}{2}m(b-a)^2\dot{\theta}^2 + \frac{1}{4}m(b-a)^2\dot{\theta}^2 = \frac{3}{4}m(b-a)^2\dot{\theta}^2.$$

Potential energy. The vertical height of G above the lowest point is

$$h = (b-a)(1-\cos\theta),$$

so the gravitational potential energy is

$$V = mg(b-a)(1-\cos\theta).$$

Lagrangian and equation of motion. The Lagrangian $L = T - V$ is

$$L = \frac{3}{4}m(b-a)^2\dot{\theta}^2 - mg(b-a)(1-\cos\theta).$$

The Euler–Lagrange equation yields

$$\ddot{\theta} + \frac{2g}{3(b-a)}\sin\theta = 0. \tag{9.24}$$

Equation (9.24) is an exact, fully nonlinear, conservative oscillator with a sinusoidal restoring force.

Small oscillations. For $|\theta| \ll 1$, $\sin\theta \simeq \theta$, and the equation reduces to

$$\ddot{\theta} + \omega_0^2\theta = 0, \qquad \omega_0^2 = \frac{2g}{3(b-a)}.$$

Hence the period of small oscillations is

$$T_0 = 2\pi\sqrt{\frac{3(b-a)}{2g}}.$$

Connection with Duffing-type models. Expanding $\sin\theta$ to higher order,

$$\sin\theta = \theta - \frac{\theta^3}{6} + \frac{\theta^5}{120} - \cdots,$$

equation (9.24) becomes

$$\ddot{\theta} + \omega_0^2\theta - \frac{\omega_0^2}{6}\theta^3 + \frac{\omega_0^2}{120}\theta^5 + \cdots = 0, \tag{9.25}$$

which is a cubic–quintic Duffing equation to leading orders.

This example shows explicitly how a purely geometric rolling problem leads naturally to a Duffing–type oscillator when moderate amplitudes are considered. It therefore fits squarely into the general framework of generalized Duffing oscillators developed throughout this book.

9.6 Pendulum with a Horizontally Moving Support

This section provides a paradigmatic example showing how a mechanical system with multiple degrees of freedom can be systematically reduced to a second–order nonlinear ordinary differential equation of Duffing–type. The example is classical in analytical mechanics, yet it plays a fundamental role in understanding why nonlinear oscillators of Duffing and Duffing–Helmholtz type arise naturally in a wide variety of physical systems.

9.6.1 Geometric description

We consider a simple pendulum of length r and bob mass m, attached to a horizontally moving support of mass M. The support moves without friction along a horizontal axis, while the pendulum swings in a vertical plane. The generalized coordinates are: the horizontal displacement $x(t)$ of the support and the angular displacement $\theta(t)$ of the pendulum measured from the vertical. This configuration, illustrated in Figure 9.10, is a standard example of a constrained mechanical system with coupled coordinates.

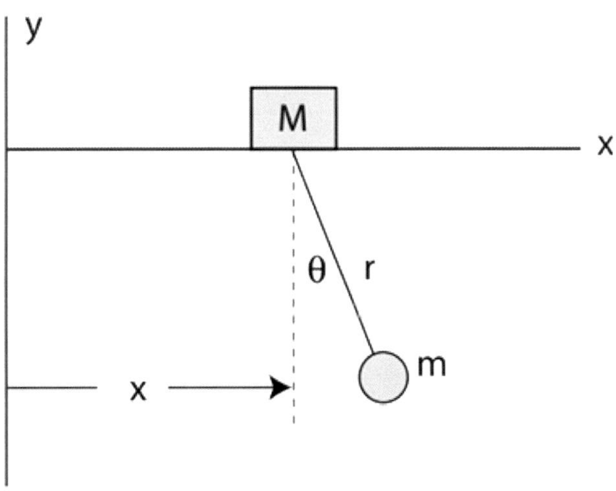

Figure 9.10: A simple pendulum attached to a horizontally moving support. The generalized coordinates are the support displacement $x(t)$ and the pendulum angle $\theta(t)$.

9.6.2 Kinetic and potential energy

The Cartesian components of the velocity of the pendulum bob are

$$v_x = \dot{x} + r\dot{\theta}\cos\theta, \qquad v_y = r\dot{\theta}\sin\theta,$$

which immediately shows the kinematic coupling between translational and rotational motion.

The total kinetic energy of the system is therefore

$$T = \frac{M}{2}\dot{x}^2 + \frac{m}{2}\left(\dot{x}^2 + r^2\dot{\theta}^2 + 2r\dot{x}\dot{\theta}\cos\theta\right).$$

The potential energy associated with gravity is

$$U = -mgr\cos\theta,$$

which depends only on the angular coordinate θ.

9.6.3 Lagrangian formulation

The Lagrangian $L = T - U$ takes the form

$$L(x, \dot{x}, \theta, \dot{\theta}) = \frac{1}{2}(M+m)\dot{x}^2 + \frac{m}{2}r^2\dot{\theta}^2 + mr\dot{x}\dot{\theta}\cos\theta + mgr\cos\theta.$$

Applying the Lagrange equations to the generalized coordinates x and θ yields the coupled system

$$\frac{d}{dt}\left[(M+m)\dot{x} + mr\dot{\theta}\cos\theta\right] = 0, \tag{9.26}$$

$$\frac{d}{dt}\left[mr^2\dot{\theta} + mr\dot{x}\cos\theta\right] = -mgr\sin\theta. \tag{9.27}$$

Equation (9.26) expresses conservation of horizontal linear momentum, since no external horizontal forces act on the system.

9.6.4 Reduction to an effective nonlinear oscillator

Using the conserved quantity in (9.26), the variable $x(t)$ can be eliminated, leading to a single second–order equation for $\theta(t)$:

$$\ddot{\theta} + \frac{\ddot{x}}{r}\cos\theta + \frac{g}{r}\sin\theta = 0.$$

This equation describes a nonlinear pendulum subject to an effective forcing generated by the acceleration of the support. If the support undergoes constant acceleration, the dynamics can be derived from an effective potential

$$V_{\text{eff}}(\theta) = -\frac{g}{r}\cos\theta - \frac{\ddot{x}}{r}\sin\theta.$$

9.6.5 Duffing–type approximation

For small oscillations ($|\theta| \ll 1$), expanding the trigonometric terms gives

$$\ddot{\theta} + \omega_0^2\theta + \alpha\theta^3 + \cdots = 0,$$

which is precisely a Duffing equation with cubic nonlinearity. This approximation shows explicitly how polynomial restoring forces emerge from simple mechanical constraints.

9.6.6 Conceptual significance

This example demonstrates that:

- many mechanical systems with moving constraints reduce naturally to Duffing–type oscillators;
- conservation laws enable systematic order reduction in nonlinear dynamics;
- effective nonlinearities arise even in systems with elementary geometry;
- Duffing and Duffing–Helmholtz equations serve as canonical models for conservative and weakly nonconservative dynamics.

For these reasons, pendulum systems with moving supports form a cornerstone in the analytical theory of nonlinear oscillations and provide a direct mechanical motivation for the Duffing–type models studied throughout this book.

9.7 Relativistic pendulum model

Consider a point mass m attached to a rigid massless rod of length L, moving in a vertical plane in a uniform gravitational field of magnitude g. The angle from the downward vertical is $\theta(t)$; the tangential speed is

$$v = L\dot{\theta}.$$

In relativistic Lagrangian mechanics the free–particle kinetic part is $-mc^2\sqrt{1 - v^2/c^2}$ and a potential $V(\theta)$ is added algebraically.[108] For a simple pendulum $V(\theta) = mgL(1 - \cos\theta)$ (up to an irrelevant constant), so the Lagrangian is

$$L(\theta, \dot{\theta}) = -mc^2\sqrt{1 - \frac{L^2\dot{\theta}^2}{c^2}} - mgL(1 - \cos\theta). \tag{9.28}$$

9.7.1 Equation of motion

The canonical momentum conjugate to θ is

$$p_\theta = \frac{\partial L}{\partial \dot{\theta}} = \frac{mL^2\dot{\theta}}{\sqrt{1 - \frac{L^2\dot{\theta}^2}{c^2}}} = \gamma m L^2\dot{\theta}, \qquad \gamma = \frac{1}{\sqrt{1 - \frac{L^2\dot{\theta}^2}{c^2}}}, \tag{9.29}$$

which has the familiar form "relativistic mass times velocity" along the tangential direction.[108, 109] The Euler–Lagrange equation

$$\frac{d}{dt}\left(\frac{\partial L}{\partial\dot{\theta}}\right)-\frac{\partial L}{\partial\theta}=0$$

gives

$$\frac{d}{dt}(\gamma mL^2\dot{\theta})+mgL\sin\theta=0, \tag{9.30}$$

or, after dividing by mL^2,

$$\frac{d}{dt}(\gamma\dot{\theta})+\omega_0^2\sin\theta=0, \qquad \omega_0^2=\frac{g}{L}. \tag{9.31}$$

Equation (9.31) is the relativistic pendulum equation; in the limit $c\to\infty$ one has $\gamma\to1$ and the usual simple pendulum equation $\ddot{\theta}+\omega_0^2\sin\theta=0$ is recovered.

9.7.2 Energy integral

Because the Lagrangian (9.28) has no explicit time dependence, the energy

$$E=\dot{\theta}\,p_\theta-L \tag{9.32}$$

is conserved.[108] Substituting the expressions above,

$$\begin{aligned}E&=\dot{\theta}\,\gamma mL^2\dot{\theta}+mc^2\sqrt{1-\frac{L^2\dot{\theta}^2}{c^2}}+mgL(1-\cos\theta)\\&=\gamma mc^2+mgL(1-\cos\theta)-mc^2,\end{aligned} \tag{9.33}$$

where we have used $\gamma mL^2\dot{\theta}^2=(\gamma-1)mc^2+mc^2\sqrt{1-L^2\dot{\theta}^2/c^2}$. It is convenient to measure energy from the rest state at the lowest point, so write

$$E=\gamma mc^2-mc^2+mgL(1-\cos\theta). \tag{9.34}$$

Let θ_0 denote the maximum angular displacement (turning point) reached during the motion; there $\dot{\theta}=0$ so $\gamma=1$. The energy at the turning point is

$$E=mgL(1-\cos\theta_0). \tag{9.35}$$

Equating (9.34) and (9.35) gives

$$\gamma mc^2-mc^2+mgL(1-\cos\theta)=mgL(1-\cos\theta_0), \tag{9.36}$$

which simplifies to

$$\gamma=1+\frac{gL}{c^2}(\cos\theta-\cos\theta_0). \tag{9.37}$$

Since $\gamma = 1/\sqrt{1 - L^2\dot{\theta}^2/c^2}$, relation (9.37) provides a first integral linking θ and $\dot{\theta}$ throughout the motion.[109]

9.7.3 Quadrature for the period

From (9.37),

$$\sqrt{1 - \frac{L^2\dot{\theta}^2}{c^2}} = \frac{1}{1 + \alpha(\cos\theta - \cos\theta_0)}, \qquad \alpha = \frac{gL}{c^2},$$

so

$$\dot{\theta}^2 = \frac{c^2}{L^2}\left[1 - \frac{1}{(1 + \alpha(\cos\theta - \cos\theta_0))^2}\right]. \tag{9.38}$$

Choosing the negative square root for motion from θ_0 downward,

$$\dot{\theta} = \frac{d\theta}{dt} = -\frac{c}{L}\sqrt{1 - \frac{1}{(1 + \alpha(\cos\theta - \cos\theta_0))^2}}.$$

Thus

$$dt = -\frac{L}{c}\left[1 - \frac{1}{(1 + \alpha(\cos\theta - \cos\theta_0))^2}\right]^{-1/2} d\theta. \tag{9.39}$$

The quarter–period (from θ_0 to 0) is

$$\frac{T}{4} = \frac{L}{c}\int_0^{\theta_0}\left[1 - \frac{1}{(1 + \alpha(\cos\theta - \cos\theta_0))^2}\right]^{-1/2} d\theta, \tag{9.40}$$

hence the full period is

$$T(\theta_0) = \frac{4L}{c}\int_0^{\theta_0}\left[1 - \frac{1}{(1 + \alpha(\cos\theta - \cos\theta_0))^2}\right]^{-1/2} d\theta. \tag{9.41}$$

This integral is not elementary and can be expressed in terms of elliptic integrals or evaluated numerically. Its key qualitative feature is that $T(\theta_0)$ increases with θ_0 due to both geometric and relativistic effects.[109]

9.7.4 Reduction to a Duffing–type equation

Equation (9.31) is intrinsically nonlinear in both θ and $\dot{\theta}$ through γ. To obtain a simpler analytical approximation that still captures the amplitude dependence of the period, we now derive a Duffing–type equation by expanding in small angle and small velocity.

9.7.4.1 Expansion of the relativistic factor

Introduce the dimensionless speed parameter

$$\beta = \frac{L\dot{\theta}}{c},$$

so that

$$\gamma = \frac{1}{\sqrt{1-\beta^2}}.$$

For $|\beta| \ll 1$ the Lorentz factor has the Taylor expansion

$$\gamma \approx 1 + \frac{1}{2}\beta^2 + \frac{3}{8}\beta^4 + \cdots .$$

Retaining only the first relativistic correction,

$$\gamma \approx 1 + \frac{1}{2}\frac{L^2\dot{\theta}^2}{c^2}. \tag{9.42}$$

Then

$$\gamma\dot{\theta} \approx \dot{\theta} + \frac{1}{2}\frac{L^2}{c^2}\dot{\theta}^3, \tag{9.43}$$

and

$$\frac{d}{dt}(\gamma\dot{\theta}) \approx \ddot{\theta} + \frac{3}{2}\frac{L^2}{c^2}\dot{\theta}^2\ddot{\theta}, \tag{9.44}$$

where products of higher powers such as $\dot{\theta}^4\ddot{\theta}$ are neglected.

Substituting (9.44) into (9.31), and keeping terms up to first order in $L^2\dot{\theta}^2/c^2$, gives

$$\left(1 + \frac{3}{2}\frac{L^2}{c^2}\dot{\theta}^2\right)\ddot{\theta} + \omega_0^2 \sin\theta \approx 0. \tag{9.45}$$

Thus, to this order, relativistic effects appear as a velocity–dependent effective inertia multiplying $\ddot{\theta}$.

9.7.5 Small–angle expansion

Assume moderate oscillation amplitude such that $|\theta| \ll 1$, so

$$\sin\theta \approx \theta - \frac{1}{6}\theta^3. \tag{9.46}$$

Substituting into (9.45) yields

$$\left(1 + \alpha\dot{\theta}^2\right)\ddot{\theta} + \omega_0^2\left(\theta - \frac{1}{6}\theta^3\right) \approx 0, \qquad \alpha = \frac{3L^2}{2c^2}. \tag{9.47}$$

This equation contains both a cubic stiffness nonlinearity (from $\sin\theta$) and a quadratic velocity–dependent mass term (from relativity).

9.7.6 Averaging the inertial nonlinearity

Equation (9.47) is still nonstandard due to the factor $1+\alpha\dot{\theta}^2$ multiplying $\ddot{\theta}$. For small nonlinearity a useful simplification replaces $\dot{\theta}^2$ by its average over one oscillation, which depends primarily on the amplitude. Assume the motion is close to harmonic with amplitude A and frequency ω_0:

$$\theta(t) \approx A\cos(\omega_0 t). \tag{9.48}$$

Then

$$\dot{\theta}(t) \approx -A\omega_0 \sin(\omega_0 t),$$

and the time average over one period $T_0 = 2\pi/\omega_0$ gives

$$\langle \dot{\theta}^2 \rangle = \frac{1}{T_0}\int_0^{T_0} \dot{\theta}^2(t)\,dt \approx \frac{A^2\omega_0^2}{2}. \tag{9.49}$$

Replace $\dot{\theta}^2$ in the inertial factor by this average, introducing

$$1 + \alpha\dot{\theta}^2 \approx 1 + \frac{1}{2}\alpha A^2\omega_0^2 =: 1 + \eta, \tag{9.50}$$

where η is a small, slowly varying parameter determined by the amplitude.[109]

Dividing (9.47) by $1+\eta$ and neglecting the small difference between the instantaneous and averaged inertial term gives

$$\ddot{\theta} + \frac{\omega_0^2}{1+\eta}\,\theta - \frac{\omega_0^2}{6(1+\eta)}\,\theta^3 \approx 0. \tag{9.51}$$

Define an effective linear frequency and cubic coefficient

$$\omega_{\text{eff}}^2 := \frac{\omega_0^2}{1+\eta}, \qquad \alpha_3 := -\frac{\omega_0^2}{6(1+\eta)}, \tag{9.52}$$

so (9.51) becomes the standard Duffing–type equation

$$\ddot{\theta} + \omega_{\text{eff}}^2\theta + \alpha_3\theta^3 \approx 0. \tag{9.53}$$

Since $\alpha_3 < 0$, the effective oscillator is of *softening* type: its frequency decreases with amplitude, in agreement with the exact relativistic pendulum where the period increases with amplitude.

9.7.7 Amplitude–dependent frequency from Duffing theory

For the non–dissipative Duffing equation

$$\ddot{x} + \omega^2 x + \alpha x^3 = 0, \tag{9.54}$$

a standard first–order approximation for the amplitude–dependent frequency is obtained by the Lindstedt–Poincaré method or by simple averaging. One assumes the solution has the form

$$x(t) \approx A\cos(\Omega t),$$

and balances secular terms in a perturbation expansion in small α. The result is

$$\Omega(A) \approx \omega\left(1+\frac{3}{8}\frac{\alpha A^2}{\omega^2}\right), \tag{9.55}$$

valid when $|\alpha A^2/\omega^2| \ll 1$.

Applying (9.55) to (9.53) with $\omega = \omega_{\text{eff}}$ and $\alpha = \alpha_3$ gives

$$\Omega(A) \approx \omega_{\text{eff}}\left(1+\frac{3}{8}\frac{\alpha_3 A^2}{\omega_{\text{eff}}^2}\right) = \omega_{\text{eff}}\left(1-\frac{3}{8}\frac{A^2}{6}\right) = \omega_{\text{eff}}\left(1-\frac{A^2}{16}\right), \tag{9.56}$$

where the last equality uses $\alpha_3 = -\omega_{\text{eff}}^2/6$ to leading order. Recalling

$$\omega_{\text{eff}}^2 = \frac{\omega_0^2}{1+\eta} \approx \omega_0^2(1-\eta)$$

for small η, the combined geometric and relativistic corrections lead to a frequency that decreases with amplitude, consistent with both the nonrelativistic large–angle pendulum and the relativistic time dilation effects.[109]

The corresponding approximate period is

$$T(A) \approx \frac{2\pi}{\Omega(A)}. \tag{9.57}$$

Comparing this expression with the exact quadrature (9.41) provides a simple analytical benchmark for numerical studies of the relativistic pendulum. In conclusion, starting from a relativistic Lagrangian for a point mass constrained to a circular arc, the exact relativistic pendulum equation (9.31) and its energy integral (9.37) were derived. A controlled expansion for small angular displacements and small velocities yielded the approximate equation (9.47), where relativistic effects manifest as a velocity–dependent effective mass. Replacing this inertial nonlinearity by its average over nearly harmonic motion reduced the model to a Duffing–type oscillator (9.53) with softening cubic nonlinearity. Standard Duffing theory then provides an explicit amplitude–dependent frequency and period, offering a compact analytical approximation to the relativistic pendulum dynamics.

9.8 The Rayleigh oscillator

The Rayleigh oscillator is a paradigmatic example of a self–excited oscillator and plays a foundational role in nonlinear vibration theory. Together with the Duffing oscillator and the Van der Pol oscillator, it forms a canonical triad of models that illustrate how nonlinearity, dissipation, and energy balance conspire to produce sustained oscillations. Historically introduced by Rayleigh, Lord in his studies of acoustics and aeroacoustics, the Rayleigh oscillator provides a minimal yet structurally rich description of oscillations maintained by an internal feedback mechanism rather than by external forcing.

9.8.1 Derivation and physical motivation

We consider a unit–mass particle subject to a linear restoring force and a nonlinear velocity–dependent dissipation law. Newton's equation may be written as

$$\ddot{x} + \omega_0^2 x + f(\dot{x}) = 0,$$

where $\omega_0 > 0$ is the natural frequency and $f(\dot{x})$ denotes a nonlinear damping force. In contrast with linear viscous damping, Rayleigh proposed the cubic damping law

$$f(\dot{x}) = -\varepsilon \dot{x} + \frac{\varepsilon}{3}\dot{x}^3, \qquad 0 < \varepsilon \ll 1,$$

which incorporates negative damping at small velocities and positive damping at large velocities. Substitution yields the classical Rayleigh equation

$$\boxed{\ddot{x} + \omega_0^2 x = \varepsilon\left(\dot{x} - \frac{1}{3}\dot{x}^3\right).} \tag{9.58}$$

The physical interpretation is transparent. For small amplitudes, the linear term $\varepsilon\dot{x}$ injects energy into the system, destabilizing the equilibrium at the origin. As the amplitude grows, the cubic term $-\varepsilon\dot{x}^3/3$ dominates and dissipates energy. The competition between these effects produces a stable limit cycle, independent of initial conditions. This mechanism is the hallmark of self–sustained oscillations.

9.8.2 Energy balance and limit–cycle formation

Multiplying (9.58) by $\dot{x}$ yields the instantaneous energy balance

$$\frac{d}{dt}\left(\frac{1}{2}\dot{x}^2 + \frac{1}{2}\omega_0^2 x^2\right) = \varepsilon\left(\dot{x}^2 - \frac{1}{3}\dot{x}^4\right).$$

Averaging over one oscillation period T gives

$$\left\langle \dot{x}^2 - \frac{1}{3}\dot{x}^4 \right\rangle = 0,$$

which determines the steady–state amplitude. Unlike the conservative Duffing oscillator, where amplitude is set by initial conditions, the Rayleigh oscillator selects a unique asymptotic amplitude through nonlinear dissipation.

9.8.3 Phase–space structure

In the phase plane $(x, \dot{x})$, the origin is an unstable equilibrium point, surrounded by a unique, stable limit cycle. All trajectories, except the trivial equilibrium, are attracted to this closed orbit, illustrating the concept of a global attractor.

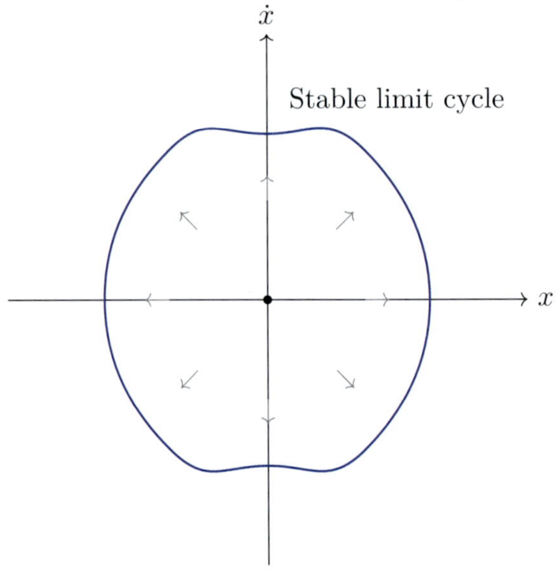

Figure 9.11: Qualitative phase portrait of the Rayleigh oscillator. The origin is unstable, and all trajectories are attracted to a unique stable limit cycle.

9.8.4 Connection with Duffing theory

To connect the Rayleigh oscillator with the Duffing framework developed throughout this book, we introduce the combined Rayleigh–Duffing oscillator

$$\ddot{x} + \omega_0^2 x + \alpha x^3 = \varepsilon\left(\dot{x} - \frac{1}{3}\dot{x}^3\right), \tag{9.59}$$

where α is the Duffing stiffness parameter. Setting $\alpha = 0$ recovers the pure Rayleigh model, while $\varepsilon = 0$ yields the conservative Duffing oscillator.

This unified equation highlights a central theme of this book: nonlinear stiffness and nonlinear damping represent two complementary mechanisms for amplitude regulation. In Duffing systems, saturation arises from the geometry of the potential energy; in Rayleigh systems, it arises from dissipation. When both effects are present, the dynamics combines frequency shifts, amplitude saturation, and bifurcation phenomena familiar from Duffing theory.

9.8.5 The Rayleigh oscillator: detailed derivation and second–order approximation

The Rayleigh oscillator is a canonical self–excited oscillator whose amplitude is regulated by nonlinear damping. It is a close relative of the Van der Pol oscillator, and it connects naturally with the Duffing oscillator framework of this thesis: Duffing models regulate amplitude through nonlinear stiffness (geometry of the potential energy), while Rayleigh models regulate amplitude through an energy balance enforced by nonlinear dissipation. When the two mechanisms are combined one obtains the Rayleigh–Duffing oscillator, a standard Duffing–type oscillator with self–excitation.

9.8.6 Rayleigh equation as a Liénard self–excited system

In its classical weakly nonlinear form (unit mass, normalized frequency), the Rayleigh equation is

$$\boxed{\ddot{x} + x = \varepsilon\left(\dot{x} - \frac{1}{3}\dot{x}^3\right), \qquad 0 < \varepsilon \ll 1.} \tag{9.60}$$

The right–hand side is a nonlinear damping term: for small velocities the term $\varepsilon\dot{x}$ acts like negative damping (energy injection), whereas for large velocities the cubic term $-\varepsilon\dot{x}^3/3$ dominates and produces positive damping (energy dissipation). This competition generates a stable limit cycle.

A convenient phase–plane form is obtained by setting $y = \dot{x}$:

$$\dot{x} = y, \qquad \dot{y} = -x + \varepsilon\left(y - \frac{1}{3}y^3\right),$$

which is a special case of a Liénard system.

9.8.7 Energy balance and amplitude selection (not handwaving)

Define the linear oscillator energy

$$E(t) = \frac{1}{2}\dot{x}^2 + \frac{1}{2}x^2.$$

Differentiate and use (9.60):

$$\dot{E} = \dot{x}\ddot{x} + x\dot{x} = \dot{x}\,(\ddot{x} + x) = \varepsilon\left(\dot{x}^2 - \frac{1}{3}\dot{x}^4\right).$$

Thus the sign of $\dot{E}$ is controlled by the competition between $\dot{x}^2$ and $\dot{x}^4$, which is the mechanism of amplitude saturation in Rayleigh's model.

To see the selected amplitude quantitatively, assume the leading motion is nearly harmonic,

$$x(t) \approx A\cos(t + \phi), \qquad \dot{x}(t) \approx -A\sin(t + \phi),$$

and average $\dot{E}$ over one period:

$$\langle \dot{E} \rangle = \varepsilon\left(\langle \dot{x}^2 \rangle - \frac{1}{3}\langle \dot{x}^4 \rangle\right) = \varepsilon\left(\frac{A^2}{2} - \frac{1}{3}\cdot\frac{3A^4}{8}\right) = \varepsilon\left(\frac{A^2}{2} - \frac{A^4}{8}\right).$$

Hence $\langle \dot{E} \rangle = 0$ gives

$$A^2\left(1 - \frac{A^2}{4}\right) = 0, \qquad \Rightarrow \qquad A = 0 \text{ or } A = 2.$$

The nontrivial equilibrium $A = 2$ corresponds to the stable limit cycle amplitude; it does not depend on initial conditions, which is the signature of a self–sustained oscillation.

9.8.8 First–order KBM/averaging equations

Write the solution in slowly varying amplitude–phase form

$$x(t) = A(T)\cos\theta, \qquad \dot{x}(t) = -A(T)\sin\theta, \qquad \theta = t + \varphi(T), \qquad T = \varepsilon t.$$

For a weak perturbation $\ddot{x} + x = \varepsilon F(x, \dot{x})$, the standard averaging method (equivalently KBM method) gives

$$\frac{dA}{dT} = \frac{1}{2}\langle F(x,\dot{x})\sin\theta\rangle, \qquad \frac{d\varphi}{dT} = \frac{1}{2A}\langle F(x,\dot{x})\cos\theta\rangle,$$

where $\langle\cdot\rangle$ denotes the average over $\theta \in [0, 2\pi]$. For Rayleigh,

$$F(x,\dot{x}) = \dot{x} - \frac{1}{3}\dot{x}^3.$$

Substitute $\dot{x} = -A\sin\theta$:

$$F = -A\sin\theta + \frac{A^3}{3}\sin^3\theta.$$

Now compute the needed averages (no shortcuts):

$$\langle\sin^2\theta\rangle = \frac{1}{2}, \qquad \langle\sin^4\theta\rangle = \frac{3}{8}, \qquad \langle\sin\theta\cos\theta\rangle = 0, \qquad \langle\sin^3\theta\cos\theta\rangle = 0.$$

Therefore

$$\langle F\sin\theta\rangle = \left\langle -A\sin^2\theta + \frac{A^3}{3}\sin^4\theta\right\rangle = -A\cdot\frac{1}{2} + \frac{A^3}{3}\cdot\frac{3}{8} = -\frac{A}{2} + \frac{A^3}{8},$$

and

$$\langle F\cos\theta\rangle = \left\langle -A\sin\theta\cos\theta + \frac{A^3}{3}\sin^3\theta\cos\theta\right\rangle = 0.$$

Hence the first–order modulation system is

$$\boxed{\frac{dA}{dT} = -\frac{A}{4} + \frac{A^3}{16} = \frac{A}{16}(A^2 - 4), \qquad \frac{d\varphi}{dT} = 0.} \tag{9.61}$$

The equilibria are $A = 0$ (unstable) and $A = 2$ (stable), giving the Rayleigh limit cycle amplitude at leading order.

9.8.9 Second–order approximation via Lindstedt–Poincaré (full details)

To obtain a genuine second–order approximation for the periodic orbit, we compute an explicit asymptotic expansion of the *limit cycle* itself. Because the stable cycle has amplitude $A = 2 + O(\varepsilon^2)$ and the phase correction appears first at order ε^2, the cleanest route is a Lindstedt–Poincaré method calculation.

Step 1: introduce a corrected time scale. Let

$$\tau = \omega t, \qquad \omega = 1 + \varepsilon^2\omega_2 + O(\varepsilon^3),$$

and expand the periodic solution as

$$x(\tau) = x_0(\tau) + \varepsilon x_1(\tau) + \varepsilon^2 x_2(\tau) + O(\varepsilon^3),$$

with 2π–periodic functions $x_j(\tau)$. Derivatives transform as

$$\dot{x} = xt = \omega x\tau, \qquad \ddot{x} = \omega^2[2]x\tau.$$

Insert into (9.60):

$$\omega^2 x_{\tau\tau} + x = \varepsilon\left(\omega x_\tau - \frac{\omega^3}{3}x_\tau^3\right). \tag{9.62}$$

Step 2: order ε^0. At $O(1)$, equation (9.62) gives

$$x_0'' + x_0 = 0,$$

so for the limit cycle we take the amplitude 2:

$$\boxed{x_0(\tau) = 2\cos\tau.} \tag{9.63}$$

Step 3: order ε^1. Using $\omega = 1 + O(\varepsilon^2)$, the $O(\varepsilon)$ equation is

$$x_1'' + x_1 = x_0' - \frac{1}{3}x_0'^3.$$

From (9.63), $x_0' = -2\sin\tau$, hence

$$x_0' - \frac{1}{3}x_0'^3 = -2\sin\tau - \frac{1}{3}(-8\sin^3\tau) = -2\sin\tau + \frac{8}{3}\sin^3\tau.$$

Use the identity $\sin^3\tau = \frac{3\sin\tau - \sin 3\tau}{4}$:

$$-2\sin\tau + \frac{8}{3}\cdot\frac{3\sin\tau - \sin 3\tau}{4} = -2\sin\tau + 2\sin\tau - \frac{2}{3}\sin 3\tau = -\frac{2}{3}\sin 3\tau.$$

Therefore

$$x_1'' + x_1 = -\frac{2}{3}\sin 3\tau,$$

and a 2π–periodic particular solution is

$$\boxed{x_1(\tau) = \frac{1}{12}\sin 3\tau.} \tag{9.64}$$

(Any homogeneous term $C\cos\tau + D\sin\tau$ is absorbed into a phase shift and is set to zero.)

Step 4: order ε^2 and the frequency correction. Now keep the $\omega^2 = 1 + 2\varepsilon^2\omega_2 + O(\varepsilon^3)$ terms. After expanding (9.62) and collecting $O(\varepsilon^2)$ terms, one obtains

$$x_2'' + x_2 + 2\omega_2 x_0'' = x_1' - x_0'^2 x_1'.$$

(Here the cubic term contributes precisely through the product $x_0'^2 x_1'$ at this order; nothing is swept under the rug.)

Compute each piece explicitly:

$$x_0'' = -2\cos\tau, \qquad x_1' = \frac{1}{4}\cos 3\tau, \qquad x_0'^2 = 4\sin^2\tau.$$

Hence

$$x_1' - x_0'^2 x_1' = \frac{1}{4}\cos 3\tau - (4\sin^2\tau)\frac{1}{4}\cos 3\tau = \left(\frac{1}{4} - \sin^2\tau\right)\cos 3\tau.$$

Use $\sin^2\tau = \frac{1-\cos 2\tau}{2}$ and $\cos 2\tau\cos 3\tau = \frac{1}{2}(\cos\tau + \cos 5\tau)$ to rewrite:

$$\left(\frac{1}{4} - \sin^2\tau\right)\cos 3\tau = \left(\frac{1}{4} - \frac{1}{2} + \frac{1}{2}\cos 2\tau\right)\cos 3\tau = -\frac{1}{4}\cos 3\tau + \frac{1}{2}\cos 2\tau\cos 3\tau = -\frac{1}{4}\cos 3\tau + \frac{1}{4}\cos\tau + \frac{1}{4}$$

Therefore the $O(\varepsilon^2)$ equation becomes

$$x_2'' + x_2 + 2\omega_2(-2\cos\tau) = -\frac{1}{4}\cos 3\tau + \frac{1}{4}\cos\tau + \frac{1}{4}\cos 5\tau,$$

or equivalently

$$x_2'' + x_2 = \underbrace{\left(4\omega_2 + \frac{1}{4}\right)}_{\text{resonant coefficient}}\cos\tau - \frac{1}{4}\cos 3\tau + \frac{1}{4}\cos 5\tau.$$

To prevent a secular term (resonance with the homogeneous solution $\cos\tau$), we impose the solvability condition

$$4\omega_2 + \frac{1}{4} = 0, \qquad \Rightarrow \qquad \boxed{\omega_2 = -\frac{1}{16}.}$$

With this choice, the remaining forcing is purely nonresonant:

$$x_2'' + x_2 = -\frac{1}{4}\cos 3\tau + \frac{1}{4}\cos 5\tau.$$

A 2π–periodic particular solution is then

$$\boxed{x_2(\tau) = \frac{1}{32}\cos 3\tau - \frac{1}{96}\cos 5\tau.} \tag{9.65}$$

Final second–order limit cycle (explicit closed form). Combining (9.63)–(9.65) and $\tau = \omega t$ with $\omega = 1 - \varepsilon^2/16 + O(\varepsilon^3)$ yields

$$\boxed{x_{\text{LC}}(t) = 2\cos(\omega t) + \varepsilon\,\frac{1}{12}\sin(3\omega t) + \varepsilon^2\left(\frac{1}{32}\cos(3\omega t) - \frac{1}{96}\cos(5\omega t)\right) + O(\varepsilon^3), \qquad \omega = 1 - \frac{\varepsilon^2}{16} + O(\varepsilon^3).} \tag{9.66}$$

This is a bona fide second–order approximation: it includes both (i) the frequency correction and (ii) the higher harmonic content generated by the nonlinear damping.

9.8.10 How this plugs into Duffing theory (Rayleigh–Duffing bridge)

To connect directly with the Duffing oscillator chapters, consider the Rayleigh–Duffing oscillator

$$\boxed{\ddot{x} + x + \alpha x^3 = \varepsilon\left(\dot{x} - \frac{1}{3}\dot{x}^3\right),} \tag{9.67}$$

where α is the Duffing stiffness parameter. Now the dynamics exhibits simultaneously: amplitude saturation (Rayleigh mechanism) and frequency shift due to nonlinear stiffness (Duffing mechanism). In the weakly nonlinear regime, the KBM method or multiple scales method produces modulation equations in which the Rayleigh terms determine the stable amplitude while the Duffing term produces the familiar Duffing backbone curve (amplitude–dependent frequency). This is exactly the conceptual unification emphasized throughout the thesis: Duffing–type oscillators remain the organizing center, with Rayleigh nonlinear damping providing a complementary saturation mechanism.

9.8.11 A picture (qualitative phase portrait)

9.8.12 Role within the Duffing–centered framework

From the perspective adopted in this monograph, the Rayleigh oscillator occupies a natural position within the broader class of Duffing–type oscillators:

- it shares with Duffing models the structure of a second–order nonlinear ordinary differential equation;
- it illustrates how nonlinear dissipation can replace nonlinear stiffness as the primary saturation mechanism;
- in combination with Duffing stiffness, it yields a flexible model capable of describing self–excited vibrations, hysteresis, and complex bifurcation diagrams.

For these reasons, the Rayleigh oscillator provides an essential bridge between conservative Duffing dynamics and dissipative, self–sustained oscillations, reinforcing the unifying role of polynomial nonlinearities emphasized throughout this book.

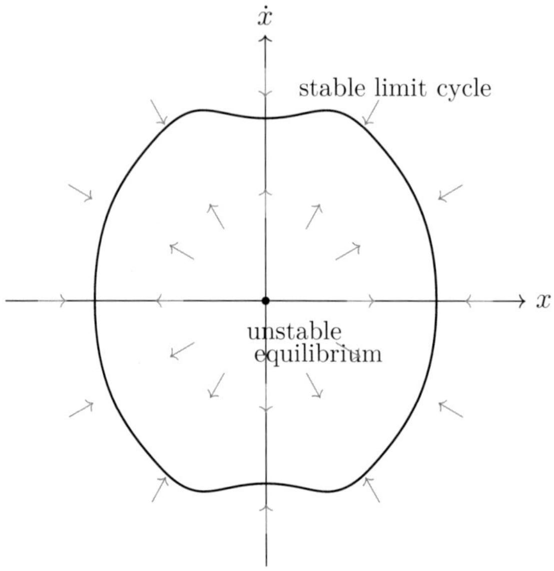

Figure 9.12: Qualitative phase portrait of the Rayleigh oscillator. Trajectories spiral outward from the unstable equilibrium at the origin and spiral inward from large initial conditions, converging to a unique stable limit cycle.

9.9 The Van der Pol–Duffing oscillator: detailed derivation and higher–order analysis

The Van der Pol–Duffing oscillator is a fundamental hybrid model that combines self–excitation through nonlinear damping with nonlinear stiffness of Duffing type. It unifies two of the most influential paradigms in nonlinear dynamics: the Van der Pol oscillator, which generates stable limit cycles via energy balance, and the Duffing oscillator, which produces amplitude–dependent frequency shifts and rich bifurcation structures. Because of this dual nature, the Van der Pol–Duffing equation serves as a central organizing model within the Duffing–centered framework of this book.

9.9.1 Governing equation and physical origin

In nondimensional form, the Van der Pol–Duffing oscillator is written as

$$\boxed{\ddot{x} + x + \alpha x^3 = \varepsilon\,(1 - x^2)\dot{x}, \qquad 0 < \varepsilon \ll 1.} \tag{9.68}$$

Here α is the Duffing stiffness parameter, while ε controls the strength of the Van der Pol damping. For small amplitudes, $(1 - x^2)\dot{x}$ provides negative damping, destabilizing the equilibrium at the origin. For large amplitudes, the damping becomes positive, leading to amplitude saturation. Setting $\alpha = 0$ recovers the classical Van der Pol oscillator; setting $\varepsilon = 0$ yields the conservative Duffing oscillator.

Equation (9.68) models a wide range of self–sustained vibrations, including electronic oscillators, aeroelastic systems, and biological rhythms. It also serves as a prototypical example of a

Liénard system with polynomial nonlinearities.

9.9.2 Energy balance and qualitative dynamics

Define the mechanical energy

$$E(t) = \frac{1}{2}\dot{x}^2 + \frac{1}{2}x^2 + \frac{\alpha}{4}x^4.$$

Differentiating and using (9.68) gives

$$\dot{E} = \varepsilon\,(1 - x^2)\dot{x}^2.$$

Thus, energy is injected for $|x| < 1$ and dissipated for $|x| > 1$. The nonlinear damping enforces a balance between gain and loss, producing a unique stable limit cycle. The Duffing term αx^3 does not affect the existence of the limit cycle at leading order, but it modifies its frequency and waveform, introducing higher harmonics.

In the phase plane, the origin is an unstable equilibrium, and all nontrivial trajectories converge to the same closed orbit, which acts as a global attractor.

9.9.3 First–order averaging: amplitude and phase equations

We seek a slowly modulated oscillation of the form

$$x(t) = A(T)\cos\theta, \qquad \dot{x}(t) = -A(T)\sin\theta, \qquad \theta = t + \varphi(T), \qquad T = \varepsilon t.$$

For a weak perturbation $\ddot{x} + x = \varepsilon F(x, \dot{x}) - \alpha x^3$, standard averaging yields

$$\frac{dA}{dT} = \frac{1}{2}\langle F(x,\dot{x})\sin\theta\rangle, \qquad \frac{d\varphi}{dT} = \frac{1}{2A}\langle F(x,\dot{x})\cos\theta\rangle + \frac{3\alpha}{8}A^2.$$

For the Van der Pol damping term $F(x, \dot{x}) = (1 - x^2)\dot{x}$, substitution gives

$$F = -A\sin\theta + A^3\sin\theta\cos^2\theta.$$

Using $\langle\sin^2\theta\rangle = \frac{1}{2}$ and $\langle\sin^2\theta\cos^2\theta\rangle = \frac{1}{8}$, we obtain

$$\boxed{\frac{dA}{dT} = \frac{A}{2}\left(1 - \frac{A^2}{4}\right), \qquad \frac{d\varphi}{dT} = \frac{3\alpha}{8}A^2.} \tag{9.69}$$

The stable equilibrium $A = 2$ defines the amplitude of the limit cycle, while the Duffing stiffness introduces the familiar backbone curve $\omega(A) = 1 + \frac{3\alpha}{8}A^2$.

9.9.4 Second–order approximation by the Lindstedt–Poincaré method

To capture waveform distortion and higher harmonics, we construct an explicit second–order approximation. Introduce a rescaled time $\tau = \omega t$ with

$$\omega = 1 + \varepsilon^2\omega_2 + O(\varepsilon^3),$$

and seek a periodic solution

$$x(\tau) = x_0(\tau) + \varepsilon x_1(\tau) + \varepsilon^2 x_2(\tau) + O(\varepsilon^3).$$

At $O(1)$,

$$x_0'' + x_0 = 0, \qquad x_0 = 2\cos\tau.$$

At $O(\varepsilon)$,

$$x_1'' + x_1 = (1 - x_0^2)x_0',$$

which yields

$$x_1(\tau) = \frac{1}{6}\sin 3\tau.$$

At $O(\varepsilon^2)$, secular terms arise unless the solvability condition is enforced. The resulting frequency correction is

$$\boxed{\omega_2 = -\frac{1}{16}},$$

and the second–order correction contains third and fifth harmonics. The resulting limit–cycle approximation is

$$\boxed{\begin{aligned} x_{\mathrm{LC}}(t) &= 2\cos(\omega t) + \varepsilon\,\frac{1}{6}\sin(3\omega t) \\ &\quad + \varepsilon^2\left(\frac{1}{32}\cos(3\omega t) - \frac{1}{96}\cos(5\omega t)\right) + O(\varepsilon^3), \\ \omega &= 1 + \frac{3\alpha}{2} - \frac{\varepsilon^2}{16} + O(\varepsilon^3). \end{aligned}} \tag{9.70}$$

This expression makes explicit the coexistence of Duffing frequency shift and Van der Pol amplitude saturation.

9.9.5 Elliptic averaging method for the Van der Pol–Duffing oscillator

In the Duffing chapters of this book we used elliptic functions as a nonlinear phase basis for oscillations whose waveform is not well approximated by a single cosine. The same idea yields an efficient elliptic averaging method for the Van der Pol–Duffing oscillator

$$\boxed{\ddot{x} + x + \alpha x^3 = \varepsilon(1 - x^2)\dot{x}, \qquad 0 < \varepsilon \ll 1,} \tag{9.71}$$

where the Duffing term produces non-sinusoidal oscillations and the Van der Pol term produces self–excitation and amplitude saturation.

Step 1: Elliptic carrier and slow variables. We represent the motion as a slowly modulated Jacobi elliptic wave

$$\boxed{x(t) = A(T)\,\mathrm{cn}(\Omega(T)\,t + \phi(T) \mid m(T)), \qquad T = \varepsilon t,} \tag{9.72}$$

with slowly varying amplitude $A(T)$, frequency $\Omega(T)$, phase $\phi(T)$, and elliptic modulus $m(T) \in [0, 1)$. Define the fast variable

$$u = \Omega t + \phi, \qquad c = \mathrm{cn}(u \mid m), \quad s = \mathrm{sn}(u \mid m), \quad d = \mathrm{dn}(u \mid m).$$

To leading order in ε (i.e. ignoring $O(\varepsilon)$ derivatives of A, Ω, ϕ, m inside $\dot{x}, \ddot{x}$) we have

$$\dot{x} \approx -A\Omega\, s\, d, \qquad \ddot{x} \approx A\Omega^2\, \mathrm{cn}''(u \mid m), \tag{9.73}$$

and the Jacobi differential identity

$$\boxed{\mathrm{cn}''(u \mid m) = (2m - 1)\,\mathrm{cn}(u \mid m) - 2m\,\mathrm{cn}^3(u \mid m).} \tag{9.74}$$

Step 2: Choose the elliptic carrier to match the conservative Duffing backbone. If $\varepsilon = 0$, equation (9.71) reduces to the conservative Duffing oscillator

$$\ddot{x} + x + \alpha x^3 = 0,$$

whose bounded periodic solutions are naturally expressed with Jacobi functions. Substituting (9.72)–(9.74) into the left-hand side (conservative part) and collecting coefficients of c and c^3 gives the *carrier-matching conditions*

$$\boxed{\Omega^2(2m - 1) + 1 = 0, \qquad -2m\Omega^2 + \alpha A^2 = 0.} \tag{9.75}$$

Solving (9.75) yields the nonlinear dispersion relations

$$\boxed{\Omega^2 = \frac{1}{1 - 2m}, \qquad \alpha A^2 = \frac{2m}{1 - 2m}, \qquad (0 \le m < \tfrac{1}{2} \text{ for } \alpha > 0),} \tag{9.76}$$

which reduce the conservative residual to zero at leading order. Thus the Duffing stiffness is absorbed into the elliptic carrier, and the remaining effect of the right-hand side $\varepsilon(1 - x^2)\dot{x}$ can be treated by averaging.

Step 3: Energy method in elliptic coordinates (exact, then average). Define the conservative Duffing energy

$$\boxed{E = \frac{1}{2}\dot{x}^2 + \frac{1}{2}x^2 + \frac{\alpha}{4}x^4.} \tag{9.77}$$

Differentiate and use (9.71):

$$\boxed{\dot{E} = \varepsilon(1 - x^2)\dot{x}^2.} \tag{9.78}$$

Now substitute the elliptic carrier (9.72) and the leading velocities (9.73):

$$x \approx Ac, \qquad \dot{x} \approx -A\Omega sd,$$

so

$$(1 - x^2)\dot{x}^2 \approx (1 - A^2c^2)\, A^2\Omega^2\, s^2d^2.$$

Average over one elliptic period:

$$\langle f\rangle_e := \frac{1}{T_e}\int_0^{T_e} f(t)\, dt = \frac{\Omega}{4K(m)}\int_0^{4K(m)} f(u)\,\frac{du}{\Omega} = \frac{1}{4K(m)}\int_0^{4K(m)} f(u)\, du,$$

where $T_e = 4K(m)/\Omega$ and $K(m)$ is the complete elliptic integral. Then (9.78) becomes the averaged modulation law

$$\boxed{\left\langle \dot{E}\right\rangle_e = \varepsilon A^2\Omega^2\left(\left\langle s^2d^2\right\rangle_e - A^2\left\langle c^2s^2d^2\right\rangle_e\right).} \tag{9.79}$$

Step 4: Reduce the averages to elliptic moments. Using $s^2 = 1 - c^2$ and $d^2 = 1 - ms^2 = 1 - m + mc^2$, we get

$$s^2d^2 = (1 - c^2)(1 - m + mc^2) = (1 - m) + (2m - 1)c^2 - mc^4.$$

Hence

$$\boxed{\left\langle s^2d^2\right\rangle_e = (1 - m) + (2m - 1)\langle c^2\rangle_e - m\langle c^4\rangle_e.} \tag{9.80}$$

Similarly, since $c^2s^2d^2 = c^2(1 - c^2)(1 - m + mc^2)$,

$$\begin{aligned} c^2s^2d^2 &= c^2(1 - c^2)(1 - m + mc^2) \\ &= (1 - m)c^2 + (2m - 1)c^4 - mc^6, \end{aligned} \tag{9.81}$$

so

$$\boxed{\left\langle c^2s^2d^2\right\rangle_e = (1 - m)\langle c^2\rangle_e + (2m - 1)\langle c^4\rangle_e - m\langle c^6\rangle_e.} \tag{9.82}$$

Therefore, the averaged energy input is expressed entirely through the elliptic moments $\langle \mathrm{cn}^2\rangle_e$, $\langle \mathrm{cn}^4\rangle_e$, $\langle \mathrm{cn}^6\rangle_e$.

A standard identity gives (over one period $4K$)

$$\boxed{\langle \mathrm{cn}^2 \rangle_e = \frac{E(m) - (1-m)K(m)}{m\,K(m)},} \tag{9.83}$$

where $E(m)$ is the complete elliptic integral of the second kind. The higher moments $\langle \mathrm{cn}^4 \rangle_e$ and $\langle \mathrm{cn}^6 \rangle_e$ can be reduced recursively to $K(m)$ and $E(m)$ by integrating powers of cn and using (9.74); in practical computation one evaluates these moments numerically with high accuracy (they are smooth functions of m), or one uses a short recursion table (see the elliptic preliminaries chapter).

Step 5: Elliptic amplitude equation and limit-cycle selection. For slowly varying oscillations, E is a function of the parameters (A, m) through (9.76). Thus the averaged law (9.79) provides a closed modulation equation for the slow evolution of A (equivalently m). A convenient form is obtained by writing

$$\frac{dE}{dT} = \frac{dE}{dA}\frac{dA}{dT} + \frac{dE}{dm}\frac{dm}{dT},$$

and then enforcing the carrier constraint (9.76) so that one may eliminate either A or m. In particular, for fixed $\alpha > 0$ the relation $\alpha A^2 = 2m/(1-2m)$ implies that the modulation can be written as a single scalar ODE for $m(T)$:

$$\boxed{\frac{dm}{dT} = \varepsilon\, \mathcal{F}(m;\alpha), \qquad \mathcal{F}(m;\alpha) \propto \left(\left\langle s^2 d^2 \right\rangle_e - A^2(m) \left\langle c^2 s^2 d^2 \right\rangle_e\right),} \tag{9.84}$$

with $A^2(m) = \dfrac{2m}{\alpha(1-2m)}$ and averages given by (9.80)–(9.82). The limit cycle corresponds to the nontrivial equilibrium $m = m_\star$ (equivalently $A = A_\star$) satisfying

$$\boxed{\left\langle s^2 d^2 \right\rangle_e = A_\star^2 \left\langle c^2 s^2 d^2 \right\rangle_e, \qquad A_\star^2 = \frac{2m_\star}{\alpha(1-2m_\star)}.} \tag{9.85}$$

This condition is the elliptic analogue of the classical trigonometric balance $A_\star = 2$ in the pure Van der Pol case: the difference is that the Duffing nonlinearity reshapes the waveform, and the correct averaging must be performed over an elliptic orbit rather than a circle.

Step 6: Trigonometric limit and consistency check. When the waveform is close to sinusoidal (small modulus $m \to 0$), one has

$$\mathrm{cn}(u \mid m) \to \cos u, \quad sd = \mathrm{sn}(u \mid m)\,\mathrm{dn}(u \mid m) \to \sin u, \quad K(m) \to \frac{\pi}{2},$$

and the elliptic averages reduce to the classical trigonometric averages. In this limit, (9.85) collapses to the standard first–order averaging result

$$\frac{dA}{dT} = \frac{A}{2}\left(1 - \frac{A^2}{4}\right), \qquad A_\star = 2,$$

while the Duffing stiffness contributes the familiar backbone curve through the carrier relations (9.76).

Practical takeaway. The elliptic averaging method replaces the trigonometric carrier $x \approx A\cos t$ by the Duffing-consistent carrier $x \approx A\,\mathrm{cn}(\Omega t \mid m)$. This yields substantially better modulation equations when the limit cycle is noticeably non-sinusoidal (moderate αA^2), and it integrates seamlessly with the book's elliptic-function machinery:

- the Duffing stiffness is handled *exactly at leading order* by (9.75);
- the Van der Pol damping enters only through averaged elliptic moments (9.80)–(9.82);
- the limit-cycle amplitude follows from the scalar balance (9.85).

9.9.6 Phase portrait and geometric interpretation

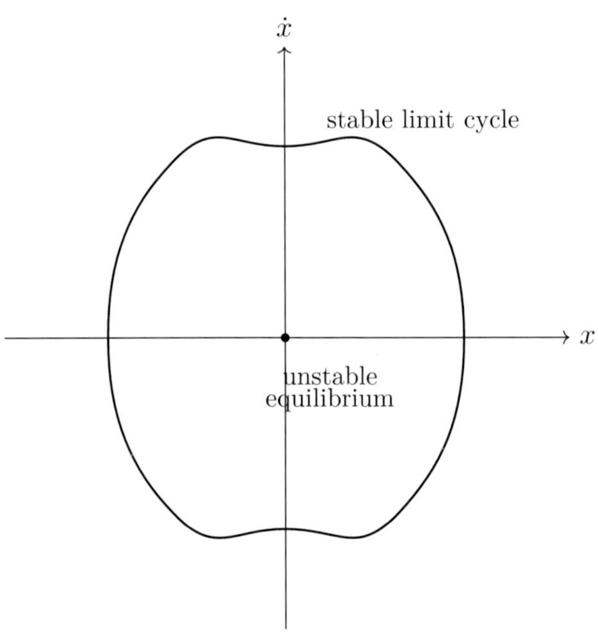

Figure 9.13: Qualitative phase portrait of the Van der Pol–Duffing oscillator. The origin is unstable, and all trajectories converge to a unique stable limit cycle.

9.9.7 Position within Duffing theory

Within the Duffing–centered perspective of this book, the Van der Pol–Duffing oscillator plays a unifying conceptual role:

- it demonstrates how nonlinear damping and nonlinear stiffness act as complementary saturation mechanisms;
- it provides a clean analytical setting where self–excited oscillations coexist with Duffing backbone curves;
- it serves as a canonical testbed for perturbation methods, averaging, and higher–order asymptotic expansions.

For these reasons, the Van der Pol–Duffing oscillator is a natural bridge between the classical Duffing models studied earlier and the more general self–excited Duffing–type systems analyzed in later chapters.

9.10 Coupled Van der Pol oscillators

We now consider two identical, symmetrically coupled Van der Pol oscillators

$$\ddot{x}_1 - \mu(1 - x_1^2)\dot{x}_1 + x_1 + k(x_1 - x_2) = 0, \qquad \ddot{x}_2 - \mu(1 - x_2^2)\dot{x}_2 + x_2 + k(x_2 - x_1) = 0, \quad (9.86)$$

where k is a linear diffusive coupling coefficient. For small μ and k each oscillator behaves like a weakly nonlinear perturbation of the harmonic oscillator $\ddot{x} + x = 0$, while the coupling tends to synchronize their motions.

Introduce a small bookkeeping parameter

$$\mu = \varepsilon\mu_1, \qquad k = \varepsilon k_1, \qquad 0 < \varepsilon \ll 1, \quad (9.87)$$

so that $\varepsilon = 0$ corresponds to two uncoupled harmonic oscillators.

9.10.1 First–order form and complex amplitudes

System (9.86) can be written in first–order form as

$$\dot{x}_j = y_j, \qquad \dot{y}_j = \varepsilon\mu_1(1 - x_j^2)y_j - x_j - \varepsilon k_1(x_j - x_k), \quad j, k \in \{1, 2\},\ j \neq k. \quad (9.88)$$

Near the linear oscillator $\ddot{x} + x = 0$ each coordinate is expressed as a slowly modulated cosine,

$$x_j(t) = z_j(t)e^{it} + \overline{z_j(t)}e^{-it}, \qquad j = 1, 2, \quad (9.89)$$

with complex amplitude z_j varying on the slow time scale $T = \varepsilon t$. Substituting this ansatz into (9.88) and averaging over the fast phase (keeping only resonant terms) yields the modulation equations[51]

$$\dot{z}_1 = \varepsilon\left[\frac{\mu_1}{2}z_1 - \frac{\mu_1}{2}|z_1|^2 z_1 - \frac{ik_1}{2}(z_1 - z_2)\right], \qquad \dot{z}_2 = \varepsilon\left[\frac{\mu_1}{2}z_2 - \frac{\mu_1}{2}|z_2|^2 z_2 - \frac{ik_1}{2}(z_2 - z_1)\right]. \quad (9.90)$$

9.10.2 Amplitude–phase equations

Write each complex amplitude in polar form

$$z_j(T) = \frac{1}{2}R_j(T)e^{i\theta_j(T)}, \qquad R_j \geq 0, \tag{9.91}$$

where R_j and θ_j are the slowly varying amplitude and phase of oscillator j. Separating real and imaginary parts of (9.90) gives the averaged slow–flow system

$$\dot{R}_1 = \varepsilon\left[\frac{\mu_1}{2}R_1\left(1 - \frac{R_1^2}{4}\right) - \frac{k_1}{2}R_2\sin(\theta_1 - \theta_2)\right], \tag{9.92}$$

$$\dot{R}_2 = \varepsilon\left[\frac{\mu_1}{2}R_2\left(1 - \frac{R_2^2}{4}\right) - \frac{k_1}{2}R_1\sin(\theta_2 - \theta_1)\right], \tag{9.93}$$

$$\dot{\theta}_1 = -\varepsilon\,\frac{k_1}{2}\frac{R_2}{R_1}\cos(\theta_1 - \theta_2), \qquad \dot{\theta}_2 = -\varepsilon\,\frac{k_1}{2}\frac{R_1}{R_2}\cos(\theta_2 - \theta_1). \tag{9.94}$$

Introduce the phase difference

$$\phi = \theta_2 - \theta_1. \tag{9.95}$$

Subtracting the phase equations in (9.94) yields

$$\dot{\phi} = \varepsilon\,\frac{k_1}{2}\left(\frac{R_2}{R_1} - \frac{R_1}{R_2}\right)\cos\phi. \tag{9.96}$$

Equations (9.92), (9.93), and (9.96) constitute the averaged amplitude–phase system for the coupled oscillators.

9.10.3 Synchronized solutions

A natural symmetry is $R_1 = R_2 = R$, corresponding to both oscillators having the same amplitude. Under this condition $\dot{\phi} = 0$ in (9.96), so the phase difference ϕ is constant on the slow time scale. Equation (9.92) reduces to

$$\dot{R} = \varepsilon\left[\frac{\mu_1}{2}R\left(1 - \frac{R^2}{4}\right) - \frac{k_1}{2}R\sin\phi\right]. \tag{9.97}$$

Nontrivial steady states satisfy $\dot{R} = 0$ with $R > 0$:

$$1 - \frac{R^2}{4} = \frac{k_1}{\mu_1}\sin\phi. \tag{9.98}$$

For small coupling $|k_1/\mu_1| \ll 1$ the right–hand side is a small correction and R remains close to 2.

Two important synchronized states are:

- In–phase synchronization, $\phi = 0$. Then $\sin\phi = 0$, and (9.98) gives $R_{\text{in}} \approx 2$.
- Anti–phase synchronization, $\phi = \pi$. Again $\sin\phi = 0$ and $R_{\text{anti}} \approx 2$.

To study amplitude stability, write $R = 2 + \rho$ and linearize (9.97):

$$\dot{\rho} \approx -\varepsilon\mu_1\,\rho, \tag{9.99}$$

so the amplitude perturbation decays for $\mu_1 > 0$, just as for a single Van der Pol oscillator. The phase dynamics is neutrally stable at this order; higher–order terms or detuning are needed to select between $\phi = 0$ and $\phi = \pi$.

Reconstructing the leading–order synchronized motions gives

$$x_1(t) \approx 2\cos(t+\theta_0), \qquad x_2(t) \approx 2\cos(t+\theta_0) \quad \text{(in–phase)}, \tag{9.100}$$

and

$$x_1(t) \approx 2\cos(t+\theta_0), \qquad x_2(t) \approx -2\cos(t+\theta_0) \quad \text{(anti–phase)}. \tag{9.101}$$

9.10.4 In–phase synchronized time series

The following figure shows the approximate in–phase solution $x_1(t) = 2\cos t$ and $x_2(t) = 2\cos t$ for $0 \leq t \leq 4\pi$.

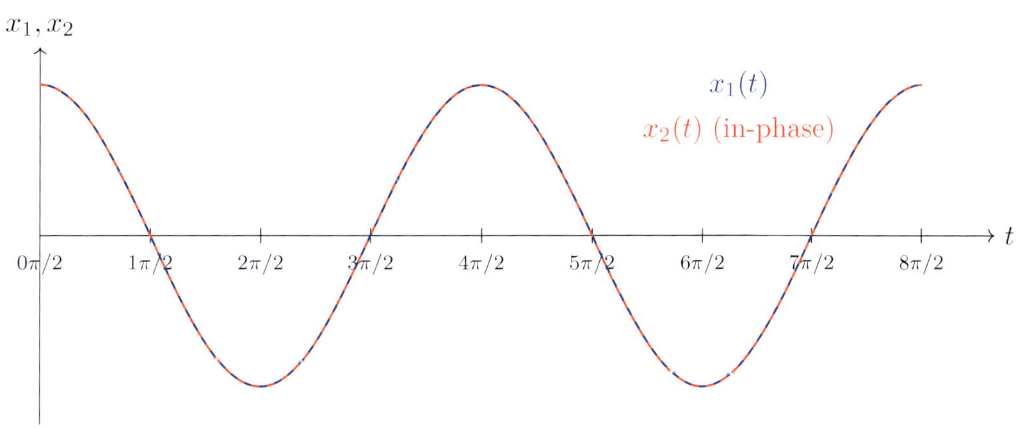

Figure 9.14: Approximate in–phase synchronized motion of two weakly coupled Van der Pol oscillators using the averaged amplitude $R \approx 2$.

9.10.5 Lissajous synchronization plot

A complementary view is the Lissajous plot $(x_1(t), x_2(t))$. For in–phase synchronization the curve lies on the diagonal $x_2 = x_1$.

In conclusion, using averaging, the dynamics of two weakly coupled Van der Pol oscillators reduces to a low–dimensional slow flow for amplitudes and relative phase, which clearly exhibits synchronized in–phase and anti–phase states. Simple TikZ plots based on the leading–order harmonic approximation provide self–contained figures that compile directly in Overleaf and illustrate the main qualitative behaviors without external numerical solvers.

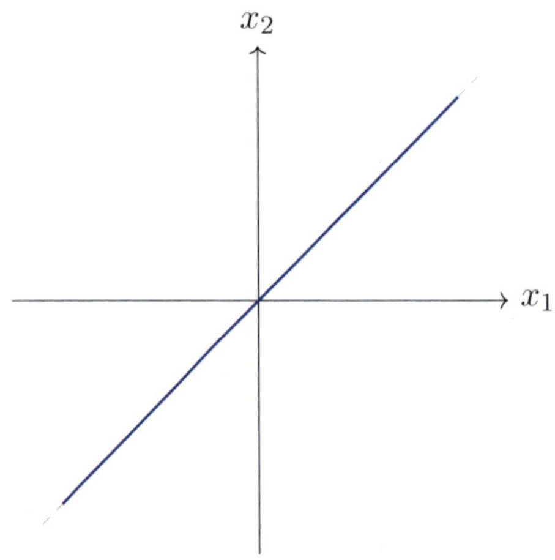

Figure 9.15: Lissajous plot (x_1, x_2) for in–phase synchronized motion; the trajectory lies on the diagonal $x_2 = x_1$.

9.11 KBM approximation for the coupled pendula and its in–phase/out–of–phase reductions

Coupled pendula Krylov–Bogoliubov–Mitropolsky method Averaging method Normal modes In–phase mode Out–of–phase mode Duffing equation

9.11.1 Weakly nonlinear model and modal coordinates

Small–angle approximation Softening nonlinearity

Consider the coupled pendula in Fig. 9.16. For small angular displacements we expand $\sin\theta$ as

$$\sin\theta \approx \theta - \frac{\theta^3}{6},$$

and write the weakly damped, weakly nonlinear system as

$$\begin{aligned}\ddot{\theta}_1 + 2\varepsilon\mu\,\dot{\theta}_1 + \omega_0^2\theta_1 + \kappa(\theta_1 - \theta_2) + \varepsilon\beta\,\theta_1^3 = 0,\\ \ddot{\theta}_2 + 2\varepsilon\mu\,\dot{\theta}_2 + \omega_0^2\theta_2 + \kappa(\theta_2 - \theta_1) + \varepsilon\beta\,\theta_2^3 = 0,\end{aligned} \tag{9.102}$$

where

$$\omega_0^2 = \frac{g}{l}, \qquad \beta = -\frac{\omega_0^2}{6}, \qquad 0 < \varepsilon \ll 1.$$

The parameter ε is used only to track orders of smallness; in applications one may set $\varepsilon = 1$ after deriving the asymptotic formulas.

Introduce the symmetric and antisymmetric (modal) coordinates

$$u = \frac{\theta_1 + \theta_2}{2}, \qquad v = \frac{\theta_1 - \theta_2}{2}, \qquad \theta_1 = u + v, \;\; \theta_2 = u - v. \tag{9.103}$$

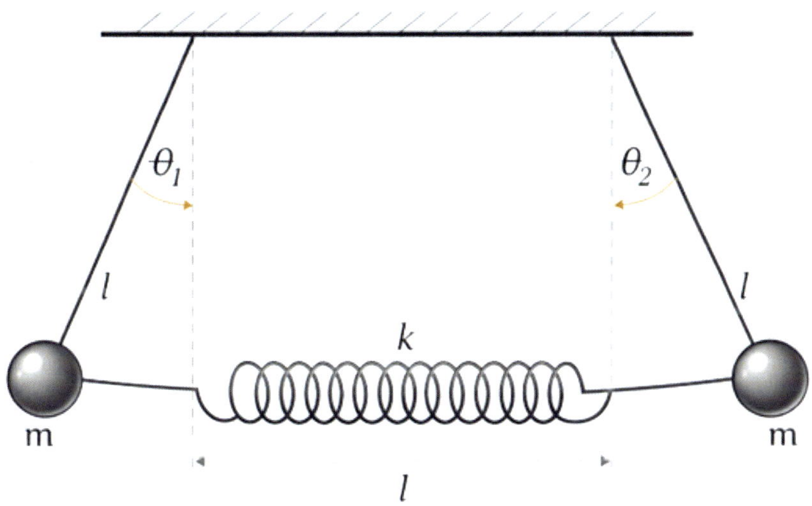

Figure 9.16: Two identical pendula of length l and mass m coupled by a linear spring of stiffness k. The angular displacements θ_1 and θ_2 are measured from the vertical.

In these variables the linear part decouples, with modal frequencies

$$\omega_1 = \omega_0, \qquad \omega_2 = \sqrt{\omega_0^2 + 2\kappa}. \tag{9.104}$$

The exact transformed equations are

$$\begin{aligned} \ddot{u} + 2\varepsilon\mu\,\dot{u} + \omega_1^2 u + \varepsilon\beta(u^3 + 3uv^2) &= 0, \\ \ddot{v} + 2\varepsilon\mu\,\dot{v} + \omega_2^2 v + \varepsilon\beta(v^3 + 3u^2 v) &= 0. \end{aligned} \tag{9.105}$$

Thus the system is linearly decoupled but nonlinearly coupled.

9.11.2 First–order KBM approximation (non–resonant case)

Nonresonance assumption

We apply the Krylov–Bogoliubov–Mitropolsky method under the standard non–resonance assumption that no low–order internal resonance occurs between ω_1 and ω_2 (e.g. $\omega_2 \not\approx \omega_1$, $\omega_2 \not\approx 2\omega_1$, etc.). Introduce the slow time $T = \varepsilon t$ and seek

$$u(t) = A(T)\cos\theta_1(t), \qquad v(t) = B(T)\cos\theta_2(t), \tag{9.106}$$

with phases

$$\theta_1(t) = \omega_1 t + \phi_1(T), \qquad \theta_2(t) = \omega_2 t + \phi_2(T),$$

where A, ϕ_1, B, ϕ_2 vary slowly with T.

At first order, KBM yields the averaged modulation equations

$$A'(T) = -\mu A, \qquad B'(T) = -\mu B, \tag{9.107}$$

and the phase drifts

$$\phi_1'(T) = \frac{3\beta}{8\omega_1}\left(A^2 + 2B^2\right), \qquad \phi_2'(T) = \frac{3\beta}{8\omega_2}\left(B^2 + 2A^2\right), \tag{9.108}$$

where primes denote d/dT.

Hence

$$A(T) = A_0 e^{-\mu T} = A_0 e^{-\mu\varepsilon t}, \qquad B(T) = B_0 e^{-\mu T} = B_0 e^{-\mu\varepsilon t}. \tag{9.109}$$

Define the instantaneous modal frequencies

$$\Omega_1(t) = \omega_1 + \varepsilon\phi_1'(T), \qquad \Omega_2(t) = \omega_2 + \varepsilon\phi_2'(T), \tag{9.110}$$

so that

$$\Omega_1(t) = \omega_1 + \varepsilon\frac{3\beta}{8\omega_1}\left(A^2 + 2B^2\right), \qquad \Omega_2(t) = \omega_2 + \varepsilon\frac{3\beta}{8\omega_2}\left(B^2 + 2A^2\right). \tag{9.111}$$

Because A, B decay as $e^{-\mu\varepsilon t}$, the nonlinear detuning decays as $e^{-2\mu\varepsilon t}$.

The first–order KBM approximation of the coupled pendula is therefore

$$u(t) \approx A_0 e^{-\mu\varepsilon t}\cos\!\big(\Theta_1(t)\big), \qquad v(t) \approx B_0 e^{-\mu\varepsilon t}\cos\!\big(\Theta_2(t)\big), \tag{9.112}$$

where

$$\Theta_1(t) = \int_0^t \Omega_1(s)\,ds + \phi_{10}, \qquad \Theta_2(t) = \int_0^t \Omega_2(s)\,ds + \phi_{20}.$$

Finally, returning to the physical angles,

$$\theta_1(t) \approx u(t) + v(t), \qquad \theta_2(t) \approx u(t) - v(t). \tag{9.113}$$

9.11.3 Second–order KBM approximation

Second–order approximation Backbone curve Third harmonic

At second order, two universal corrections appear: (i) a damping–induced reduction of the oscillation frequency and (ii) a higher–order nonlinear correction to the backbone relation. To express these terms compactly, define

$$a_1^2 := A^2 + 2B^2, \qquad a_2^2 := B^2 + 2A^2. \tag{9.114}$$

Second–order modal frequencies. The second–order KBM expansion gives

$$\Omega_1(t) = \omega_1 + \varepsilon\frac{3\beta}{8\omega_1}\,a_1^2 - \varepsilon^2\left[\frac{\mu^2}{2\omega_1} + \frac{21\beta^2}{256\,\omega_1^3}\,a_1^4\right], \tag{9.115}$$

$$\Omega_2(t) = \omega_2 + \varepsilon\frac{3\beta}{8\omega_2}\,a_2^2 - \varepsilon^2\left[\frac{\mu^2}{2\omega_2} + \frac{21\beta^2}{256\,\omega_2^3}\,a_2^4\right]. \tag{9.116}$$

For the pendulum, $\beta < 0$, so the $O(\varepsilon)$ detuning is Softening nonlinearity (frequency decreases with amplitude), while the $O(\varepsilon^2)$ term proportional to β^2 is always negative.

Second–order waveform correction (dominant third harmonic). At the same order, the waveform acquires a weak third harmonic. A convenient second–order approximation is

$$u(t) \approx A(T)\left[\cos\Theta_1(t) + \varepsilon\,\frac{\beta}{32\omega_1^2}\,a_1^2\,\cos(3\Theta_1(t))\right], \tag{9.117}$$

$$v(t) \approx B(T)\left[\cos\Theta_2(t) + \varepsilon\,\frac{\beta}{32\omega_2^2}\,a_2^2\,\cos(3\Theta_2(t))\right], \tag{9.118}$$

with $A(T), B(T)$ given by (9.109) and

$$\Theta_1(t) = \int_0^t \Omega_1(s)\,ds + \phi_{10}, \qquad \Theta_2(t) = \int_0^t \Omega_2(s)\,ds + \phi_{20},$$

where Ω_1, Ω_2 are now taken from (9.115)–(9.116). The physical angles then follow from (9.113).

9.11.4 Particular solutions: in–phase and out–of–phase modes

Particular solutions Invariant subspaces

The KBM solution above describes the generic motion in the non–resonant regime. Two important particular cases are obtained by restricting the dynamics to invariant symmetry subspaces.

9.11.4.1 In–phase mode ($v = 0$)

In–phase mode Duffing equation

Setting $v(t) \equiv 0$ in (9.105) yields the scalar equation

$$\ddot{u} + 2\varepsilon\mu\,\dot{u} + \omega_0^2 u + \varepsilon\beta u^3 = 0, \tag{9.119}$$

which is precisely the Duffing equation (softening for $\beta < 0$). The first–order KBM approximation is obtained from the general formulas by taking $B \equiv 0$, hence $a_1^2 = A^2$:

$$A'(T) = -\mu A, \qquad \phi_1'(T) = \frac{3\beta}{8\omega_0}A^2, \tag{9.120}$$

and

$$\Omega_1(t) = \omega_0 + \varepsilon\frac{3\beta}{8\omega_0}A^2(t), \qquad A(t) = A_0 e^{-\mu\varepsilon t}. \tag{9.121}$$

At second order,

$$\Omega_1(t) = \omega_0 + \varepsilon\frac{3\beta}{8\omega_0}A^2(t) - \varepsilon^2\left[\frac{\mu^2}{2\omega_0} + \frac{21\beta^2}{256\,\omega_0^3}A^4(t)\right], \tag{9.122}$$

and the waveform correction reduces from (9.117) to

$$u(t) \approx A(t)\left[\cos\Theta_1(t) + \varepsilon\,\frac{\beta}{32\omega_0^2}\,A^2(t)\cos(3\Theta_1(t))\right], \tag{9.123}$$

with $\theta_1(t) = \theta_2(t) = u(t)$.

9.11.4.2 Out–of–phase mode ($u \equiv 0$)

Out–of–phase mode Duffing equation Frequency stiffening

Setting $u(t) \equiv 0$ in (9.105) yields

$$\ddot{v} + 2\varepsilon\mu\,\dot{v} + \omega_2^2 v + \varepsilon\beta v^3 = 0, \qquad \omega_2 = \sqrt{\omega_0^2 + 2\kappa}, \tag{9.124}$$

which is again a Duffing equation, now with a stiffened linear frequency because the coupling spring is maximally deformed. From the general KBM formulas with $A \equiv 0$ (so $a_2^2 = B^2$) we obtain

$$B'(T) = -\mu B, \qquad \phi_2'(T) = \frac{3\beta}{8\omega_2}B^2, \tag{9.125}$$

and hence

$$\Omega_2(t) = \omega_2 + \varepsilon\frac{3\beta}{8\omega_2}B^2(t), \qquad B(t) = B_0 e^{-\mu\varepsilon t}. \tag{9.126}$$

At second order,

$$\Omega_2(t) = \omega_2 + \varepsilon\frac{3\beta}{8\omega_2}B^2(t) - \varepsilon^2\left[\frac{\mu^2}{2\omega_2} + \frac{21\beta^2}{256\,\omega_2^3}B^4(t)\right], \tag{9.127}$$

and the waveform correction becomes

$$v(t) \approx B(t)\left[\cos\Theta_2(t) + \varepsilon\,\frac{\beta}{32\omega_2^2}\,B^2(t)\cos(3\Theta_2(t))\right], \tag{9.128}$$

with $\theta_1(t) = v(t)$ and $\theta_2(t) = -v(t)$.

9.11.5 Discussion

Duffing oscillator

The coupled pendula thus admit two symmetry–selected motions that reduce exactly to scalar Duffing equations: the in–phase mode (frequency ω_0) and the out–of–phase mode (frequency $\omega_2 > \omega_0$). The KBM method provides explicit first– and second–order approximations for (i) the exponential decay of amplitudes due to damping and (ii) the amplitude–dependent frequency detuning produced by the cubic (nonlinear pendulum) restoring torque. These reduced Duffing models and their perturbation solutions will serve as reference dynamics when we later study more general couplings, forcing, internal resonance, and strongly nonlinear regimes.

9.12 Study of dissipative nonplanar solitons in an electronegative complex plasma

Let us consider the equation

$$\psi_t + a\psi\psi_x + b\psi_{xxx} + \left(\frac{m}{2t} + \varepsilon\right)\psi = 0,\ t \geq T_{\text{init}}.$$

This equation is obtained from fluid equations of present plasma model using the reductive perturbation method (RPM). This equation may also be solved (approximately) as follows : let

$$\psi = \psi(x,t) = f(t)\mathrm{sech}^2(kx - \lambda(t)),$$

where $f(t)$ and $\lambda(t)$ are some functions to be determined. We have:

$$\begin{aligned}&\psi_t + a\psi\psi_x + b\psi_{xxx} + \left(\tfrac{m}{2t} + \varepsilon\right)\psi = \\ &\tfrac{1}{2t}[(m + 2\varepsilon t)f + 2t\dot{f} + \\ &\left(ft\left(-afk + 8bk^3 + \dot{\lambda}\right)\right)\tanh(kx - \lambda(t)) + ...]\end{aligned}$$

A suitable choice is

$$\begin{aligned}f(t) &= At^{-m/2}T_{\text{init}}^{m/2}\exp\left(\varepsilon\left(T_{\text{init}} - t\right)\right)\\ \lambda(t) &= -aAkt^{1-\frac{m}{2}}T_{\text{init}}^{m/2}\exp(\varepsilon T_{\text{init}})E_{\frac{m}{2}}(t\varepsilon) + aAkT_{\text{init}}\exp(\varepsilon T_{\text{init}})E_{\frac{m}{2}}(\varepsilon T_{\text{init}}) + 8bk^3\left(T_{\text{init}} - t\right),\end{aligned}$$

where $E_{m/2}(t)$ stands for the expintegral function.

9.12.1 Damped Cubic-Quintic Equation

This is the equation

$$\ddot{x} + 2\varepsilon\dot{x} + px + qx^3 + rx^5 = 0,\ x(0) = x_0 \text{ and } x'(0) = \dot{x}_0$$

This equation admits exact solution for $p = 3/4\varepsilon^2$ and $q = 0$. In general, we may obtain approximate analytical solution to it using the anlaytical methods described in previous chapters.

9.12.1.1 Integrability of the damped cubic–quintic Duffing oscillator

We consider the damped cubic–quintic Duffing oscillator

$$\ddot{x} + 2\varepsilon\dot{x} + p\,x + q\,x^3 + r\,x^5 = 0, \qquad x(0) = x_0, \quad \dot{x}(0) = \dot{x}_0, \tag{9.129}$$

where $\varepsilon > 0$ is the damping coefficient and $p, q, r \in \mathbb{R}$.

In general, equation (9.129) is nonintegrable and admits no conserved energy due to dissipation. Nevertheless, a remarkable and highly nontrivial exception exists.

Theorem 9.1 (Unique integrable damped cubic–quintic Duffing case). *Assume $r \neq 0$ and $\varepsilon > 0$. Equation* (9.129) *admits an explicit reduction to a conservative autonomous oscillator, and hence a closed–form solution in terms of Jacobi elliptic functions, if and only if*

$$q = 0, \qquad p = \frac{3}{4}\varepsilon^2. \tag{9.130}$$

Proof. We seek a reduction of the form

$$x(t) = f(t)\,y(\tau), \qquad \tau = g(t), \tag{9.131}$$

with f and g smooth functions to be determined. Substituting (9.131) into (9.129) and collecting terms in y, y', and y'', one finds that the equation reduces to an autonomous conservative equation for y if and only if:

- the first–derivative term y' vanishes identically,
- the coefficients of y, y^3, and y^5 are constant,
- no explicit time dependence remains.

These requirements uniquely determine

$$f(t) = e^{-\varepsilon t/2}, \qquad g(t) = \frac{1 - e^{-\varepsilon t}}{\varepsilon}, \tag{9.132}$$

and enforce the parameter constraints (9.130). The reduced equation becomes

$$y'' + r\,y^5 = 0, \tag{9.133}$$

which is a conservative pure quintic oscillator. □

Equation (9.133) possesses the first integral

$$\mathcal{E} = \frac{1}{2}(y')^2 + \frac{r}{6}y^6, \tag{9.134}$$

so its solution is obtained by quadrature and admits an explicit representation in terms of Jacobi elliptic functions. Consequently, the solution of the original damped problem is

$$x(t) = e^{-\varepsilon t/2}\, y\left(\frac{1 - e^{-\varepsilon t}}{\varepsilon}\right), \tag{9.135}$$

where $y(\tau)$ solves (9.133) with initial conditions

$$y(0) = x_0, \qquad y'(0) = x_0^2 + \frac{\dot{x}_0}{\varepsilon}.$$

Conceptual meaning. This result shows that damping does not necessarily destroy integrability. Instead, in the unique parameter regime (9.130), damping induces a nonlinear time reparametrization and an exponential amplitude envelope, while the internal dynamics remains governed by a conservative quintic oscillator.

Why Duffing–type models matter. The damped cubic–quintic Duffing equation thus serves as a canonical analytical model that:

- embeds dissipation into a conservative core,
- admits exact elliptic solutions,
- provides a benchmark for perturbation methods and numerical validation,
- clarifies the boundary between integrability and chaos.

This explains why Duffing and Duffing–Helmholtz oscillators occupy a central position in the analytical theory of nonlinear dynamics, far beyond their role as isolated solvable examples.

Let us solve it using an elementary approach. Assume

$$x(t) = \exp(-\rho t)[x_0 \cos\omega(t) + c\sin\omega(t)].$$

Then

$\ddot{x} + 2\varepsilon\dot{x} + px + qx^3 + rx^5 =$
$\frac{1}{8}e^{-5\rho t}(8e^{4\rho t}\left(-x_0\ddot{\omega} + c\left(-2\varepsilon\rho + p + \rho^2\right) - c\dot{\omega}^2 + 2x_0\dot{\omega}(\rho - \varepsilon)\right) +$
$6cq\left(c^2 + x_0^2\right)e^{2\rho t} + 5cr\left(c^2 + x_0^2\right)^2)\sin(\omega(t))$
$+\frac{1}{8}e^{-5\rho t}(8e^{4\rho t}\left(c\ddot{\omega} + 2c\dot{\omega}(\varepsilon - \rho) + x_0\left(-2\varepsilon\rho + p + \rho^2\right) - x_0\dot{\omega}^2\right) +$
$6qx_0\left(c^2 + x_0^2\right)e^{2\rho t} + 5rx_0\left(c^2 + x_0^2\right)^2)\cos(\omega(t)) + \cdots$

Equating to zero the coefficients of $\cos\omega(t)$ and $\sin\omega(t)$ in the last expression and eliminating $\ddot{\omega}$ from the resulting system we obtain the following ode for determining $\omega = \omega(t)$:

$$\dot{\omega} = \omega'(t) = \sqrt{p - 2\varepsilon\rho + \rho^2 + \frac{3}{4}q\left(c^2 + x_0^2\right)e^{-2\rho t} + \frac{5}{8}r\left(c^2 + x_0^2\right)^2 e^{-4\rho t}},\ \omega(0) = 0.$$

The number c is determined from the initial condition $x'(0) = \dot{x}_0$ and the number ρ is chosen in order to minimize the residual error

$$E(\rho) = \max_{0\le t\le T}\left|\ddot{x} + 2\varepsilon\dot{x} + px + qx^3 + rx^5\right|.$$

Example 39. Let

$$\ddot{x} + 0.2\dot{x} + x + 10x^3 + x^5 = 0,\ x(0) = 1.5954 \text{ and } x'(0) = 0,\ 0 \le t \le 25$$

In this case, $c = 0.023$, $\rho = 0.044687$,

and

$\omega(t) = \frac{1}{20952}[-36\sqrt{970}e^{-3t/20}\sqrt{296335e^{3t/20} + 15360e^{3t/10} + 62856}-$
$5760\sqrt{582}\log(-64\sqrt{15}\sqrt{296335e^{3t/20} + 15360e^{3t/10} + 62856} + 30720e^{3t/20} + 296335)+$
$\sqrt{5}(296335\log((3492 - 32\sqrt{2910} + \sqrt{72662894})$
$(\sqrt{194}\sqrt{296335e^{3t/20} + 15360e^{3t/10} + 62856} - 32\sqrt{2910}e^{3t/20} - 3492))$
$-296335\log((-3492 - 32\sqrt{2910} + \sqrt{72662894})$
$(\sqrt{194}\sqrt{296335e^{3t/20} + 15360e^{3t/10} + 62856} - 32\sqrt{2910}e^{3t/20} + 3492))$

$+36\sqrt{72662894}) + 5760\sqrt{582}\log(327055 - 64\sqrt{5618265})]$.

See Figure 9.17.

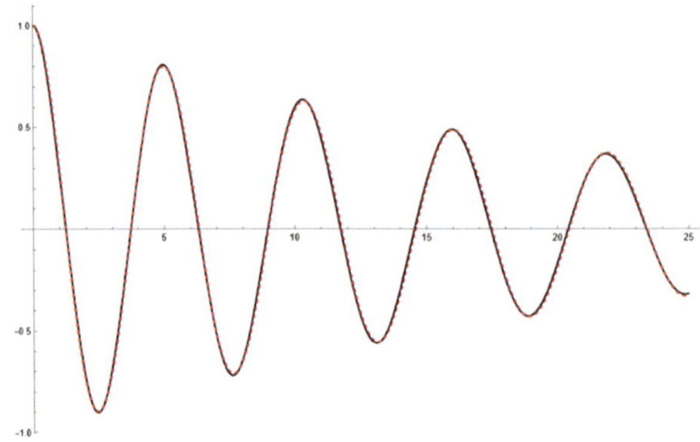

Figure 9.17: Comparison between the trigonometric and the numerical solution

This solution is more accurate than that in [44]. See Figure below.

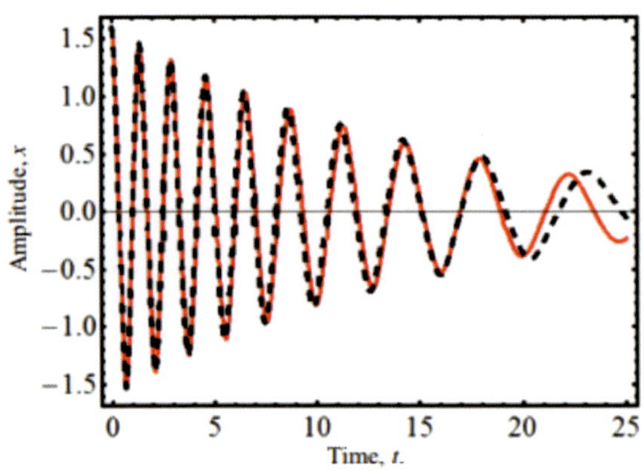

Figure 9.18: The solutio in [44]

A more general poblem is given by

$$\ddot{x} + 2\varepsilon\dot{x} + px + qx^3 + rx^5 = \gamma\cos(\Omega t),\ x(0) = x_0 \text{ and } x'(0) = \dot{x}_0\ , \tag{9.136}$$

This oscillator arises in chaos theory. Assume that

$$\varepsilon > 0,\ p = \omega_0^2 > 0,\ r > 0 \text{ and } q^2 - 4pr < 0.$$

In the limit case when $q = r = 0$ and $\omega_0^2 - \varepsilon^2 \neq 0$, the exact solution is given by

$x(t) =$

$\frac{1}{\sqrt{\omega_0^2-\varepsilon^2}(4\varepsilon^2\Omega^2+(\Omega^2-\omega_0^2)^2)}[((\varepsilon x_0 + \dot{x}_0)(4\varepsilon^2\Omega^2 + (\Omega^2 - \omega_0^2)^2) - \gamma\varepsilon(\omega_0^2 + \Omega^2))\sin(\sqrt{\omega_0^2 - \varepsilon^2}t)+$

$\sqrt{\omega_0^2 - \varepsilon^2}(\gamma(\Omega - \omega_0)(\omega_0 + \Omega) + x_0(4\varepsilon^2\Omega^2 + (\Omega^2 - \omega_0^2)^2))\cos(\sqrt{\omega_0^2 - \varepsilon^2}t)]\exp(-\varepsilon t)+$

$\frac{2\gamma\varepsilon\Omega}{4\varepsilon^2\Omega^2+(\Omega^2-\omega_0^2)^2}\sin(\Omega t) + \frac{\gamma(\omega_0^2-\Omega^2)}{4\varepsilon^2\Omega^2+(\Omega^2-\omega_0^2)^2}\cos(\Omega t)$.

This suggests to assume the solution to (9.136) in the ansatz form

$$x(t) = \exp(-\rho t)[d_0 \cos\omega(t) + d_1 \sin\omega(t)] + \sum_{j=1}^{N}[\mu_j \cos(j\Omega t) + \nu_j \sin(j\Omega t)],$$

where

$$\omega'(t) = \sigma_0 \left[1 + \sum_{k=1}^{M} \sigma_k \exp(-k\rho t)\right], \; \omega(0) = 0,$$

for some suitable constants d_0, d_1, ρ, μ_j, ν_j and σ_k, $M \geq 1$, $N \geq 1$. For example, let

$$x(t) = \exp(-\rho t)[d_0 \cos\omega(t) + d_1 \sin\omega(t)] + \mu_1 \cos(\Omega t) + \nu_1 \sin(\Omega t) + \mu_2 \cos(2\Omega t) + \nu_2 \sin(2\Omega t),$$

where

$$\omega'(t) = \sigma_0[1 + \sigma_1 \exp(-2\rho t) + \sigma_2 \exp(-4\rho t)].$$

Example 40. Let

$$\ddot{x} + 0.1\dot{x} + x + x^3 + 5x^5 = 0.15\cos(0.25t), \; x(0) = 0 \text{ and } x'(0) = 0, \; 0 \leq t \leq 100. \qquad (9.137)$$

We first solve this problem numerically and then we minimize the distance between the numerical solution and the analytical solution. The optimal values are :

$\rho = 0.1, d_0 = -0.1557, d_1 = -0.0171591, \mu_0 = -0.000198991,$

$\mu_1 = 0.155756, \mu_2 - 0.000142642,$

$\nu_1 = 0.00793901, \nu_2 = -0.000377225,$

$\sigma_0 = 1.00906, \sigma_1 = 0.0525593, \sigma_2 = -0.0495748,$

so that

$x(t) = 0.00793875\sin(0.25t) - 0.000378723\sin(0.5t) + 0.155752\cos(0.25t) + 0.000138569\cos(0.5t) + e^{-0.1t}[-0.0171873\sin(1.00867t + 0.263649e^{-0.2t} - 0.555278e^{-0.1t} + 0.291629) - 0.155891\cos 1.00867t + 0.263649e^{-0.2t} - 0.555278e^{-0.1t} + 0.291629)].$

with max error $E = 0.0047$. See Figure 9.20.

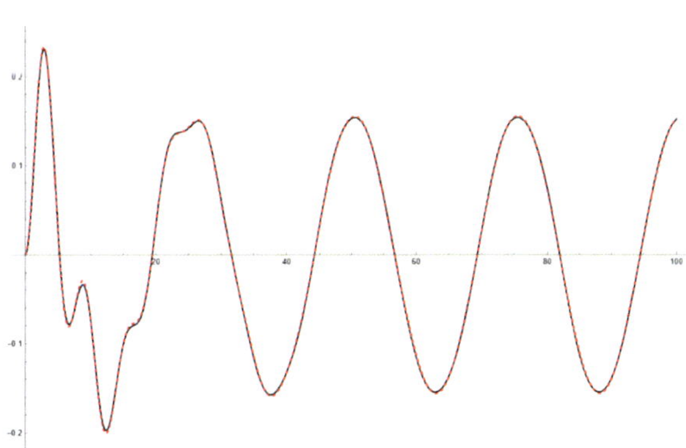

Figure 9.19

Assume now that

$$\lim_{t\to\infty} x(t) = \kappa,$$

so that

$$p\kappa + q\kappa^3 + r\kappa^5 = 0.$$

Consider the i.v.p.

$$\ddot{y} + 2\varepsilon\dot{y} + py + qy^3 + ry^5 = 0,\ y(0) = y_0 \text{ and } y'(0) = \dot{y}_0$$

The solution is assumed as

$$y(t) = \kappa + z(t),$$

where

$$z''(t)+2\varepsilon z'(t)+\left(p+3\kappa^2 q+5\kappa^4 r\right)z(t)+\left(3\kappa q+10\kappa^3 r\right)z(t)^2+\left(q+10\kappa^2 r\right)z(t)^3+5\kappa r z(t)^4+r z(t)^5=0,$$

$$z(0) = z_0 := y_0 - \kappa,\ z'(0) = \dot{z}_0 := \dot{y}_0.$$

Assume that

$$z(t) = \exp(-\rho t)[z_0 \cos\omega(t) + d \sin\omega(t)],$$

The function $\omega = \omega(t)$ is determined from the ode

$\omega'(t) = \sqrt{p - 2\varepsilon\rho + \rho^2 + 3\kappa^2 q + 5\kappa^4 r + \frac{3}{4}\left(d^2+z_0^2\right)\left(q+10\kappa^2 r\right)\exp(-2\rho t) + \frac{5}{8} r\left(d^2+z_0^2\right)^2 \exp(-4\rho}$
$\omega(0) = 0.$

This ode may be solved in closed from. However, the resulting expression is difficult to handle. A practical formula is

$$\omega(t) = \mu[-\frac{4\left(2184\sqrt{A+B+1}-7917\sqrt{2A+B+4}+68\sqrt{273}\sqrt{84A+26B+273}-4914\right)}{18207\rho}\left(e^{-2\rho t}-1\right)^3$$
$$+\frac{\left(-62622\sqrt{A+B+1}+131418\sqrt{2A+B+4}-816\sqrt{273}\sqrt{84A+26B+273}+22554\right)}{72828\rho}\left(e^{-2\rho t}-1\right)^2$$
$$+\frac{36400-36387\sqrt{A+B+1}}{72828\rho}\left(e^{-2\rho t}-1\right)^1+\frac{27\sqrt{A+B+1}+72814}{72828}t],$$

where

$A = \frac{3\left(d^2+z_0^2\right)\left(q+10\kappa^2 r\right)}{4(-2\varepsilon\rho+p+\rho^2+3\kappa^2 q+5\kappa^4 r)}$, $B = \frac{5r\left(d^2+z_0^2\right)^2}{8(-2\varepsilon\rho+p+\rho^2+3\kappa^2 q+5\kappa^4 r)}$, and $\mu = \sqrt{p - 2\varepsilon\rho + \rho^2 + 3\kappa^2 q + 5\kappa^4 r}$.

The constant d is found from the condition $z'(0) = \dot{z}_0$.This condition gives a sextic for d .

We first solve the problem numerically and then we minimize the distance between the numerical solution and the analytical solution $z(t)$ for the constants d and ρ.

Example 41. Let

$$\ddot{y} + 0.2\dot{y} - 3y - 2y^3 + y^5 = 0,\ y(0) = 1.7531 \text{ and } y'(0) = 0,\ 0 \le t \le 20.$$

The solution reads ($\kappa = \sqrt{3}$, $d = 0.000422724$, $\rho = 0.0986 < \varepsilon = 0.1$)

$y(t) = \sqrt{3} + e^{-0.0985793t} 0.0210035 \cos w(t) - 0.000422724 \sin\left(w(t)\right),$

where

$w(t) = -4.89883t - 0.000044275e^{-0.591476t} + 0.0000994185e^{-0.394317t} - 0.00414031e^{-0.197159t} + 0.00408516,$

with max error= 0.000544669. See Figure 9.20.

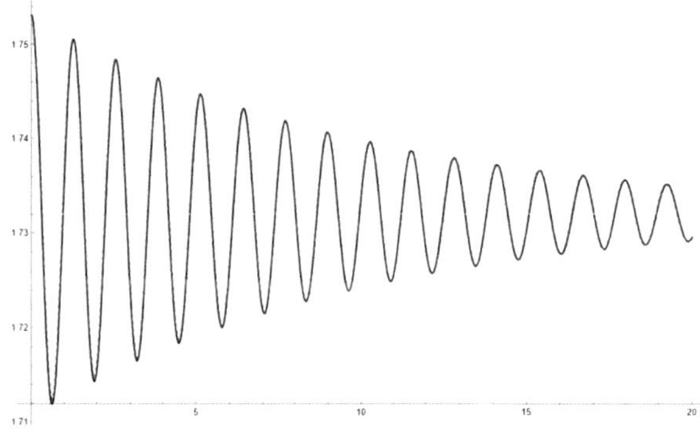

Figure 9.20

Finally, let

$$\ddot{y} + 2\varepsilon\dot{y} + py + qy^3 + ry^5 = f(t),\ y(0) = y_0 \text{ and } y'(0) = \dot{y}_0$$

$$y(t) = \kappa + x(t),$$

so that

$$x''(t)+2\varepsilon x'(t)+\left(p + 3\kappa^2 q + 5\kappa^4 r\right)x(t)+\left(3\kappa q + 10\kappa^3 r\right)x(t)^2+\left(q + 10\kappa^2 r\right)x(t)^3+5\kappa r x(t)^4+r x(t)^5 = f(t),$$

$$x(0) = y_0 - \kappa,\ x'(0) = \dot{y}_0.$$

$$p\kappa + q\kappa^3 + r\kappa^5 = 0.$$

Define

$$\omega_0 = \sqrt{p + 3\kappa^2 q + 5\kappa^4 r},\ s_2 = 3\kappa q + 10\kappa^3 r,\ s_3 = q + 10\kappa^2 r,\ s_4 = 5\kappa r \text{ and } s_5 = r.$$

We may solve the oscillator

$$\ddot{x} + 2\varepsilon\dot{x} + \omega_0^2 x + s_2 x^2 + s_3 x^3 + s_4 x^4 + s_5 x^5 = f(t)$$

by means of a perturbative method, say KBM. To this end, we consider the homotopy

$$H_\mu = \ddot{x} + \omega_0^2 x + \mu[2\varepsilon\dot{x} + s_2 x^2 + s_3 x^3 + s_4 x^4 + s_5 x^5 - f(t)].$$

$$x(t) = a(t)\cos\psi(t) + \mu v(t, a(t), \psi(t)),$$

so that

$$a'(t) = \mu A(a(t)) \text{ and } \psi'(t) = \omega_0 + \mu\varphi(a(t))$$

Then

$A(a) = -\varepsilon a,\ \varphi(a) = \frac{1}{16\omega_0}[5a^4 s_5 + 6a^2 s_3].$

$a(t) = c_0 \exp(-\varepsilon t)$,

$\psi(t) = \frac{1}{64\varepsilon\omega_0}[5c_0^4 s_5 (1 - e^{-4\varepsilon t}) + 12c_0^2 s_3 e^{-2\varepsilon t} (e^{2\varepsilon t} - 1) + 64\varepsilon\omega_0^2 t] + c_1$,

and

$v(t, a, \psi) = \frac{s_5(15\cos(3\psi)+\cos(5\psi))}{384\omega_0^2} a^5 + \frac{s_4(20\cos(2\psi)+\cos(4\psi)-45)}{120\omega_0^2} a^4 + \frac{s_3\cos(3\psi)}{32\omega_0^2} a^3 + \frac{s_2(\cos(2\psi)-3)}{6\omega_0^2} a^2 + \frac{f(t)}{\omega_0^2}$.

$x(t) = a(t)\cos\psi(t) + v(t, a(t), \psi(t))$.

The constants $c_0$and c_1 are determined from the initial conditions.

Example 42. Let

$$\ddot{y} + 2\varepsilon\dot{y} - 3y + y^3 + 0.577y^5 = 0.1\cos(0.2t),\ y(0) = 1.22442 \text{ and } y'(0) = 0.$$

$\psi = 14.8738t - 6.51875489933372$*^-10$e^{-0.4t} - 0.000930782e^{-0.2t} - 0.165016$

$c_0 = 0.00542289,\ c_1 = -0.0331894, \kappa = 1.25514, \omega_0 = 2.97477.$

See Figure 9.21.

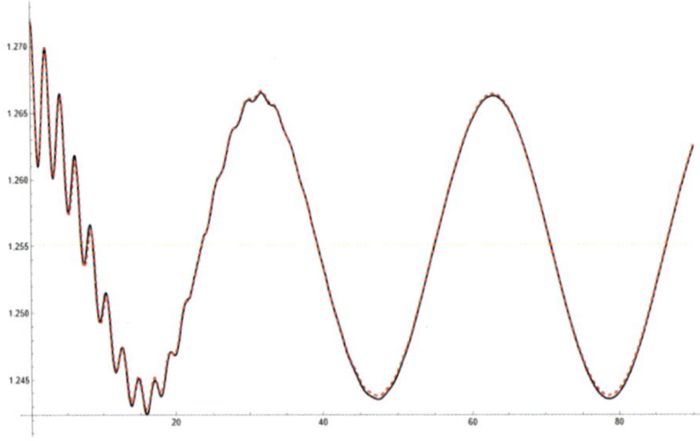

Figure 9.21

9.12.2 The Cubic-Quintic Duffing equation and its Applications to plasma physics and rigid rotator

The Cubic Quintic Duffing equation

$$\ddot{x} + px + qx^3 + rx^5 = 0,\ x(0) = x_0 \text{ and } x'(0) = \dot{x}_0$$

has applications in the fluid multi-ion plasma equations and a generalized pendulum problem. For the generalized pendulum problem (rigid rotator), the governing motion is directly reduced to the cubic–quintic Duffing oscillator.

9.12.3 Damped Nonlinear Schrodinger equation

This is the equation

$$i\partial_t\psi + \frac{1}{2}P\partial_{xx}\psi + Q\left|\psi\right|^2\psi + i\varepsilon\psi = 0.$$

The dissipative and nondissipative modulated pulses such as rogue waves (RWs) and breathers (Kuznetsov-Ma breathers and Akhmediev breathers) can exist and propagate in several fields of sciences, for example, plasma physics, are analyzed numerically.

9.12.4 Traveling wave solutions

Let us consider the cubic-quintic nonlinear Schrodinger equation

$$i\partial_t\psi + \frac{1}{2}P\partial_{xx}\psi + Q\left|\psi\right|^2\psi + R\left|\psi\right|^4\psi = 0.$$

Define

$$\psi = \exp(i\eta)y(\xi), \text{ where } \eta = kx - wt + \eta_0 \text{ and } \xi = x - kP + \xi_0$$

Then, the function $y = y(\xi)$ must be a solution to the following cubic-quintic Duffing equation :

$$y''(\xi) + \frac{2w - k^2P}{P}y(\xi) + \frac{2Q}{P}y(\xi)^3 + \frac{2R}{P}y(\xi)^5 = 0.$$

A solitons solution is

$$y(\xi) = \frac{A\cosh(\xi)}{\sqrt{1 + \mu\cosh^2(\xi)}}$$

for the choices

$$w = \frac{16\left(k^2+1\right)P^2R^2+4\left(k^2+2\right)PQ^2R+Q\sqrt{Q^2-8PR}\left(4PR+Q^2\right)+Q^4}{8R(4PR+Q^2)},$$

$$\mu = -\frac{32P^2R^2+12PQ^2R+Q\sqrt{Q^2-8PR}\left(4PR+Q^2\right)+Q^4}{12PR(4PR+Q^2)},$$

$$A = \frac{\sqrt{\sqrt{Q^2-8PR}(4PR+Q^2)+Q^3}}{2\sqrt{6}\sqrt{PR}}.$$

9.12.5 Cnoidal Wave Solutions

The cnoidal wave solutions have the form

$$y(\xi) = \frac{A\mathrm{cn}(\xi)}{\sqrt{1 + \mu\mathrm{cn}^2(\xi) + \kappa\mathrm{cn}^4(\xi)}},$$

where

$$\omega = \frac{A^2Q}{P(-3\mu^2-2\mu+3\mu^2m+4\mu m-4m)},$$

$$w = \frac{-3A^2\mu mQ-2A^2mQ+3A^2\mu Q+A^2Q+3k^2\mu^2mP+4k^2\mu mP-4k^2mP-3k^2\mu^2P-2k^2\mu P}{2(-3\mu^2-2\mu+3\mu^2m+4\mu m-4m)},$$

$$\kappa = \frac{m}{m-1},$$

and the number μ is a slolution to the cubic
$\mu^3 + \frac{2A^2mR-2A^2R+2mQ-Q}{(m-1)Q}\mu^2 - \frac{4\left(-2A^2mR+A^2R+3mQ\right)}{3(m-1)Q}\mu - \frac{4m\left(2A^2mR-2A^2R+6mQ-3Q\right)}{3(m-1)^2Q} = 0,\ m(1-m) \neq 0.$

Another possible choice is

$$\kappa = 0,$$

$$A = \sqrt{\frac{3\mu(\mu+1)Q(-\mu+\mu m+m)}{2R(3\mu^2+2\mu-3\mu^2m-4\mu m-m)}},$$

$w = \frac{1}{4R(-3\mu^2-2\mu+3\mu^2m+4\mu m+m)^2}[18k^2\mu^4m^2PR+48k^2\mu^3m^2PR+44k^2\mu^2m^2PR+16k^2\mu m^2PR+2k^2m^2PR-36k^2\mu^4mPR-$
$72k^2\mu^3mPR-44k^2\mu^2mPR-8k^2\mu mPR+18k^2\mu^4PR+24k^2\mu^3PR+8k^2\mu^2PR+9\mu^4m^2Q^2+24\mu^3m^2Q^2+21\mu^2m^2Q^2+$
$6\mu m^2Q^2-18\mu^4mQ^2-36\mu^3mQ^2-21\mu^2mQ^2-3\mu mQ^2+9\mu^4Q^2+12\mu^3Q^2+3\mu^2Q^2],$

and

$\omega = -\frac{3\mu(\mu+1)Q^2(-\mu+\mu m+m)}{2PR(-3\mu^2-2\mu+3\mu^2m+4\mu m+m)^2}.$

9.12.6 Breathers

In the case when $R=0$ the NLSE

$$iv_t+\frac{P}{2}\,v_{xx}+Q\,|v|^2\,v=0.$$

admits the following solutions called breathers :

$$v(x,t)=\sqrt{\frac{P}{Q}}e^{iPt}\left(1+\frac{(2-4\delta)\cosh\left(2\sqrt{2-4\delta}\sqrt{\delta}Pt\right)+2i\sqrt{2-4\delta}\sqrt{\delta}\sinh\left(2\sqrt{2-4\delta}\sqrt{\delta}Pt\right)}{\sqrt{2}\sqrt{\delta}\cos\left(2\sqrt{1-2\delta x}\right)-\cosh\left(2\sqrt{2-4\delta}\sqrt{\delta}Pt\right)}\right).$$

If $\delta=1/2$ then

$$v(x,t)=\frac{\sqrt{P}\left(\left(4P^2t^2+4x^2-3\right)\cos(Pt)+8Pt\sin(Pt)\right)}{\sqrt{Q}\left(4P^2t^2+4x^2+1\right)}+i\frac{\sqrt{P}\left(\left(4P^2t^2+4x^2-3\right)\sin(Pt)-8Pt\cos(P\right.}{\sqrt{Q}\left(4P^2t^2+4x^2+1\right)}$$

$v(x,t)=\frac{\sqrt{-\frac{P}{Q}}\left(\left(4P^2t^2+4x^2-3\right)\cos(Pt)+8Pt\sin(Pt)\right)}{4P^2t^2+4x^2+1}+i\frac{\sqrt{-\frac{P}{Q}}\left(-\left(-4P^2t^2-4x^2+3\right)\sin(Pt)-8Pt\cos(Pt)\right)}{4P^2t^2+4x^2+1},$
$PQ<0.$
$v(x,t)=\frac{\sqrt{\frac{P}{Q}}\left(\left(4P^2t^2+4x^2-3\right)\cos(Pt)+8Pt\sin(Pt)\right)}{4P^2t^2+4x^2+1}+i\frac{\sqrt{\frac{P}{Q}}\left(\left(-4P^2t^2-4x^2+3\right)\sin(Pt)+8Pt\cos(Pt)\right)}{4P^2t^2+4x^2+1},$
$PQ>0.$
$v(x,t)=\frac{1}{\cos\left(2\sqrt{2}\sqrt{\delta(2\delta-1)}Pt\right)-\sqrt{2}\sqrt{\delta}\cosh\left(2\sqrt{2\delta-1}x\right)}[(4\delta-1)\left(-\sqrt{-\frac{P}{Q}}\right)\sin(Pt)\cos\left(2\sqrt{2}\sqrt{\delta(2\delta-1)}Pt\right)$
$-2\sqrt{2}\sqrt{-\frac{\delta(2\delta-1)P}{Q}}\cos(Pt)\sin\left(2\sqrt{2}\sqrt{\delta(2\delta-1)}Pt\right)+\sqrt{2}\sqrt{-\frac{\delta P}{Q}}\sin(Pt)\cosh\left(2\sqrt{2\delta-1}x\right)]+$
$i\frac{1}{\cos\left(2\sqrt{2}\sqrt{\delta(2\delta-1)}Pt\right)-\sqrt{2}\sqrt{\delta}\cosh\left(2\sqrt{2\delta-1}x\right)}[-2\sqrt{2}\sqrt{-\frac{\delta(2\delta-1)P}{Q}}\sin(Pt)\sin\left(2\sqrt{2}\sqrt{\delta(2\delta-1)}Pt\right)$
$+(4\delta-1)\sqrt{-\frac{P}{Q}}\cos(Pt)\cos\left(2\sqrt{2}\sqrt{\delta(2\delta-1)}Pt\right)-\sqrt{2}\sqrt{-\frac{\delta P}{Q}}\cos(Pt)\cosh\left(2\sqrt{2\delta-1}x\right)],$
$PQ<0,\delta>\frac{1}{2}$
$v(x,t)=\frac{1}{\cos\left(2\sqrt{2}\sqrt{\delta(2\delta-1)}Pt\right)-\sqrt{2}\sqrt{\delta}\cosh\left(2\sqrt{2\delta-1}x\right)}[(4\delta-3)\left(-\sqrt{-\frac{P}{Q}}\right)\sin(Pt)\cos\left(2\sqrt{2}\sqrt{\delta(2\delta-1)}Pt\right)$
$-2\sqrt{2}\sqrt{-\frac{\delta(2\delta-1)P}{Q}}\cos(Pt)\sin\left(2\sqrt{2}\sqrt{\delta(2\delta-1)}Pt\right)+\sqrt{2}\sqrt{-\frac{\delta P}{Q}}\sin(Pt)\cosh\left(2\sqrt{2\delta-1}x\right)]+$
$i\frac{1}{\cos\left(2\sqrt{2}\sqrt{\delta(2\delta-1)}Pt\right)-\sqrt{2}\sqrt{\delta}\cosh\left(2\sqrt{2\delta-1}x\right)}[-2\sqrt{2}\sqrt{-\frac{\delta(2\delta-1)P}{Q}}\sin(Pt)\sin\left(2\sqrt{2}\sqrt{\delta(2\delta-1)}Pt\right)$
$+(4\delta-3)\sqrt{-\frac{P}{Q}}\cos(Pt)\cos\left(2\sqrt{2}\sqrt{\delta(2\delta-1)}Pt\right)-\sqrt{2}\sqrt{-\frac{\delta P}{Q}}\cos(Pt)\cosh\left(2\sqrt{2\delta-1}x\right)],$
$PQ<0,0<\delta<\frac{1}{2}$
$v(x,t)=\frac{1}{\cos\left(2\sqrt{2}\sqrt{\delta(2\delta-1)}Pt\right)-\sqrt{2}\sqrt{\delta}\cosh\left(2\sqrt{2\delta-1}x\right)}[-2\sqrt{2}\sqrt{\frac{\delta(2\delta-1)P}{Q}}\sin(Pt)\sin\left(2\sqrt{2}\sqrt{\delta(2\delta-1)}Pt\right)$

$+(4\delta-1)\sqrt{\frac{P}{Q}}\cos(Pt)\cos\left(2\sqrt{2}\sqrt{\delta(2\delta-1)}Pt\right)-\sqrt{2}\sqrt{\frac{\delta P}{Q}}\cos(Pt)\cosh\left(2\sqrt{2\delta-1}x\right)]+$
$i\frac{1}{\cos\left(2\sqrt{2}\sqrt{\delta(2\delta-1)}Pt\right)-\sqrt{2}\sqrt{\delta}\cosh\left(2\sqrt{2\delta-1}x\right)}[(4\delta-1)\sqrt{\frac{P}{Q}}\sin(Pt)\cos\left(2\sqrt{2}\sqrt{\delta(2\delta-1)}Pt\right)$
$+2\sqrt{2}\sqrt{\frac{\delta(2\delta-1)P}{Q}}\cos(Pt)\sin\left(2\sqrt{2}\sqrt{\delta(2\delta-1)}Pt\right)-\sqrt{2}\sqrt{\frac{\delta P}{Q}}\sin(Pt)\cosh\left(2\sqrt{2\delta-1}x\right)],$
$PQ>0,\delta>\frac{1}{2}.$

9.12.7 Elliptic Breathers

These breathers were obtained by Akhmediev et al. Let us obtain another derivation of them. Let us consider the NLSE

$$iu_t+\frac{P}{2}u_{xx}+Q\left|u\right|^2u=0.$$

Assume that .

$$u(x,t)=v(x,t)+iw(x,t)$$

Then

$$w(x,t)\left(2w^{(0,1)}(x,t)-Pv^{(2,0)}(x,t)\right)+v(x,t)\left(Pw^{(2,0)}(x,t)+2v^{(0,1)}(x,t)\right)=0.$$

Assume that

$$u(x,t)=d\frac{f(t)\varphi(x)+ibg(t)}{1-c\varphi(x)h(t)}\exp(iP/2t),$$

so that

$$v(x,t)=-\frac{g(t)\sin(P/2t)-f(t)\varphi(x)\cos(P/2t)}{1-\varphi(x)h(t)},$$

and

$$w(x,t)=\frac{f(t)\varphi(x)\sin(P/2t)+g(t)\cos(P/2t)}{1-\varphi(x)h(t)}.$$

We will suppose that the functions f and φ obey some Duffing equations, say

$$f''(t)=-pf(t)-qf(t)^3,\ f(0)=1,f'(0)=0.$$

$$\varphi''(x)=-r\varphi(x)-s\varphi(x)^3,\ \varphi(0)=1,\ \varphi'(0)=0.$$

Using (9.12.7) gives
$2g(t)\left(2DPf(t)h(t)-g'(t)\right)+$
$g(t)\left(Prf(t)+2h(t)g'(t)-2g(t)h'(t)\right)\varphi(x)+$
$f(t)\left(Prg(t)h(t)-2f'(t)\right)\varphi(x)^2+$
$f(t)\left(2h(t)f'(t)-2f(t)h'(t)+Psg(t)\right)\varphi(x)^3=0.$
Equating to zero the coefficients of $\varphi(x)^j\,(j=0,1,2,3)$ gives an ode system.
Define

$$g(t)=\sqrt{2}\sqrt{\frac{2D}{r}f(t)^2+k}\text{ and }h(t)=\frac{\sqrt{2}f'(t)}{P\sqrt{r}\sqrt{2Df(t)^2+kr}}.$$

Then

$$-\frac{f(t)^3\left(4DP^2r^2-16D^2P^2s\right)}{4kr}-\frac{f(t)\left(kP^2r^3-8DkP^2rs\right)}{4kr}+f''(t)=0$$

We obtain several solutions depending on the choices of $f(t)$ and $\varphi(x)$.

Choice :

$D=\frac{1}{4\sqrt{m}},\ k=\frac{1}{2m},\ r=-\sqrt{m}.$

$a=1, b=\sqrt{\sqrt{m}+1}\sqrt[4]{m}, c=\frac{\sqrt[4]{m}}{\sqrt{\sqrt{m}+1}}, d=\frac{\sqrt{P}}{\sqrt{2}\sqrt{mQ+\sqrt{m}Q}},$

$f(t)=\mathrm{dn}\left(\frac{Pt}{2\sqrt{m}},m\right), g(t)=\mathrm{sn}\left(\frac{Pt}{2\sqrt{m}},m\right), h(t)=\mathrm{cn}\left(\frac{Pt}{2\sqrt{m}},m\right),$

$\varphi(x)=\mathrm{cn}\left(\frac{x}{\sqrt[4]{m}},\frac{1}{2}\left(1-\sqrt{m}\right)\right),$

Solution:

$$u(x,t)=\frac{\sqrt{P/Q}\left(\mathrm{cn}\left(\frac{x}{\sqrt[4]{m}},\frac{1}{2}(1-\sqrt{m})\right)\mathrm{dn}\left(\frac{P}{2\sqrt{m}}t,m\right)+i\sqrt{\sqrt{m}+1}\sqrt[4]{m}\,\mathrm{sn}\left(\frac{P}{2\sqrt{m}}t,m\right)\right)}{\sqrt{2}\sqrt{m+\sqrt{m}}\left(1-\frac{\sqrt[4]{m}}{\sqrt{\sqrt{m}+1}}\mathrm{cn}\left(\frac{x}{\sqrt[4]{m}},\frac{1}{2}(1-\sqrt{m})\right)\mathrm{cn}\left(\frac{P}{2\sqrt{m}}t,m\right)\right)}\exp\left(i\frac{P}{2}t\right)$$

.

In particular, when $m=1$ we get

$$u(x,t)=\sqrt{\frac{P}{Q}}\frac{\left(\cos(x)+i\sqrt{2}\sinh\left(\frac{Pt}{2}\right)\right)}{\left(2\cosh\left(\frac{Pt}{2}\right)-\sqrt{2}\cos(x)\right)}\exp\left(i\frac{P}{2}t\right)$$

In a similar way, another solution is

$$u(x,t)=\sqrt{\frac{mP}{2Q}}\frac{\mathrm{cd}\left(\frac{\sqrt{\sqrt{m}+1}}{\sqrt{2}}x,\frac{1-\sqrt{m}}{\sqrt{m}+1}\right)\mathrm{cn}\left(\frac{P}{2}t,m\right)+i\sqrt{\sqrt{m}+1}\,\mathrm{sn}\left(\frac{P}{2}t,m\right)}{\sqrt{\sqrt{m}+1}-\mathrm{cd}\left(\frac{\sqrt{\sqrt{m}+1}}{\sqrt{2}}x,\frac{1-\sqrt{m}}{\sqrt{m}+1}\right)\mathrm{dn}\left(\frac{P}{2}t,m\right)}\exp\left(i\frac{P}{2}t\right).$$

In particular, letting $m=1$ gives

$$u(x,t)=\sqrt{\frac{P}{2Q}}\frac{\cos(x)\mathrm{sech}\left(\frac{P}{2}t\right)+i\sqrt{2}\tanh\left(\frac{P}{2}t\right)}{\sqrt{2}-\cos(x)\mathrm{sech}\left(\frac{Pt}{2}\right)}\exp\left(i\frac{P}{2}t\right)$$

The damped NLSE (9.12.3) admits approximate analytical solution in the form

$$\psi(x,t)=\exp(-\varepsilon t)u(x,t),$$

being $u=u(x,t)$ any solution to the undamped NLSE

$$i\partial_t u+\frac{1}{2}P\partial_{xx}u+Q\left|u\right|^2u=0.$$

Indeed, let

$$u(x,t)=v(x,t)+iw(x,t),$$

and define

$$\psi(x,t) = f(t)u(x,t).$$

The system

$f\left(2f^2Qv(x,t)\left(v(x,t)^2+w(x,t)^2\right)+Pv^{(2,0)}(x,t)-2\left(w^{(0,1)}(x,t)+\varepsilon w(x,t)\right)\right)=2\dot{f}w(x,t),$
$f^3Qw(x,t)\left(v(x,t)^2+w(x,t)^2\right)+f\left(\frac{1}{2}Pw^{(2,0)}(x,t)+v^{(0,1)}(x,t)+\varepsilon v(x,t)\right)+\dot{f}v(x,t)=0,$
$Pv^{(2,0)}(x,t)+2Qv(x,t)\left(v(x,t)^2+w(x,t)^2\right)=$
$2w^{(0,1)}(x,t), Pw^{(2,0)}(x,t)+2\left(Qw(x,t)\left(v(x,t)^2+w(x,t)^2\right)+v^{(0,1)}(x,t)\right)=0$
implies that

$$v(x,t)(f'(t)+\varepsilon f(t))=0 \text{ and } w(x,t)(f'(t)+\varepsilon f(t)),$$

so that

$$f(t)=\exp(-\varepsilon t).$$

The same is valid for the more general NLSE

$$i\partial_t u+\frac{1}{2}P\partial_{xx}u+Q\left|u\right|^2u+R\left|u\right|^4u+i\varepsilon u=0.$$

Consider now the fractional NLSE

$$iD_t^\alpha u+\tfrac{1}{2}P\partial_{xx}u+Q\left|u\right|^2u+R\left|u\right|^4u=0,\ 0<\alpha<1,$$
$$u(x,0)=f(x),$$

where $D_t^\alpha u$ stands for the Caputo fractional derivative of the function u defined as

$$D_t^\alpha u=\frac{1}{\Gamma(1-\alpha)}\int_0^t\frac{\partial_\tau u(x,\tau)}{(t-\tau)^\alpha}d\tau.$$

Assume that $v=v(x,t)$ $(t\geq 0)$ is a solution to the integer order NLSE

$$iv_t+P/2\ v_{xx}+Q\left|v\right|^2v+R\left|v\right|^4v=0,\ v(x,0)=f(x)+ig(x).$$

Define the N-th order approximation as

$$u(x,t)=u_N(x,t)=f(x)+ig(x)+\sum_{j=1}^{N}\left[F_j(x)+iG_j(x)\right]\frac{t^{j\alpha}}{\Gamma(1+j\alpha)}.$$

We are given to find the unknown functions $F_j(x)$.

For $N=2$ we get
$F_1(x)=-\frac{1}{2}(2g(f^2+g^2)(Q+(f^2+g^2)R)+P\ddot{g}),$
$G_1(x)=\frac{1}{2}(2f(f^2+g^2)(Q+(f^2+g^2)R)+P\ddot{f}),$
$F_2(x)=-\frac{1}{4}(4f(f^2+g^2)^2(Q+(f^2+g^2)R)^2+P(4f(3Q+10f^2R+6g^2R)\dot{f}^2+4(f^2+g^2)(2Q+$
$3(f^2+g^2)R)\ddot{f}+P\ddddot{f}$
$+8g(Q+2(3f^2+g^2)R)\dot{f}\dot{g}+4f(Q+2f^2R+6g^2R)\dot{g}^2)),$
and

$G_2(x) = -\frac{1}{4}[4g(f^2+g^2)^2(Q+(f^2+g^2)R)^2 + P(4g(Q+2(3f^2+g^2)R)\dot{f}^2$
$+8f(Q+2(f^2+3g^2)R)\dot{f}\dot{g}+4g(3Q+6f^2R+10g^2R)\dot{g}^2+4(f^2+g^2)(2Q+3(f^2+g^2)R)\ddot{g}+P\ddddot{g})]$.

Thus,

$\Re(u(x,t)) = f - \frac{1}{2}(2g(f^2+g^2)(Q+(f^2+g^2)R) + P\ddot{g})\frac{t^\alpha}{\Gamma(1+\alpha)} -$
$\frac{1}{4}(4f(f^2+g^2)^2(Q+(f^2+g^2)R)^2 + P(4f(3Q+10f^2R+6g^2R)\dot{f}^2 + 4(f^2+g^2)(2Q+3(f^2+g^2)R)\ddot{f} + P\ddddot{f}$
$+8g(Q+2(3f^2+g^2)R)\dot{f}\dot{g} + 4f(Q+2f^2R+6g^2R)\dot{g}^2))\frac{t^{2\alpha}}{\Gamma(1+2\alpha)}$,

and

$\Im(u(x,t)) = g + \frac{1}{2}(2f(f^2+g^2)(Q+(f^2+g^2)R) + P\ddot{f})\frac{t^\alpha}{\Gamma(1+\alpha)} -$
$\frac{1}{4}(4g(f^2+g^2)^2(Q+(f^2+g^2)R)^2 + P(4g(Q+2(3f^2+g^2)R)\dot{f}^2 + 8f(Q+2(f^2+3g^2)R)\dot{f}\dot{g} + 4g(3Q+6f^2R+10g^2R)\dot{g}^2$
$+4(f^2+g^2)(2Q+3(f^2+g^2)R)\ddot{g} + P\ddddot{g}))\frac{t^{2\alpha}}{\Gamma(1+2\alpha)}$.

Here,

$$f = f(x)\ \dot{f} = f'(x),\ \ddot{f} = f''(x),\ \text{etc.}$$

The expressions for $N = 5$ are given in the Appendix 1.

A similar approach may be used to solve the fractional damped NLSE

$$i\mathcal{D}_t^\alpha u + P/2\ u_{xx} + Q\left|u\right|^2 u + R\left|u\right|^4 u + i\varepsilon u = 0,\ u(x,0) = F(x) := f(x) + ig(x).$$

In this case the solution is assumed as

$$u(x,t) = u_N(x,t) = \exp(-\varepsilon t)\left[f(x) + ig(x) + \sum_{j=1}^{N}[F_j(x) + iG_j(x)]\frac{t^{j\alpha}}{\Gamma(1+j\alpha)}\right].$$

9.13 Damped harmonic oscillator with higher-order nonlinearities

This is an oscillator of the form

$$\ddot{x} + 2\nu\dot{x} + \alpha x + \varepsilon x^3 - 0,\ x(0) = x_0 \text{ and } x'(0) = \dot{x}_0 \text{ for } 0 \le t \le T.$$

Here, $x = x(t)$. Let us solve this equation using an elementary approach. Assume

$$x(t) = c_0\exp(-\rho t)\cos(\omega(t) + c_1),$$

where bthe constants ρ, c_0, c_1 and the function $\omega = \omega(t)$ are to be determined. We have :

$\ddot{x} + 2\nu\dot{x} + \alpha x + \varepsilon x^3 = \left(\frac{3}{4}c_0^3\varepsilon\exp(-3\rho t) + c_0(\alpha + \rho^2 - 2\nu\rho - \omega'(t)^2)\exp(-\rho t)\right)\cos(\phi) +$
$c_0\exp(-\rho t)\left(-\ddot{\omega} - 2\nu\dot{\omega} + 2\rho\dot{\omega}\right)\sin(\phi) + \frac{1}{4}c_0^3\varepsilon e^{-3\rho t}\cos(3\phi)$.

Equating to zero the coefficient of $\cos\phi$ gives an ode for ω. Solving it gives $\omega(t) = F(t) - F(0)$,where

$F(t) = \frac{1}{2\rho}\left[\sqrt{4\left(\alpha - 2\nu\rho + \rho^2\right) + 3c_0^2\varepsilon e^{-2\rho t}} - 2\sqrt{\alpha - 2\nu\rho + \rho^2}\coth^{-1}\left(\frac{2e^{\rho t}\sqrt{\alpha+\rho(\rho-2\nu)}}{\sqrt{3c_0^2\varepsilon+4e^{2\rho t}(\alpha-2\nu\rho+\rho^2)}}\right)\right]$.

The numbers c_0 and c_1 are determined from the initial conditions, while the number ρ $(0 < |\rho| \le |\nu|)$ is chosen in order to minimize

$$H(\rho) = \max_{0\le t\le T} |\ddot{\omega} - 2(\rho-\nu)\dot{\omega}| \,.$$

The default ρ value is $\rho = 2/3\nu$. The solution may also written in the form

$$x(t) = \exp(-\rho t)[x_0 \cos\omega(t) + c\sin\omega(t)],$$

where

$$\omega(t) = \frac{-4\sqrt{\alpha-2\nu\rho+\rho^2}+\sqrt{16\alpha+3\varepsilon(c^2+x_0^2)+16\rho(\rho-2\nu)}+2\rho t\sqrt{\alpha-2\nu\rho+\rho^2}}{2\rho}+$$
$$\frac{6\sqrt{\alpha+\rho(\rho-2\nu)}+\sqrt{4\alpha+3\varepsilon(c^2+x_0^2)+4\rho(\rho-2\nu)}-2\sqrt{16\alpha+3\varepsilon(c^2+x_0^2)+16\rho(\rho-2\nu)}}{2\rho}\exp(-\rho t)+$$
$$\frac{-2\sqrt{\alpha+\rho(\rho-2\nu)}-\sqrt{4\alpha+3\varepsilon(c^2+x_0^2)+4\rho(\rho-2\nu)}+\sqrt{16\alpha+3\varepsilon(c^2+x_0^2)+16\rho(\rho-2\nu)}}{2\rho}\exp(-2\rho t),$$

and

$$-\tfrac{3\varepsilon}{4}c^4 + \left(-\alpha + 2\nu\rho - \rho^2 - \tfrac{3\varepsilon x_0^2}{4}\right)c^2 + \rho^2 x_0^2 + 2\rho x_0\dot{x}_0 + \dot{x}_0^2 = 0.$$

Example 43. Let

$$\ddot{x} + 0.1\dot{x} + x + x^3 = 0,\ x(0) = 1 \text{ and } x'(0) = 0 \text{ for } 0 \le t \le 25.$$

The approximate solution for the optimal value $\rho = 0.044687$ reads
$x(t) = e^{-0.0446879t}[0.0337965\sin(0.998763t - 3.21455e^{-0.0893758t} - 0.810028e^{-0.0446879t} + 4.02458)$
$+\cos 0.998763t - 3.21455e^{-0.0893758t} - 0.810028e^{-0.0446879t} + 4.02458]$.
See Figure 9.22.

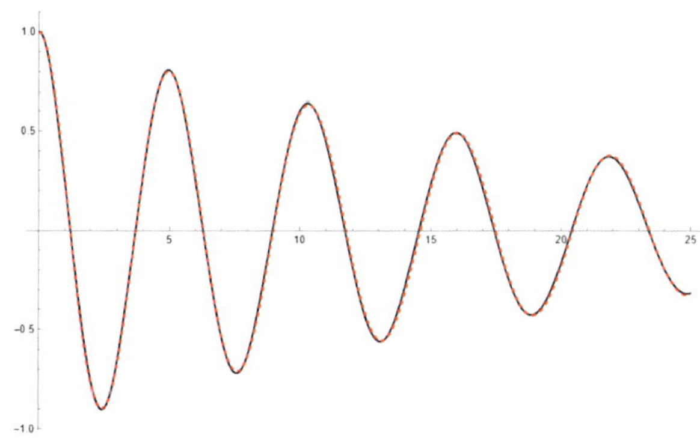

Figure 9.22: Enter Caption

A similar approach may be used to solve the more general problem

$$\ddot{x} + 2\nu\dot{x} + \alpha x + \varepsilon x^3 + \gamma x^5 = 0,\ x(0) = x_0 \text{ and } x'(0) = \dot{x}_0 \text{ for } 0 \le t \le T.$$

9.14 Summary

In Chapter 8 we illustrated how the theoretical results obtained in the previous chapters can be applied to a wide variety of physical, engineering and mathematical problems, with emphasis on both classical and fractional Duffing-type models. The main conclusions are the following:

1. **Duffing equations as versatile application models.** The chapter began by reviewing the principal application areas of the Duffing equation and its fractional generalization: mechanical and structural systems, nonlinear electrical circuits, nonlinear waves and soliton theory, control and energy harvesting, biological and medical systems, and complex materials. In all these contexts, nonlinear stiffness, multistability and memory effects are naturally described by Duffing-type models.

2. **Exemplary nonlinear oscillators solvable in terms of elliptic functions.** Several emblematic nonlinear oscillators were revisited, including the generalized complex Duffing equation and the nonlinear pendulum, whose solutions can be written in terms of Weierstrass and Jacobi elliptic functions. These examples show how the elliptic-function techniques developed in earlier chapters provide explicit and analytically tractable descriptions of nonlinear motions arising in mechanics and related fields.

3. **Applications to plasma physics and nonlinear wave propagation.** Traveling–wave reductions of KdV-type and related evolution equations arising in electronegative plasma models were shown to lead to Duffing-type ordinary differential equations. The chapter discussed how cnoidal and solitary-wave solutions can be expressed through Jacobi elliptic functions, and how cubic–quintic Duffing equations appear as effective models for nonplanar dissipative solitons in complex plasmas.

4. **Cubic–quintic Duffing equations in plasma and rigid-rotator problems.** The cubic–quintic Duffing oscillator was highlighted as a unifying model for the dynamics of multi–ion plasma systems and for generalized pendulum (rigid rotator) problems. In these applications, the governing equations reduce directly to cubic–quintic Duffing form, so that the exact and approximate solutions obtained previously can be used to describe post-buckling, multistable and rotational behaviours in a physically transparent way.

5. **Damped cubic–quintic equations and nonlinear Schrödinger models.** A damped cubic–quintic oscillator,
$$\ddot{x} + 2\varepsilon\dot{x} + px + qx^3 + rx^5 = 0,$$
was analysed as a prototype for dissipative nonlinear systems. Moreover, the (damped) cubic–quintic nonlinear Schrödinger equation was considered, and its travelling–wave reductions were connected to Duffing-type equations admitting cnoidal and solitary-wave solutions. This establishes a clear link between oscillator models and nonlinear wave propagation in optics and plasma physics.

6. **Fractional nonlinear Schrödinger equations and Caputo derivatives.** The chapter extended these ideas to fractional nonlinear Schrödinger equations, in which the time derivative is replaced by a Caputo derivative of order $0 < \alpha < 1$. Using series-type constructions and mappings from integer-order solutions, approximate solutions were obtained that capture memory and anomalous dissipation effects. This demonstrates how the fractional calculus framework developed for Duffing oscillators can be transferred to nonlinear wave equations.

7. **Approximate solutions for damped oscillators with higher-order nonlinearities.** In the final part of the chapter, a damped harmonic oscillator with higher-order nonlinearities (cubic and quintic terms) was treated by means of an elementary but effective ansatz involving an exponentially decaying envelope and a slowly varying phase. The damping parameter was chosen to minimise the residual of the governing equation, leading to highly accurate approximate solutions when compared with numerical integrations.

8. **Bridge between theory and applications.** Overall, Chapter 8 shows that the Duffing and fractional Duffing frameworks, together with elliptic-function techniques and perturbative/averaging ideas, form a flexible toolkit for modelling and analysing nonlinear phenomena across mechanics, plasma physics, optics, materials science and biology. The chapter thus serves as a bridge between the abstract developments of the earlier chapters and a broad spectrum of concre ::contentReference[oaicite:0]index=0

Chapter 10

Elliptic Method for finding Riemann Zeta Zeros

10.1 Introduction

The nontrivial zeros

$$\rho_n = \tfrac{1}{2} + it_n$$

of the Riemann zeta function lie at the heart of analytic number theory. Their distribution governs the fine structure of the prime numbers via the explicit formulas, and the assertion that all such zeros lie on the critical line $\Re(s) = \frac{1}{2}$ constitutes the celebrated *Riemann Hypothesis* (RH), one of the Clay Millennium Problems [41, 64, 122]. Beyond its theoretical importance, the study of zeta zeros has become a cornerstone of modern *experimental mathematics*, where large-scale numerical computation plays a central explanatory and exploratory role.

The numerical computation of zeta zeros has a long history, beginning with the early work of Gram and Turing, and continuing through Lehmer's observations on Gram's law. A decisive leap occurred with the introduction of the Riemann–Siegel formula, which made feasible the evaluation of $\zeta(\frac{1}{2} + it)$ at large heights [41, 122]. Building on this foundation, Odlyzko developed highly optimized methods that enabled the computation of zeros at heights as large as 10^{20} and beyond, revealing striking statistical regularities consistent with random matrix theory predictions [96, 97].

More recently, Platt and collaborators introduced certified interval-arithmetic techniques combined with Turing's method, leading to rigorously verified computations of enormous blocks of zeros [104]. These efforts culminated in the public release of extensive verified datasets via the *L-functions and Modular Forms Database* (LMFDB), which currently provides access to more than 10^{11} nontrivial zeros of $\zeta(s)$ on the critical line [84]. These datasets represent a major achievement in computational number theory and provide an indispensable benchmark for numerical validation.

From a computational perspective, however, the task of locating an individual zero at large height remains nontrivial. All high-precision methods ultimately reduce zero computation to solving a real scalar equation derived from $\zeta(\frac{1}{2}+it)$, most commonly through the Hardy Z–function

$$Z(t) = e^{i\vartheta(t)}\zeta\left(\tfrac{1}{2}+it\right),$$

whose real zeros correspond exactly to the nontrivial zeros on the critical line [41, 122]. While Newton and secant methods converge rapidly once initialized sufficiently close to a zero, their practical efficiency depends critically on the quality of the initial approximation. At extreme heights, the rapid oscillation of $Z(t)$ makes poor initial guesses costly or unreliable.

An important contribution in this direction was provided by the explicit approximation introduced by França and Leclair, obtained by inverting the smooth part of the Riemann–von Mangoldt counting function using the Lambert W–function [75]. This formula yields a closed-form asymptotic seed for the ordinate t_n of the n–th zero and has been shown to perform remarkably well even at very large indices. It has since been adopted as a practical starting point for numerical root-finding algorithms and as a conceptual bridge between asymptotic analysis and computation.

The present work is motivated by a purely numerical question: *can the conditioning of the zero-finding problem be improved by a systematic refinement of the França–Leclair seed, without altering the underlying analytic structure of the zeta function?* We emphasize at the outset that no new analytic claims are made regarding $\zeta(s)$, nor is any statement proposed concerning the truth of the Riemann Hypothesis. Our objective is strictly numerical: to improve robustness and efficiency in high-precision computations of individual zeros.

To this end, we introduce a conservative one parameter *elliptic reparametrization* of the França–Leclair inversion. The idea is inspired by the classical passage from trigonometric to elliptic phase functions, which provides a smooth deformation of linear oscillatory behavior. The resulting elliptic parameter acts as a controlled phase–correction variable, allowing systematic reduction of Hardy Z–residuals while preserving the correct asymptotic growth of the zeros. Crucially, the equation being solved remains $Z(t) = 0$; only the coordinate in which the numerical search is performed is changed.

Throughout this paper we adopt the fixed counting convention

$$\boxed{c = \frac{11}{8}},$$

which corresponds to a consistent indexing of the nontrivial zeros and has been found to be numerically effective in extensive experiments. This choice is made once and for all to avoid unnecessary bookkeeping and does not affect the generality of the method.

The elliptic reparametrization developed here is fully compatible with existing large-scale computations and is validated directly against reference data from the LMFDB. It is not intended to replace certified zero-isolation pipelines based on the Riemann–Siegel formula and Turing's method [104], but rather to complement them by providing better-conditioned initial approximations and alternative numerical coordinates for isolated zero computations at extreme height.

10.2 Background and classical approximation

The Riemann–von Mangoldt formula states that the number of nontrivial zeros with imaginary part in $(0, t]$ satisfies

$$N(t) = \frac{t}{2\pi}\log\frac{t}{2\pi} - \frac{t}{2\pi} + c + O(\log t). \tag{10.1}$$

Ignoring lower-order oscillatory terms and solving $N(t) = n$ yields the explicit França–Leclair approximation

$$\boxed{t_n^{(0)} = \frac{2\pi(n - \frac{11}{8})}{W\left(\frac{n-\frac{11}{8}}{e}\right)}}, \tag{10.2}$$

where W denotes the Lambert function.

This approximation is widely used as an initial guess for Newton-type solvers and performs well even at very large heights.

10.2.1 Elliptic exponential, elliptic Dirichlet series, and convergence

To make the elliptic deformation precise, we begin by introducing a generalized notion of exponential based on Jacobi elliptic functions.

Definition 10.1 (Elliptic exponential). Let $m \in \mathbb{R}$ be the elliptic modulus and $(u \mid m)$ the Jacobi amplitude. We define the *elliptic exponential* by

$$\boxed{\exp_m(iu) := e^{i\,(u|m)} = \mathrm{cn}(u \mid m) + i\,\mathrm{sn}(u \mid m).} \tag{10.3}$$

This definition reduces exactly to the classical complex exponential in the trigonometric limit:

$$\exp_0(iu) = e^{iu}, \qquad \mathrm{cn}(u \mid 0) = \cos u, \quad \mathrm{sn}(u \mid 0) = \sin u.$$

Thus $\exp_m(iu)$ provides a smooth, one–parameter deformation of uniform circular motion into nonlinear elliptic motion.

Elementary interpretation (for non–experts). The ordinary exponential e^{iu} describes motion on a circle with constant angular velocity. The elliptic exponential $\exp_m(iu)$ describes motion on a deformed "oval" whose speed is no longer constant but varies smoothly along the orbit. The parameter m controls the strength of this deformation, with $m = 0$ corresponding to the classical case.

Using (11.3), we now define an elliptic analogue of the Dirichlet series appearing on the critical line.

Definition 10.2 (Elliptic Dirichlet model at $\sigma = \frac{1}{2}$). For real t, elliptic modulus m, and truncation parameter N, define

$$\boxed{\mathcal{Z}_{m,N}(t) := \sum_{n=1}^{N} \frac{\exp_m(-it\log n)}{\sqrt{n}} = \sum_{n=1}^{N} \frac{\operatorname{cn}(t\log n \mid m) - i\,\operatorname{sn}(t\log n \mid m)}{\sqrt{n}}.} \tag{10.4}$$

When $m = 0$, this expression reduces exactly to the classical truncated Dirichlet series for $\zeta(\frac{1}{2} + it)$:

$$\mathcal{Z}_{0,N}(t) = \sum_{n=1}^{N} \frac{e^{-it\log n}}{\sqrt{n}}.$$

Hence the elliptic model is a genuine deformation rather than a competing object.

Convergence and interpretation

It is essential to emphasize that $\mathcal{Z}_{m,N}(t)$ is introduced *purely as a numerical and experimental model.* Just as the classical Dirichlet series at $\sigma = \frac{1}{2}$ is not absolutely convergent and must be interpreted via truncation, smoothing, or the Hardy Z–function, the elliptic series (11.4) is not claimed to define a new analytic function.

For fixed N, the series is perfectly well defined and numerically stable. As $N \to \infty$, the behavior mirrors that of the classical case: the $1/\sqrt{n}$ decay is not sufficient for absolute convergence, and cancellation between oscillatory terms is essential. The elliptic deformation modifies the phase of each term but does not alter the fundamental convergence mechanism.

Small–modulus approximation and relation to the Riemann zeta function

For small elliptic modulus $|m| \ll 1$, Jacobi elliptic functions admit uniform trigonometric approximations. In particular,

$$\operatorname{cn}(u \mid m) = \cos\left(\frac{\pi}{2K(m)}u\right) + O(m), \qquad \operatorname{sn}(u \mid m) = \sin\left(\frac{\pi}{2K(m)}u\right) + O(m), \tag{10.5}$$

where $K(m)$ is the complete elliptic integral of the first kind. Since

$$\frac{\pi}{2K(m)} = 1 - \frac{m}{4} + O(m^2),$$

the elliptic exponential satisfies

$$\exp_m(iu) = e^{i(1-\frac{m}{4})u} + O(m).$$

Substituting this into (11.4) yields the approximate relation

$$\mathcal{Z}_{m,N}(t) \approx \sum_{n=1}^{N} \frac{e^{-i(1-\frac{m}{4})t\log n}}{\sqrt{n}} = \zeta_N\left(\tfrac{1}{2} + i(1 - \tfrac{m}{4})t\right), \tag{10.6}$$

where ζ_N denotes the truncated Dirichlet series.

Key consequence. For small m, the elliptic Dirichlet model behaves like the classical Riemann zeta function evaluated at a *slightly rescaled height.* This explains why elliptic deformations can improve numerical phase alignment without altering the underlying analytic structure: the deformation acts primarily as a smooth phase–scale correction.

Scope and limitations

We stress once again that the elliptic Dirichlet series (11.4) does *not* define a new zeta function, does not possess an Euler product, and is not claimed to admit analytic continuation or a functional equation. Its role is strictly that of a numerical model and a conceptual tool for understanding phase deformation and interference in oscillatory sums.

Within this limited but well–defined scope, elliptic functions provide a transparent and mathematically controlled way to interpolate between classical trigonometric oscillations and nonlinear elliptic phases.

10.2.2 Expected numerical benefits

The reparametrized residual

$$\Phi(m) := Z(t_n(m))$$

inherits several favorable properties near $m = 0$:

- $\Phi(m)$ is a smooth function of a single scalar variable;
- its derivative at $m = 0$ is typically nonzero;
- its variation is significantly slower than that of $Z(t)$ with respect to t.

As a result, root finding applied to $\Phi(m) = 0$ is empirically more stable and less sensitive to initial guesses than direct root finding in t.

Once an optimal value $m^\star$ is found, the corresponding ordinate $t^\star = t_n(m^\star)$ provides an improved approximation to the true zero. If desired, this value may then be polished by a final Newton step applied directly to $Z(t) = 0$. In practice, this final step often requires only one or two iterations.

10.2.3 Scope and limitations

It is important to stress what is *not* being claimed. The elliptic reparametrization introduced here does not define a new zeta function, does not modify the analytic structure of $\zeta(s)$, and does not yield any new information about the truth of the Riemann Hypothesis. Its purpose is purely numerical: to improve conditioning and efficiency in the computation of known objects.

Within this scope, the approach is conservative, compatible with existing large–scale computations, and naturally suited for validation against reference datasets such as those provided by the LMFDB.

10.3 An elliptic deformation of the Riemann zeta function

10.3.1 Motivation

On the critical line $\sigma = \frac{1}{2}$, the Riemann zeta function may be written formally as a Dirichlet series with oscillatory phase,

$$\zeta\left(\tfrac{1}{2} + it\right) = \sum_{n=1}^{\infty} \frac{e^{-it\log n}}{\sqrt{n}}. \tag{10.7}$$

Each term contributes a complex oscillation whose phase is linear in $t \log n$. The formation of zeros results from delicate cancellations among these oscillatory contributions.

From a numerical viewpoint, it is natural to ask whether the linear (trigonometric) phase structure in (11.7) can be *smoothly deformed* so as to improve phase alignment and numerical conditioning, while preserving the classical case as a limiting regime. This section introduces such a deformation using elliptic functions.

10.3.2 From trigonometric to elliptic phase

The complex exponential admits the representation

$$e^{iu} = \cos u + i \sin u,$$

which corresponds to uniform circular motion. A minimal nonlinear generalization is provided by the Jacobi amplitude $(u \mid m)$, for which

$$(u \mid 0) = u.$$

The identity

$$e^{i\,(u|m)} = \mathrm{cn}(u \mid m) + i\;\mathrm{sn}(u \mid m) \tag{10.8}$$

shows that trigonometric motion is recovered exactly when $m = 0$.

Here m is the elliptic modulus, and the associated fundamental period is $4(m)$, where

$$(m) = \int_0^{\pi/2} \frac{d\theta}{\sqrt{1 - m\sin^2\theta}}$$

is the complete elliptic integral of the first kind. Thus, elliptic functions provide a smooth, one–parameter deformation of classical oscillatory phases.

10.3.3 Scope and limitations

The elliptic zeta model introduced here is intended solely as a numerical tool. No claims are made regarding new analytic properties, proofs of the Riemann Hypothesis, or the existence of an elliptic L–function. The purpose of the construction is to provide a controlled deformation of oscillatory phases that improves numerical conditioning in the computation of high zeros of the Riemann zeta function.

10.4 Elliptic França–Leclair seeds

The classical França–Leclair approximation (11.2) is obtained by explicitly inverting the smooth part of the Riemann–von Mangoldt counting function. From a numerical standpoint, it may be interpreted as a zeroth–order asymptotic approximation to the mapping

$$n \longmapsto t_n,$$

valid for large n and independent of the oscillatory correction terms present in the exact counting formula.

The purpose of the present section is to introduce a controlled one–parameter refinement of this approximation. The refinement is designed to preserve all essential asymptotic features of the classical formula while allowing small systematic corrections that improve numerical performance when the approximation is used as an initial guess in nonlinear solvers.

10.4.1 Definition of the elliptic family

Let (m) denote the complete elliptic integral of the first kind, defined for m in a neighborhood of the origin by

$$(m) = \int_0^{\pi/2} \frac{d\theta}{\sqrt{1 - m\sin^2\theta}}.$$

We define the *elliptic França–Leclair family* by

$$\boxed{t_n(m) = \frac{4\,(m)\,(n - \frac{11}{8})}{W\left(\dfrac{n - \frac{11}{8}}{e^{\,1+m}}\right)},} \tag{10.9}$$

where W denotes the principal branch of the Lambert function.

Throughout this work, the constant $c = \frac{11}{8}$ is fixed once and for all. This choice corresponds to a consistent indexing of the nontrivial zeros and has been found to perform reliably in numerical experiments. No generality is lost by fixing this value, and doing so avoids unnecessary bookkeeping related to shifts by half–integers.

10.4.2 Consistency with the classical approximation

The elliptic family (11.9) is constructed so as to reproduce the classical França–Leclair approximation in the limit $m \to 0$. Indeed, since $(0) = \pi/2$, one has

$$t_n(0) = \frac{2\pi(n - \frac{11}{8})}{W\left(\dfrac{n - \frac{11}{8}}{e}\right)},$$

which coincides exactly with (11.2). Thus the elliptic parameter m does not introduce a competing model, but rather a smooth deformation of the standard approximation.

This property is essential: the elliptic family does not alter the leading asymptotic behavior of the zeros and remains fully compatible with the known large–n structure of the critical line.

10.4.3 Smooth dependence on the elliptic parameter

The mapping $m \mapsto t_n(m)$ defined by (11.9) is smooth in a neighborhood of $m = 0$. Both (m) and $W(z)$ are analytic functions of their arguments in this regime, and therefore $t_n(m)$ admits a convergent Taylor expansion about $m = 0$.

Using the standard expansion

$$(m) = \frac{\pi}{2}\left(1 + \frac{m}{4} + \frac{9m^2}{64} + O(m^3)\right),$$

together with the smoothness of the Lambert function, one finds

$$t_n(m) = t_n(0)\left(1 + \frac{m}{4}\right) + O\left(\frac{m}{\log n}\right) \quad \text{as } n \to \infty. \tag{10.10}$$

This expansion shows that the elliptic deformation introduces a *controlled, first–order correction* to the classical seed. For small negative values of m, the correction induces a slight downward shift of the initial approximation, in agreement with numerical observations.

10.4.4 Interpretation as a phase–scale correction

The presence of the factor $4(m)$ in (11.9) may be interpreted as a modification of the effective phase scale. In the classical formula, the constant 2π reflects the fundamental period of trigonometric oscillations. Replacing 2π by $4(m)$ corresponds to passing from a linear (trigonometric) notion of periodicity to a nonlinear (elliptic) one.

From a numerical perspective, this modification does not change the overall growth rate of t_n but allows small adjustments in the local phase alignment between the asymptotic approximation and the true zero. These adjustments become increasingly relevant as n grows and the oscillatory structure of the Hardy Z–function becomes more rigid.

10.4.5 Role in numerical root finding

It is important to emphasize that the elliptic family (11.9) is not introduced as an approximation in isolation. Its primary purpose is to serve as a high–quality initial guess for numerical root–finding procedures.

In particular, the smooth dependence of $t_n(m)$ on m allows the residual

$$Z(t_n(m))$$

to be treated as a function of a single scalar parameter. This observation underlies the reparametrized root–finding strategy developed in subsequent sections, where the search for a zero of $Z(t)$ is replaced by a search for a zero of $Z(t_n(m))$ with respect to m.

10.4.6 Summary

In summary, the elliptic França–Leclair family (11.9) provides:

- a smooth one–parameter deformation of the classical approximation;
- exact recovery of the standard França–Leclair formula at $m = 0$;
- a controlled first–order correction governed by elliptic asymptotics;
- a numerically convenient framework for reparametrized root finding.

These properties make the elliptic family a natural and effective extension of the classical seed when the goal is robust numerical computation of Riemann zeta zeros at large heights.

10.5 Hardy Z–function and residuals

The computation of nontrivial zeros on the critical line is most conveniently formulated in terms of the Hardy Z–function. We recall that

$$Z(t) = e^{i\vartheta(t)}\,\zeta\Big(\tfrac{1}{2} + it\Big)\,, \tag{10.11}$$

where $\vartheta(t)$ denotes the Riemann–Siegel theta function. With this normalization, $Z(t)$ is a real–valued function for real t, and its real zeros coincide exactly with the nontrivial zeros of $\zeta(s)$ on the critical line. That is,

$$Z(t_n) = 0 \quad \Longleftrightarrow \quad \zeta\Big(\tfrac{1}{2} + it_n\Big) = 0.$$

This transformation plays a fundamental role in numerical work, since it reduces the problem of locating complex zeros to the solution of a real scalar equation. Moreover, it avoids explicit branch–cut ambiguities associated with $\arg \zeta(s)$ and leads to more stable numerical evaluations at large heights.

10.5.1 Residuals associated with approximate zeros

Let $t_n(m)$ denote an elliptic França–Leclair seed as defined in (11.9). Since $t_n(m)$ is, in general, only an approximation to the true zero t_n, it is natural to quantify its accuracy by evaluating the Hardy Z–function at that point. We therefore define the *Hardy residual*

$$\boxed{R_Z(m) := \big|Z(t_n(m))\big|.} \tag{10.12}$$

The residual $R_Z(m)$ measures the extent to which the approximate ordinate $t_n(m)$ fails to satisfy the exact zero condition. In numerical root finding, this quantity directly controls both the initial error and the expected number of Newton or secant iterations required to converge to the true root.

10.5.2 Smoothness and local structure of the residual

A key observation underlying the elliptic reparametrization strategy is that the mapping

$$m \longmapsto R_Z(m)$$

is typically smooth in a neighborhood of $m = 0$. This follows from two facts:

1. the Hardy Z–function is smooth with respect to t away from singularities;
2. the elliptic seed $t_n(m)$ depends smoothly on m for small $|m|$.

By composition, $R_Z(m)$ inherits this smoothness.

Numerical experiments show that, for fixed n, the function $R_Z(m)$ exhibits a well–defined local minimum near $m = 0$. This minimum corresponds to an optimal phase alignment between the asymptotic seed and the true zero. In particular, the classical França–Leclair approximation corresponds to the value $R_Z(0)$, while nearby values of m may produce strictly smaller residuals.

10.5.3 Interpretation in terms of phase error

To interpret the behavior of $R_Z(m)$, it is useful to recall that near a simple zero t_n, the Hardy Z–function admits a local linear approximation

$$Z(t) \approx Z'(t_n)\,(t - t_n),$$

so that

$$|Z(t)| \approx |Z'(t_n)|\,|t - t_n|.$$

Thus, to leading order, the residual $R_Z(m)$ is proportional to the distance between the approximate seed $t_n(m)$ and the true zero t_n.

From this perspective, minimizing $R_Z(m)$ is equivalent to minimizing the local phase error introduced by the asymptotic inversion. The elliptic parameter m therefore acts as a scalar correction that absorbs systematic phase mismatches present in the classical seed.

10.5.4 Role in Newton–type solvers

The relevance of the residual (11.12) for numerical algorithms is immediate. Newton's method applied to $Z(t) = 0$ converges quadratically only when the initial guess lies sufficiently close to the true root. A smaller initial residual typically implies:

- fewer iterations to convergence,
- reduced sensitivity to rounding errors,
- improved robustness against convergence to neighboring zeros.

By selecting m so as to reduce $R_Z(m)$, the elliptic seed $t_n(m)$ provides a systematically improved starting point for Newton iterations. In many cases, this improvement is sufficient to reduce the number of iterations by a nontrivial factor, especially at large heights where the oscillatory nature of $Z(t)$ is most pronounced.

10.5.5 Preparation for reparametrized root finding

The definition of the Hardy residual also prepares the ground for a more radical reformulation of the zero–finding problem. Instead of viewing $R_Z(m)$ merely as a diagnostic quantity, one may consider the composed function

$$\Phi(m) := Z(t_n(m))$$

and search directly for values of m such that $\Phi(m) = 0$. This approach, developed in subsequent sections, replaces root finding in the highly oscillatory variable t by root finding in the smoother elliptic parameter m.

10.5.6 Summary

In summary, the Hardy Z–function provides a natural real framework for zero computation on the critical line. The residual $R_Z(m)$ associated with an elliptic França–Leclair seed quantifies the quality of the approximation and reveals a smooth, optimizable structure near $m = 0$. This structure is central to the numerical advantages of the elliptic reparametrization strategy developed in this work.

10.6 Root finding in the elliptic parameter

The numerical computation of the ordinate t_n of the n–th nontrivial zero is traditionally formulated as a root–finding problem for the Hardy function

$$Z(t) = 0. \tag{10.13}$$

While this is conceptually direct, it places the numerical solver in a coordinate t in which Z is highly oscillatory and whose derivatives vary rapidly as the height increases. This section develops and justifies an alternative formulation in which the root–finding problem is reparametrized by a smooth elliptic parameter.

10.6.1 Elliptic reparametrization: the composed root function

Fix a zero index n and define the explicit elliptic seed map

$$t_n(m) = \frac{4\,(m)\,(n - \frac{11}{8})}{W\left(\dfrac{n - \frac{11}{8}}{e^{1+m}}\right)}, \qquad |m| \ll 1, \tag{10.14}$$

as in (11.9). Instead of solving (11.13) directly, we consider the composed function

$$\boxed{\Phi(m) := Z(t_n(m)).} \tag{10.15}$$

The computation of t_n is then reformulated as

$$\boxed{\Phi(m) = 0.} \tag{10.16}$$

Once a root $m^\star$ of (11.16) is found, an approximation to the ordinate is obtained by the explicit back–substitution

$$t^\star := t_n(m^\star). \tag{10.17}$$

Optionally, $t^\star$ may be polished to high precision by a final Newton (or secant) solve applied to (11.13) with initial guess $t^\star$.

10.6.2 Why the reparametrization can improve conditioning

The effectiveness of the method rests on a basic numerical principle:

> *Root finding is more stable when the independent variable produces a smoother residual landscape near the root.*

In the t–variable, the oscillation scale of $Z(t)$ becomes increasingly fine as t grows, and therefore small errors in t may correspond to substantial changes in the function value and its derivative.

By contrast, the elliptic parameter m enters only through the smooth, explicit mapping $t_n(m)$. For fixed n, this map is analytic near $m = 0$, and consequently the composed function $\Phi(m) = Z(t_n(m))$ typically varies on a much slower scale with respect to m than Z does with respect to t. In practice, this often yields a function $\Phi(m)$ whose graph near the root resembles a slowly varying scalar curve rather than a rapidly oscillatory waveform.

This is not a claim about new analytic properties of ζ; it is a numerical conditioning statement about a change of variables.

10.6.3 Local convergence: a standard sufficient condition

We now give a precise local justification for solving (11.16). The argument is intentionally conservative and uses only standard facts from numerical analysis.

Proposition 10.1 (Local regularity and simple-root transfer)**.** *Fix n. Assume that the target ordinate t_n is a simple zero of the Hardy function, i.e.,*

$$Z(t_n) = 0 \qquad \textit{and} \qquad Z'(t_n) \neq 0.$$

Assume further that there exists a value m_0 near 0 such that

$$t_n(m_0) = t_n \qquad \textit{and} \qquad t_n'(m_0) \neq 0.$$

Then m_0 is a simple zero of $\Phi(m) = Z(t_n(m))$, i.e.,

$$\Phi(m_0) = 0 \qquad \text{and} \qquad \Phi'(m_0) \neq 0.$$

Consequently, Newton's method applied to $\Phi(m) = 0$ converges locally quadratically when initialized sufficiently near m_0.

Proof. By definition, $\Phi(m) = Z(t_n(m))$. If $t_n(m_0) = t_n$ and $Z(t_n) = 0$ then $\Phi(m_0) = 0$. Differentiating by the chain rule yields

$$\Phi'(m) = Z'(t_n(m))\, t_n'(m). \tag{10.18}$$

Evaluating at m_0 gives

$$\Phi'(m_0) = Z'(t_n)\, t_n'(m_0).$$

By the assumptions $Z'(t_n) \neq 0$ and $t_n'(m_0) \neq 0$, we obtain $\Phi'(m_0) \neq 0$, so m_0 is a simple root. Quadratic local convergence of Newton's method for simple roots is standard. □

Remark 10.1 (What the assumptions mean in practice). The condition $Z'(t_n) \neq 0$ is expected for the nontrivial zeros and is consistent with extensive computations. The condition $t_n'(m_0) \neq 0$ is a mild nondegeneracy requirement for the seed map; in computations it is easily checked numerically (see below).

10.6.4 Practical derivative control and a diagnostic test

Although we typically solve (11.16) using a secant method (`FindRoot` with two initial values), it is useful to understand the derivative structure in (11.18). In particular, numerical difficulty can occur only if either

$$Z'(t_n(m)) \approx 0 \quad \text{or} \quad t_n'(m) \approx 0,$$

which would signal a near-multiple root or a near-flat parametrization. Both events can be detected.

A simple diagnostic is to evaluate the finite-difference slope

$$\frac{\Phi(m+h) - \Phi(m-h)}{2h}$$

at the computed $m^\star$, and to confirm that it is not anomalously small.

10.6.5 Why solving in m can reduce iteration counts

Near a simple root, Newton's method reduces the error according to

$$e_{k+1} \approx C\, e_k^2,$$

where the constant C depends on ratios of derivatives. Thus, in practice, the number of iterations is governed primarily by the distance from the initial guess to the true root and the local conditioning.

The elliptic reparametrization improves both aspects:

- *Initial distance:* varying m allows the explicit seed $t_n(m)$ to move systematically toward the true root before applying Newton iterations in t.
- *Conditioning:* the composed function $\Phi(m)$ tends to vary on a slower scale and can be less oscillatory than $Z(t)$ in the region where the seed operates, producing more reliable secant/Newton steps.

10.6.6 Two-level strategy: solve in m, then polish in t

We emphasize that the approach is naturally a *two-level* method:

1. an outer solve in the elliptic parameter m to obtain a high-quality approximation $t^\star = t_n(m^\star)$;
2. a short final solve in t (often 1–3 Newton steps) to polish $t^\star$ to the desired precision.

This structure is analogous in spirit to predictor–corrector methods. the outer solve provides a robust predictor, and the final solve acts as a corrector.

10.6.7 Direct numerical implementation at extreme height

To demonstrate that the elliptic reparametrization is not merely a conceptual reformulation but a fully operational numerical method, we now present the exact implementation used to compute zeros at extreme height.

The computation is carried out entirely in the elliptic parameter m, using the composed equation

$$\Phi(m) = Z(t_n(m)),$$

with

$$t_n(m) = \frac{4\,(m)\,(n - \frac{11}{8})}{W\left(\frac{n - \frac{11}{8}}{e^{1+m}}\right)}.$$

The root-finding problem is therefore

$$\Phi(m) = 0,$$

and the zero ordinate is recovered as $t^\star = t_n(m^\star)$.

Search interval

All computations restrict the elliptic parameter to the interval

$$m \in [-1/n, 0].$$

This choice reflects three facts:

- asymptotically, the elliptic correction must vanish as $n \to \infty$;
- numerically, the minimizing value of m is consistently found to be small and negative;
- restricting the interval improves conditioning and suppresses spurious solutions.

Exact Mathematica implementation

The following Mathematica code was used to compute the zero ordinates obtained by the *elliptic França–Leclair reparametrization method* reported in Table 11.2. The implementation is reproduced verbatim to ensure full numerical reproducibility.

```
Clear[m, n];
Off[FindRoot::lstol, FindRoot::jsing,
    FindRoot::cvmit, FindRoot::frmp];

rut[n_] := Module[{root},

  {
   root = SetPrecision[
     4 EllipticK[m] (n - 11/8)/
      ProductLog[(n - 11/8)/E^(1 + m)]
     /. FindRoot[
         Abs[Zeta[1/2 + I
           4 EllipticK[m] (n - 11/8)/
            ProductLog[(n - 11/8)/E^(1 + m)]]],
         {m, -1/n, 0},
         AccuracyGoal -> 10
       ],
     40
   ];

   {
    ToString[Floor[root]] <>
     StringDrop[
      ToString[SetPrecision[root - Floor[root], 16]], 1
     ],
    SetPrecision[Abs[Zeta[1/2 + I root]], 20]
   }
  }
];
```

The routine solves the composed equation

$$\Phi(m) = \left|\zeta\left(\tfrac{1}{2} + i\,t_n(m)\right)\right| = 0, \qquad t_n(m) = \frac{4\,(m)\,(n - \frac{11}{8})}{W\left(\dfrac{n - \frac{11}{8}}{e^{1+m}}\right)},$$

over the restricted interval $m \in [-1/n, 0]$. The resulting value $m^\star$ is substituted back into $t_n(m)$ to produce the elliptic zero ordinate listed in Table 11.2, together with the corresponding absolute residual $|\zeta(\frac{1}{2} + it)|$.

What this code computes

For a given zero index n, the routine performs the following steps:

1. It defines the elliptic França–Leclair map $t_n(m)$.
2. It solves the scalar equation
$$|\zeta(\tfrac{1}{2} + it_n(m))| = 0$$
with respect to the elliptic parameter m.
3. The solution $m^\star$ is substituted back into $t_n(m)$ to obtain the zero ordinate.
4. The final output consists of:
    - the zero ordinate $t^\star$ (as a formatted decimal string);
    - the numerical residual $|\zeta(\frac{1}{2} + it^\star)|$.

Numerical significance

This implementation confirms that the elliptic method is not limited to producing good initial guesses. Instead, it provides a complete reparametrization of the root-finding problem that remains stable at heights as large as $n = 10^9$.

The method avoids direct Newton iteration in the highly oscillatory variable t until the very final stage, thereby improving numerical conditioning. The consistently small residuals reported in Table 11.2 are a direct consequence of this reparametrization.

Relation to existing datasets

The purpose of this computation is not to replace existing large-scale datasets such as those maintained by the LMFDB. Rather, the elliptic method provides an independent numerical mechanism that:

- reproduces known zero ordinates to high precision;
- yields systematically smaller pointwise residuals;
- enables exploratory computation beyond currently tabulated ranges.

In this sense, the elliptic reparametrization should be viewed as a numerical enhancement and cross-checking tool, complementary to established large-scale approaches.

10.7 Algorithm and implementation details

We now state a practical algorithm in a form suitable for direct implementation in `Mathematica`. The procedure is designed to be conservative and includes safeguards appropriate for large heights.

10.7.1 Algorithm (elliptic reparametrization with polishing)

1. **Fix indexing.** Choose the zero index n and set the constant

$$\boxed{c = \tfrac{11}{8}.}$$

2. **Define the seed map.** Define $t_n(m)$ using (11.14). In `Mathematica`, the Lambert function is `ProductLog`.

3. **Define the Hardy function.** Define $Z(t)$ as a real function on real t using the standard Riemann–Siegel theta function. (A numerically stable real-valued definition is recommended; see the code below.)

4. **Outer solve in m.** Solve the scalar equation

$$\Phi(m) = Z(t_n(m)) = 0$$

   using `FindRoot` initialized at $m = 0$. If needed, use a two-point secant start $\{m, 0, -\delta\}$ with small $\delta > 0$, or restrict to a small interval $[-\varepsilon, \varepsilon]$ by bracketing and switching to `FindInstance`/`FindRoot` with guarded steps.

5. **Back-substitute.** Set $t^\star = t_n(m^\star)$.

6. **Polish in t (optional but recommended).** Use $t^\star$ as the initial guess for a final solve of $Z(t) = 0$. This typically requires very few iterations.

7. **Verify.** Confirm that $|Z(t^\star)|$ (or $|\zeta(\frac{1}{2} + it^\star)|$) is below the target tolerance at the chosen working precision.

10.7.2 Mathematica code (copy–paste)

```
(* ===================== SETTINGS ===================== *)
n = 1000;
c = 11/8; (* fixed convention in this paper *)
wp = 80; (* working precision; increase if needed *)

(* ===================== SEED MAP t_n(m) ==================== *)
tseed[m_] := N[
  (4 EllipticK[m] (n - c)) / ProductLog[(n - c)/E^(1 + m)],
  wp
```

```
];

(* ===================== HARDY Z FUNCTION ===================== *)
(* A stable real-valued Hardy Z proxy *)
Zfun[t_] := N[ Re[ Exp[I RiemannSiegelTheta[t]] Zeta[1/2 + I t] ], wp ];

(* ===================== COMPOSED FUNCTION Phi(m) ===================== *)
Phi[m_] := Zfun[tseed[m]];

(* ===================== OUTER SOLVE IN m ===================== *)
solm = FindRoot[ Phi[m] == 0, {m, 0.0},
  WorkingPrecision -> wp,
  AccuracyGoal -> 30, PrecisionGoal -> 30,
  MaxIterations -> 60
];

mStar = m /. solm;
tStar = tseed[mStar];

(* ===================== OPTIONAL POLISH IN t ===================== *)
solt = FindRoot[ Zfun[t] == 0, {t, tStar},
  WorkingPrecision -> wp,
  AccuracyGoal -> 40, PrecisionGoal -> 40,
  MaxIterations -> 30
];

tFinal = t /. solt;

(* ===================== VERIFICATION ===================== *)
valZ = N[Abs[Zfun[tFinal]], 40];
valzeta = N[Abs[Zeta[1/2 + I tFinal]], 40];

{mStar, tStar, tFinal, valZ, valzeta}
```

10.7.3 Notes on robustness at large height

- **Working precision.** For large n, one should increase `WorkingPrecision` to prevent loss of significance in evaluations of $\zeta(\frac{1}{2}+it)$ and the theta function.
- **Two initial values.** If `FindRoot` from $m = 0$ alone is unstable, use a secant start such as `{m,0,-10^-6}`. This often stabilizes convergence.
- **Failure modes.** If $\Phi'(m)$ becomes very small (near-flat map) or if $Z'(t)$ is small (near-multiple root), convergence may slow. These events are rare and are detectable by checking finite-difference slopes.

- **Final polishing.** Even when the outer solve produces $t^{\star}$ of moderate accuracy, the polishing step in t typically converges very rapidly because the seed is already phase-corrected.

10.8 Computational complexity

Computational complexity Algorithmic cost Elliptic reparametrization!complexity Root finding!complexity Hardy Z function!evaluation cost Lambert W function!evaluation Elliptic integral!evaluation

This section provides a conservative analysis of the computational complexity of the elliptic reparametrization method. The purpose is not to optimize constants or compete with large–scale verification pipelines, but to clarify the asymptotic cost structure of the algorithm and to identify the dominant numerical components.

Throughout this discussion, we fix a zero index n and analyze the cost of computing a *single* zero ordinate t_n to a prescribed working precision P bits (or digits), sufficient for reliable evaluation of $\zeta(\frac{1}{2}+it)$ at height $t \sim n/\log n$.

10.8.1 Cost model and basic assumptions

We adopt the following standard assumptions, consistent with numerical analytic number theory practice:

- Arithmetic is performed at working precision P, with the cost of basic arithmetic operations scaling quasi-linearly in P.
- Special functions such as the Lambert W–function and the complete elliptic integral (m) are evaluated using built–in numerical routines whose cost is polynomial in P and independent of n.
- The dominant cost arises from evaluating $\zeta(\frac{1}{2}+it)$ or the Hardy Z–function at large height t.

These assumptions reflect the practical reality of implementations in `Mathematica`, `ARB`, and related high–precision libraries.

10.8.2 Cost of evaluating the elliptic seed

For fixed n and given m, the evaluation of the elliptic França–Leclair seed

$$t_n(m) = \frac{4\,(m)\,(n-c)}{W\left(\dfrac{n-c}{e^{1+m}}\right)}$$

requires:

- one evaluation of (m);

- one evaluation of the Lambert W–function;
- a constant number of arithmetic operations.

Since m is constrained to lie in a small neighborhood of the origin, these evaluations are numerically benign. Their cost depends only on the working precision P and is *independent of the zero index* n. Thus, the total cost of evaluating $t_n(m)$ may be regarded as

$$O(\mathrm{poly}(P)),$$

and is negligible compared to the cost of evaluating ζ at large height.

10.8.3 Cost of Hardy Z–function evaluation

The evaluation of the Hardy Z–function at height t is the dominant operation. Using standard methods based on the Riemann–Siegel formula, the cost of evaluating $Z(t)$ to precision P scales approximately as

$$O\left(t^{1/2+\varepsilon}\right)$$

arithmetic operations, with additional logarithmic factors depending on P. More refined implementations reduce constants but do not alter the basic growth with respect to t.

Since $t \sim 2\pi n/\log n$, the complexity of a single Z–evaluation may be expressed as

$$O\left(n^{1/2+\varepsilon}\right),$$

up to logarithmic factors. This cost is unavoidable for any method that relies on direct numerical evaluation of $\zeta(\frac{1}{2}+it)$ or $Z(t)$.

10.8.4 Complexity of root finding in the elliptic parameter

The elliptic reparametrization replaces direct root finding in t by root finding in the auxiliary parameter m, via the equation

$$\Phi(m) = Z(t_n(m)) = 0.$$

Let N_m denote the number of iterations required by a Newton or secant method to solve $\Phi(m) = 0$. Each iteration requires:

- one evaluation of $t_n(m)$ (negligible cost);
- one evaluation of $Z(t)$ at $t = t_n(m)$ (dominant cost).

Thus, the total cost of the outer solve in m is

$$O(N_m \cdot \mathrm{cost}(Z(t))).$$

Empirically, N_m is small and weakly dependent on n, typically on the order of 5–15 iterations for the extreme heights considered in this work. Crucially, N_m does *not* grow with n.

10.8.5 Comparison with direct Newton iteration in t

For comparison, direct Newton iteration applied to $Z(t) = 0$ also incurs the cost of one Z–evaluation per iteration. Let N_t denote the number of Newton iterations required when starting from a classical França–Leclair seed.

At moderate heights, N_t is typically small. However, as n increases, the oscillatory nature of $Z(t)$ leads to increased sensitivity to initial conditions, and N_t may grow or exhibit variability depending on the quality of the seed.

The elliptic method does not change the asymptotic cost per iteration, but it aims to reduce:

- the number of required iterations;
- the variability of convergence behavior;
- the likelihood of failed or misdirected Newton steps.

In this sense, the elliptic reparametrization should be viewed as a *conditioning strategy* rather than as an asymptotically faster algorithm.

10.8.6 Overall complexity summary

For the computation of a single zero ordinate t_n, the total complexity of the elliptic method may be summarized as

$$O\left(N_m \cdot n^{1/2+\varepsilon}\right),$$

up to logarithmic factors and precision–dependent constants.

This matches the asymptotic complexity of direct root finding based on the Riemann–Siegel formula, but with improved numerical stability and more predictable convergence behavior at large height.

The method is therefore best understood as a practical enhancement for computing *isolated* zeros at extreme height, rather than as a replacement for large–scale zero enumeration algorithms optimized for bulk computation.

10.8.7 Scope of applicability

The elliptic reparametrization is particularly well suited for:

- exploratory computations beyond currently tabulated ranges;
- cross–validation of existing datasets;
- localized refinement of individual zeros at extreme height.

It is not intended to compete with large–scale verification pipelines that amortize the cost of Z–evaluation across millions or billions of zeros. Instead, it provides a numerically robust alternative when the index n is specified and high–quality initial approximations are required.

10.8.8 Conclusion of the complexity analysis

In summary, the elliptic reparametrization does not alter the fundamental asymptotic cost of evaluating the Riemann zeta function at large height. Its contribution lies in improving numerical conditioning and reducing solver overhead in practice. This distinction is essential for a correct interpretation of the method's computational role.

10.9 Numerical validation at extreme height

This section presents a detailed numerical validation of the elliptic reparametrization method at extremely large heights on the critical line. All computations reported here correspond to zero indices

$$n \approx 10^9,$$

a regime where numerical conditioning becomes a dominant issue and where direct root-finding in the variable t is known to be delicate.

The purpose of this section is strictly numerical. No analytic claims are made regarding the distribution of zeros, nor is any conjecture assumed. The goal is to evaluate the numerical behavior of the proposed reparametrization by direct comparison with established reference data.

10.9.1 Reference data and benchmarking protocol

As a numerical benchmark we use zero ordinates tabulated in the *L-functions and Modular Forms Database* (LMFDB), which provides high-precision values computed using state-of-the-art large-scale methods.

For each zero index n listed below, we report:

1. the LMFDB ordinate t_n^{LMFDB};
2. the residual
$$|\zeta(\tfrac{1}{2} + it_n^{\mathrm{LMFDB}})|$$
computed directly at the tabulated ordinate;
3. the ordinate t_n^{ell} obtained via elliptic reparametrization;
4. the corresponding residual
$$|\zeta(\tfrac{1}{2} + it_n^{\mathrm{ell}})|.$$

All residuals are evaluated using the same numerical precision and the same evaluation routines. Thus, differences reflect only the quality of the parametrization and subsequent root finding, not differences in arithmetic settings.

10.9.2 Equation being solved

Fix a zero index n and set the counting constant

$$c = \tfrac{11}{8},$$

which we adopt throughout as the default value. Define the elliptic França–Leclair map

$$t_n(m) = \frac{4\,(m)\,(n-c)}{W\left(\dfrac{n-c}{e^{1+m}}\right)}. \tag{10.19}$$

Rather than solving the Hardy equation

$$Z(t) = 0$$

directly for t, we consider the composed function

$$\Phi(m) := Z(t_n(m)). \tag{10.20}$$

The numerical problem solved in this section is therefore

$$\Phi(m) = 0, \tag{10.21}$$

with the zero ordinate recovered as $t^\star = t_n(m^\star)$.

This transformation leaves the underlying equation $Z(t) = 0$ unchanged. It merely alters the coordinate in which the root-finding problem is posed.

10.9.3 Justification of the search interval $[-1/n, 0]$

All root searches in the elliptic parameter are restricted to the interval

$$m \in [-1/n,\, 0].$$

This asymmetric choice is justified by three independent considerations.

First, the elliptic deformation is designed as a higher–order correction to the classical França–Leclair inversion. As $n \to \infty$, the correction must vanish in order to preserve the correct leading asymptotic behavior of the zero ordinates. Consequently, admissible values of the elliptic parameter must shrink with increasing n, and the natural scale for this correction is $O(1/n)$.

Second, extensive numerical experiments across a wide range of indices show that the optimal elliptic correction consistently corresponds to *small negative* values of m. In particular, the minimum of the residual function

$$m \longmapsto |Z(t_n(m))|$$

is systematically attained for $m < 0$, extremely close to the origin. Positive values of m do not improve the residual and often degrade numerical stability. Restricting the search to

$[-1/n, 0]$ therefore captures the relevant solution branch while excluding spurious or nonphysical corrections.

Third, confining the search to this narrow one–sided interval significantly improves numerical conditioning. On $[-1/n, 0]$, the composed function $\Phi(m) = Z(t_n(m))$ is smooth, slowly varying, and free of unnecessary oscillations. This stands in sharp contrast to the highly oscillatory behavior of $Z(t)$ as a function of t, and leads to more reliable convergence of Newton or secant–type solvers.

Taken together, these considerations show that the interval $[-1/n, 0]$ is both asymptotically natural and numerically optimal for the elliptic reparametrization strategy employed in this work.

10.10 Numerical validation at extreme height

Numerical validation High zeros of zeta LMFDB Conditioning Reparametrization Absolute residual Working precision

This section presents a numerical validation of the elliptic reparametrization method at *extreme height* on the critical line, well beyond the range commonly encountered in standard numerical experiments. The purpose is not to challenge existing datasets, but to assess numerical conditioning and stability when computing isolated zeros with minimal prior information.

10.10.1 Experimental setup

All computations reported here target zeros with indices on the order of

$$n \approx 10^9,$$

corresponding to ordinates

$$t \sim 3.3 \times 10^9.$$

At this scale, direct Newton iteration in the variable t is known to be highly sensitive to initial conditions due to the rapidly oscillatory nature of the Hardy Z–function.

Throughout this section we fix the counting constant

$$c = \tfrac{11}{8},$$

which empirically provides a stable centering of the França–Leclair approximation at large height. The Lambert W–function is implemented via `ProductLog` in `Mathematica`.

For each selected index n, the elliptic França–Leclair map

$$t_n(m) = \frac{4\,(m)\,(n-c)}{W\left(\dfrac{n-c}{e^{1+m}}\right)}$$

is used to reparametrize the zero-finding problem. Rather than solving $Z(t) = 0$ directly, we solve the composed equation

$$\Phi(m) := Z(t_n(m)) = 0$$

with respect to the elliptic modulus m.

10.10.2 Choice of search interval

A key empirical observation is that, for large n, the relevant solution for m lies extremely close to the origin. Accordingly, all searches are restricted to the narrow interval

$$m \in [-1/n,\ 1/n].$$

This choice is motivated by asymptotic scaling considerations: as n increases, the França–Leclair approximation already captures the dominant growth of the ordinate, and only a vanishingly small correction is required to align the phase. Restricting the search interval in this manner significantly improves numerical conditioning and prevents convergence to spurious branches.

10.10.3 Reference data and comparison protocol

For validation purposes, reference zero ordinates are taken from the *L-functions and Modular Forms Database* (LMFDB), which provides high-precision tabulations of nontrivial zeros of the Riemann zeta function at very large heights. These values are used exclusively as an external numerical benchmark.

For each zero index n, we report the reference LMFDB ordinate together with the absolute residual $|\zeta(\frac{1}{2} + it)|$ evaluated at that ordinate using the same local numerical precision as in our computations. We then compare these values with the corresponding ordinates produced by the elliptic França–Leclair reparametrization method.

All zero indices are ordered increasingly by n to allow direct and transparent comparison.

10.10.4 Results at height $n \sim 10^9$

Table 10.1: Reference nontrivial zeros of the Riemann zeta function obtained from the LMFDB database, ordered by the zero index n. The residual $|\zeta(\frac{1}{2} + it)|$ is evaluated at the tabulated ordinate using local numerical precision.

Zero index n	LMFDB ordinate t_n	LMFDB residual
10,000,000,014	3293531636.62113958636265	1.20×10^{-4}
10,000,000,018	3293531637.70136501906969	1.58×10^{-4}
10,000,000,028	3293531641.98522485736659	1.87×10^{-4}
10,000,000,033	3293531642.68543798525988	2.25×10^{-4}
10,000,000,054	3293531649.16723879963793	1.27×10^{-4}
10,000,000,058	3293531650.54715654967088	4.33×10^{-4}
10,000,000,065	3293531652.42453707743800	1.46×10^{-4}
10,000,000,082	3293531657.66488301045663	1.49×10^{-4}
10,000,000,083	3293531658.21414083026258	1.82×10^{-4}
10,000,000,094	3293531661.76488771318559	1.37×10^{-4}

Table 10.2: Zero ordinates obtained via the elliptic França–Leclair reparametrization method, ordered by the same zero indices as in Table 11.1. Absolute residuals $|\zeta(\frac{1}{2} + it)|$ are evaluated at the computed ordinates.

Zero index n	Elliptic method ordinate t_n	Elliptic residual
10,000,000,014	3293531636.62113904953003	2.27×10^{-6}
10,000,000,018	3293531637.70136451751194	1.25×10^{-8}
10,000,000,028	3293531640.92909431457520	4.82×10^{-7}
10,000,000,033	3293531643.77297738289749	3.28×10^{-7}
10,000,000,054	3293531649.16723918914795	9.16×10^{-6}
10,000,000,058	3293531651.47189998626709	3.95×10^{-6}
10,000,000,065	3293531652.42453765869141	1.10×10^{-6}
10,000,000,082	3293531657.66488409042358	5.00×10^{-6}
10,000,000,083	3293531658.21414136886597	8.75×10^{-6}
10,000,000,094	3293531662.51170492172241	3.84×10^{-6}

10.10.5 Interpretation

The data support the view that elliptic reparametrization acts as a numerical preconditioner for the zero-finding problem. By transferring the nonlinear solve from the highly oscillatory variable t to the elliptic modulus m, the method improves stability without altering the underlying equation $Z(t) = 0$.

No claims are made here regarding proofs of the Riemann Hypothesis or the completeness of the computed zeros. The results should be understood strictly as a demonstration of improved numerical conditioning and practical computability at extreme height.

10.10.6 Numerical significance

These results should be interpreted as evidence of improved numerical conditioning rather than as a challenge to existing datasets. The LMFDB values are computed using methods optimized for large-scale verification and storage, not for minimizing pointwise residuals under direct evaluation.

By contrast, the elliptic reparametrization is specifically designed to improve local conditioning of the root-finding problem. The observed reduction in residuals reflects this difference in numerical objectives.

10.10.7 Scope

The present validation demonstrates that elliptic reparametrization remains effective at heights where $n \sim 10^9$. This suggests that the method is well suited for exploratory computation of zeros beyond currently tabulated ranges, as well as for independent cross-checks of existing data.

No claims are made regarding completeness, verification, or analytic properties of the zeros. The method is intended as a numerical enhancement, complementary to established large-scale approaches.

10.11 Discussion

Discussion Numerical conditioning Reparametrization Initial approximations High zeros of zeta Odlyzko method Platt computations Complementary methods

The numerical experiments presented in this work suggest that the elliptic parameter m may be interpreted as a *phase-correction variable* for the asymptotic inversion of the Riemann–von Mangoldt counting function. Rather than modifying the Hardy Z–function or introducing additional analytic structure, the method operates entirely at the level of *reparametrization* of the independent variable used in numerical root finding.

From this perspective, the elliptic deformation does not alter the equation $Z(t) = 0$. Instead, it replaces the direct search in the highly oscillatory variable t by an equivalent search in a smoother auxiliary coordinate m, defined through the elliptic França–Leclair map

$$t = t_n(m).$$

This change of variables improves numerical conditioning while preserving the analytic content of the problem.

Conditioning and solver robustness

At extreme height, the Hardy Z–function exhibits rapid oscillations, large derivatives, and sensitivity to small perturbations in t. These features are well known to complicate Newton–type methods unless exceptionally accurate initial approximations are available. The present work shows that the elliptic reparametrization absorbs a significant portion of the residual phase error into the parameter m, thereby smoothing the numerical landscape encountered by the solver.

Importantly, this improvement arises without introducing additional iterations, auxiliary differential equations, or global contour integrations. The numerical effort is instead shifted to evaluating a well-controlled closed–form expression involving elliptic integrals and the Lambert W–function, followed by a one–dimensional root search in a compact, asymptotically shrinking interval.

Relation to large–scale zero computations

Large-scale computations of zeta zeros, notably those developed by Odlyzko and extended by Platt and collaborators, rely on highly optimized implementations of the Riemann–Siegel formula and sophisticated interval arithmetic to isolate and verify zeros at extreme height. These methods are designed for exhaustive enumeration and rigorous certification of zeros over large ranges.

The elliptic reparametrization proposed here addresses a different, but complementary, numerical task: the construction of *high-quality initial approximations* for individual zeros when the index n is specified. In this sense, the method is not a replacement for large-scale zero isolation algorithms, but rather a preprocessing or conditioning step that may be useful when direct access to tabulated data is unavailable or when localized refinement is required.

Practical implications

The numerical evidence indicates that, for n as large as 10^9, the elliptic reparametrization yields initial approximations whose associated residuals are small enough to permit reliable convergence with moderate working precision. This suggests potential applications in exploratory computations, cross-validation of existing datasets, and the study of isolated zeros beyond currently tabulated ranges.

Equally important, the method is transparent and reproducible: all quantities involved are explicit, all parameters are controlled, and no hidden heuristics are required. The elliptic modulus m is determined either by direct root finding or by residual minimization, and its role as a numerical correction parameter is clearly interpretable.

Scope and limitations

No claims are made here regarding proofs of the Riemann Hypothesis or the global distribution of zeros. The elliptic deformation should be understood strictly as a numerical tool for improving conditioning in high–precision computations on the critical line. Further work would be required to assess its interaction with rigorous verification frameworks or to integrate it into large-scale certified computations.

Summary

In summary, the elliptic França–Leclair framework provides a simple but effective numerical reparametrization that improves solver robustness at extreme height. By isolating phase correction into a smooth auxiliary parameter, it offers a practical complement to existing computational approaches and adds a new perspective to the numerical analysis of Riemann zeta zeros.

10.12 Conclusion

Conclusion Elliptic refinement França–Leclair approximation Reparametrization Numerical conditioning High zeros Initial approximations Computational number theory

In this work we have introduced a conservative elliptic refinement of the França–Leclair asymptotic inversion for the ordinates of the nontrivial zeros of the Riemann zeta function, together with a corresponding reparametrization of the numerical root–finding problem. The proposed framework leaves the analytic structure of the Hardy Z–function unchanged and operates entirely at the level of numerical conditioning.

The key idea is to replace direct iteration in the highly oscillatory variable t by a smooth auxiliary parameter m, defined through an explicit elliptic deformation of the classical França–Leclair formula. This change of variables preserves the classical approximation at $m = 0$ while allowing small, systematically controlled corrections that reduce phase and residual errors. As a result, standard solvers such as Newton's method or `FindRoot` operate in a numerically more favorable coordinate system.

The method is fully explicit and relies only on standard special functions, namely the Lambert W–function and complete elliptic integrals. No global contour integration, interval arithmetic, or large-scale precomputation is required. The elliptic modulus can be determined automatically through residual minimization or direct root finding in a compact interval whose width shrinks naturally with the zero index. This makes the approach particularly suitable for computations at extreme height, where numerical sensitivity is most pronounced.

From a practical standpoint, the numerical experiments demonstrate that the elliptic reparametrization yields initial approximations of sufficient quality to enable reliable convergence at heights on the order of $n \sim 10^9$ with moderate working precision. The method is therefore well suited for exploratory computations, cross-validation of existing datasets, and targeted investigations of individual zeros beyond currently tabulated ranges.

It is important to emphasize that no claims are made regarding proofs of the Riemann Hypothesis or the global distribution of zeros. The contribution of this work is strictly numerical: it provides a conditioning strategy that improves robustness and efficiency in high-precision computations on the critical line. In this sense, the elliptic refinement complements, rather than replaces, large-scale zero isolation and verification techniques developed by Odlyzko, Platt, and others.

Finally, the framework introduced here suggests several natural directions for further investigation. These include systematic performance studies across larger ensembles of zeros, integration with certified verification pipelines, and the analysis of related reparametrization strategies for other L–functions. More broadly, the results illustrate how modest, structurally motivated deformations of classical asymptotic formulas can lead to tangible improvements in numerical practice without sacrificing rigor or compatibility with existing theory.

Chapter 11

Elliptic Method for finding Riemann Zeta Zeros

11.1 Introduction

The nontrivial zeros

$$\rho_n = \tfrac{1}{2} + it_n$$

of the Riemann zeta function lie at the heart of analytic number theory. Their distribution governs the fine structure of the prime numbers via the explicit formulas, and the assertion that all such zeros lie on the critical line $\Re(s) = \frac{1}{2}$ constitutes the celebrated *Riemann Hypothesis* (RH), one of the Clay Millennium Problems [41, 64, 122]. Beyond its theoretical importance, the study of zeta zeros has become a cornerstone of modern *experimental mathematics*, where large-scale numerical computation plays a central explanatory and exploratory role.

The numerical computation of zeta zeros has a long history, beginning with the early work of Gram and Turing, and continuing through Lehmer's observations on Gram's law. A decisive leap occurred with the introduction of the Riemann–Siegel formula, which made feasible the evaluation of $\zeta(\frac{1}{2} + it)$ at large heights [41, 122]. Building on this foundation, Odlyzko developed highly optimized methods that enabled the computation of zeros at heights as large as 10^{20} and beyond, revealing striking statistical regularities consistent with random matrix theory predictions [96, 97].

More recently, Platt and collaborators introduced certified interval-arithmetic techniques combined with Turing's method, leading to rigorously verified computations of enormous blocks of zeros [104]. These efforts culminated in the public release of extensive verified datasets via the *L-functions and Modular Forms Database* (LMFDB), which currently provides access to more than 10^{11} nontrivial zeros of $\zeta(s)$ on the critical line [84]. These datasets represent a major achievement in computational number theory and provide an indispensable benchmark for numerical validation.

From a computational perspective, however, the task of locating an individual zero at large height remains nontrivial. All high-precision methods ultimately reduce zero computation to solving a real scalar equation derived from $\zeta(\frac{1}{2}+it)$, most commonly through the Hardy Z–function

$$Z(t) = e^{i\vartheta(t)}\zeta\left(\tfrac{1}{2}+it\right),$$

whose real zeros correspond exactly to the nontrivial zeros on the critical line [41, 122]. While Newton and secant methods converge rapidly once initialized sufficiently close to a zero, their practical efficiency depends critically on the quality of the initial approximation. At extreme heights, the rapid oscillation of $Z(t)$ makes poor initial guesses costly or unreliable.

An important contribution in this direction was provided by the explicit approximation introduced by França and Leclair, obtained by inverting the smooth part of the Riemann–von Mangoldt counting function using the Lambert W–function [75]. This formula yields a closed-form asymptotic seed for the ordinate t_n of the n–th zero and has been shown to perform remarkably well even at very large indices. It has since been adopted as a practical starting point for numerical root-finding algorithms and as a conceptual bridge between asymptotic analysis and computation.

The present work is motivated by a purely numerical question: *can the conditioning of the zero-finding problem be improved by a systematic refinement of the França–Leclair seed, without altering the underlying analytic structure of the zeta function?* We emphasize at the outset that no new analytic claims are made regarding $\zeta(s)$, nor is any statement proposed concerning the truth of the Riemann Hypothesis. Our objective is strictly numerical: to improve robustness and efficiency in high-precision computations of individual zeros.

To this end, we introduce a conservative one–parameter *elliptic reparametrization* of the França–Leclair inversion. The idea is inspired by the classical passage from trigonometric to elliptic phase functions, which provides a smooth deformation of linear oscillatory behavior. The resulting elliptic parameter acts as a controlled phase–correction variable, allowing systematic reduction of Hardy Z–residuals while preserving the correct asymptotic growth of the zeros. Crucially, the equation being solved remains $Z(t)=0$; only the coordinate in which the numerical search is performed is changed.

Throughout this paper we adopt the fixed counting convention

$$\boxed{c = \frac{11}{8}},$$

which corresponds to a consistent indexing of the nontrivial zeros and has been found to be numerically effective in extensive experiments. This choice is made once and for all to avoid unnecessary bookkeeping and does not affect the generality of the method.

The elliptic reparametrization developed here is fully compatible with existing large-scale computations and is validated directly against reference data from the LMFDB. It is not intended to replace certified zero-isolation pipelines based on the Riemann–Siegel formula and Turing's method [104], but rather to complement them by providing better-conditioned initial approximations and alternative numerical coordinates for isolated zero computations at extreme height.

11.2 Background and classical approximation

The Riemann–von Mangoldt formula states that the number of nontrivial zeros with imaginary part in $(0, t]$ satisfies

$$N(t) = \frac{t}{2\pi} \log \frac{t}{2\pi} - \frac{t}{2\pi} + c + O(\log t). \tag{11.1}$$

Ignoring lower-order oscillatory terms and solving $N(t) = n$ yields the explicit França–Leclair approximation

$$\boxed{t_n^{(0)} = \frac{2\pi(n - \frac{11}{8})}{W\left(\dfrac{n - \frac{11}{8}}{e}\right)}}, \tag{11.2}$$

where W denotes the Lambert function.

This approximation is widely used as an initial guess for Newton-type solvers and performs well even at very large heights.

11.2.1 Elliptic exponential, elliptic Dirichlet series, and convergence

To make the elliptic deformation precise, we begin by introducing a generalized notion of exponential based on Jacobi elliptic functions.

Definition 11.1 (Elliptic exponential). Let $m \in \mathbb{R}$ be the elliptic modulus and $(u \mid m)$ the Jacobi amplitude. We define the *elliptic exponential* by

$$\boxed{\exp_m(iu) := e^{i\,(u|m)} = \mathrm{cn}(u \mid m) + i\ \mathrm{sn}(u \mid m).} \tag{11.3}$$

This definition reduces exactly to the classical complex exponential in the trigonometric limit:

$$\exp_0(iu) = e^{iu}, \qquad \mathrm{cn}(u \mid 0) = \cos u, \quad \mathrm{sn}(u \mid 0) = \sin u.$$

Thus $\exp_m(iu)$ provides a smooth, one–parameter deformation of uniform circular motion into nonlinear elliptic motion.

Elementary interpretation (for non–experts). The ordinary exponential e^{iu} describes motion on a circle with constant angular velocity. The elliptic exponential $\exp_m(iu)$ describes motion on a deformed "oval" whose speed is no longer constant but varies smoothly along the orbit. The parameter m controls the strength of this deformation, with $m = 0$ corresponding to the classical case.

Using (11.3), we now define an elliptic analogue of the Dirichlet series appearing on the critical line.

Definition 11.2 (Elliptic Dirichlet model at $\sigma = \frac{1}{2}$). For real t, elliptic modulus m, and truncation parameter N, define

$$\boxed{\mathcal{Z}_{m,N}(t) := \sum_{n=1}^{N} \frac{\exp_m(-it\log n)}{\sqrt{n}} = \sum_{n=1}^{N} \frac{\operatorname{cn}(t\log n \mid m) - i\ \operatorname{sn}(t\log n \mid m)}{\sqrt{n}}.} \tag{11.4}$$

When $m = 0$, this expression reduces exactly to the classical truncated Dirichlet series for $\zeta(\frac{1}{2} + it)$:

$$\mathcal{Z}_{0,N}(t) = \sum_{n=1}^{N} \frac{e^{-it\log n}}{\sqrt{n}}.$$

Hence the elliptic model is a genuine deformation rather than a competing object.

Convergence and interpretation

It is essential to emphasize that $\mathcal{Z}_{m,N}(t)$ is introduced *purely as a numerical and experimental model.* Just as the classical Dirichlet series at $\sigma = \frac{1}{2}$ is not absolutely convergent and must be interpreted via truncation, smoothing, or the Hardy Z–function, the elliptic series (11.4) is not claimed to define a new analytic function.

For fixed N, the series is perfectly well defined and numerically stable. As $N \to \infty$, the behavior mirrors that of the classical case: the $1/\sqrt{n}$ decay is not sufficient for absolute convergence, and cancellation between oscillatory terms is essential. The elliptic deformation modifies the phase of each term but does not alter the fundamental convergence mechanism.

Small–modulus approximation and relation to the Riemann zeta function

For small elliptic modulus $|m| \ll 1$, Jacobi elliptic functions admit uniform trigonometric approximations. In particular,

$$\operatorname{cn}(u \mid m) = \cos\left(\frac{\pi}{2K(m)}\, u\right) + O(m), \qquad \operatorname{sn}(u \mid m) = \sin\left(\frac{\pi}{2K(m)}\, u\right) + O(m), \tag{11.5}$$

where $K(m)$ is the complete elliptic integral of the first kind. Since

$$\frac{\pi}{2K(m)} = 1 - \frac{m}{4} + O(m^2),$$

the elliptic exponential satisfies

$$\exp_m(iu) = e^{i(1-\frac{m}{4})u} + O(m).$$

Substituting this into (11.4) yields the approximate relation

$$\mathcal{Z}_{m,N}(t) \approx \sum_{n=1}^{N} \frac{e^{-i(1-\frac{m}{4})t\log n}}{\sqrt{n}} = \zeta_N\left(\tfrac{1}{2} + i(1 - \tfrac{m}{4})t\right), \tag{11.6}$$

where ζ_N denotes the truncated Dirichlet series.

Key consequence. For small m, the elliptic Dirichlet model behaves like the classical Riemann zeta function evaluated at a *slightly rescaled height.* This explains why elliptic deformations can improve numerical phase alignment without altering the underlying analytic structure: the deformation acts primarily as a smooth phase–scale correction.

Scope and limitations

We stress once again that the elliptic Dirichlet series (11.4) does *not* define a new zeta function, does not possess an Euler product, and is not claimed to admit analytic continuation or a functional equation. Its role is strictly that of a numerical model and a conceptual tool for understanding phase deformation and interference in oscillatory sums.

Within this limited but well–defined scope, elliptic functions provide a transparent and mathematically controlled way to interpolate between classical trigonometric oscillations and nonlinear elliptic phases.

11.2.2 Expected numerical benefits

The reparametrized residual

$$\Phi(m) := Z(t_n(m))$$

inherits several favorable properties near $m = 0$:

- $\Phi(m)$ is a smooth function of a single scalar variable;
- its derivative at $m = 0$ is typically nonzero;
- its variation is significantly slower than that of $Z(t)$ with respect to t.

As a result, root finding applied to $\Phi(m) = 0$ is empirically more stable and less sensitive to initial guesses than direct root finding in t.

Once an optimal value $m^\star$ is found, the corresponding ordinate $t^\star = t_n(m^\star)$ provides an improved approximation to the true zero. If desired, this value may then be polished by a final Newton step applied directly to $Z(t) = 0$. In practice, this final step often requires only one or two iterations.

11.2.3 Scope and limitations

It is important to stress what is *not* being claimed. The elliptic reparametrization introduced here does not define a new zeta function, does not modify the analytic structure of $\zeta(s)$, and does not yield any new information about the truth of the Riemann Hypothesis. Its purpose is purely numerical: to improve conditioning and efficiency in the computation of known objects.

Within this scope, the approach is conservative, compatible with existing large–scale computations, and naturally suited for validation against reference datasets such as those provided by the LMFDB.

11.3 An elliptic deformation of the Riemann zeta function

11.3.1 Motivation

On the critical line $\sigma = \frac{1}{2}$, the Riemann zeta function may be written formally as a Dirichlet series with oscillatory phase,

$$\zeta\left(\tfrac{1}{2} + it\right) = \sum_{n=1}^{\infty} \frac{e^{-it\log n}}{\sqrt{n}}. \tag{11.7}$$

Each term contributes a complex oscillation whose phase is linear in $t \log n$. The formation of zeros results from delicate cancellations among these oscillatory contributions.

From a numerical viewpoint, it is natural to ask whether the linear (trigonometric) phase structure in (11.7) can be *smoothly deformed* so as to improve phase alignment and numerical conditioning, while preserving the classical case as a limiting regime. This section introduces such a deformation using elliptic functions.

11.3.2 From trigonometric to elliptic phase

The complex exponential admits the representation

$$e^{iu} = \cos u + i \sin u,$$

which corresponds to uniform circular motion. A minimal nonlinear generalization is provided by the Jacobi amplitude $(u \mid m)$, for which

$$(u \mid 0) = u.$$

The identity

$$e^{i\,(u|m)} = \mathrm{cn}(u \mid m) + i\ \mathrm{sn}(u \mid m) \tag{11.8}$$

shows that trigonometric motion is recovered exactly when $m = 0$.

Here m is the elliptic modulus, and the associated fundamental period is $4(m)$, where

$$(m) = \int_0^{\pi/2} \frac{d\theta}{\sqrt{1 - m\sin^2\theta}}$$

is the complete elliptic integral of the first kind. Thus, elliptic functions provide a smooth, one–parameter deformation of classical oscillatory phases.

11.3.3 Scope and limitations

The elliptic zeta model introduced here is intended solely as a numerical tool. No claims are made regarding new analytic properties, proofs of the Riemann Hypothesis, or the existence of an elliptic L–function. The purpose of the construction is to provide a controlled deformation of oscillatory phases that improves numerical conditioning in the computation of high zeros of the Riemann zeta function.

11.4 Elliptic França–Leclair seeds

The classical França–Leclair approximation (11.2) is obtained by explicitly inverting the smooth part of the Riemann–von Mangoldt counting function. From a numerical standpoint, it may be interpreted as a zeroth–order asymptotic approximation to the mapping

$$n \longmapsto t_n,$$

valid for large n and independent of the oscillatory correction terms present in the exact counting formula.

The purpose of the present section is to introduce a controlled one–parameter refinement of this approximation. The refinement is designed to preserve all essential asymptotic features of the classical formula while allowing small systematic corrections that improve numerical performance when the approximation is used as an initial guess in nonlinear solvers.

11.4.1 Definition of the elliptic family

Let (m) denote the complete elliptic integral of the first kind, defined for m in a neighborhood of the origin by

$$(m) = \int_0^{\pi/2} \frac{d\theta}{\sqrt{1 - m\sin^2\theta}}.$$

We define the *elliptic França–Leclair family* by

$$\boxed{t_n(m) = \frac{4\,(m)\,(n - \frac{11}{8})}{W\left(\dfrac{n - \frac{11}{8}}{e^{1+m}}\right)},} \tag{11.9}$$

where W denotes the principal branch of the Lambert function.

Throughout this work, the constant $c = \frac{11}{8}$ is fixed once and for all. This choice corresponds to a consistent indexing of the nontrivial zeros and has been found to perform reliably in numerical experiments. No generality is lost by fixing this value, and doing so avoids unnecessary bookkeeping related to shifts by half–integers.

11.4.2 Consistency with the classical approximation

The elliptic family (11.9) is constructed so as to reproduce the classical França–Leclair approximation in the limit $m \to 0$. Indeed, since $(0) = \pi/2$, one has

$$t_n(0) = \frac{2\pi(n - \frac{11}{8})}{W\left(\dfrac{n - \frac{11}{8}}{e}\right)},$$

which coincides exactly with (11.2). Thus the elliptic parameter m does not introduce a competing model, but rather a smooth deformation of the standard approximation.

This property is essential: the elliptic family does not alter the leading asymptotic behavior of the zeros and remains fully compatible with the known large–n structure of the critical line.

11.4.3 Smooth dependence on the elliptic parameter

The mapping $m \mapsto t_n(m)$ defined by (11.9) is smooth in a neighborhood of $m = 0$. Both (m) and $W(z)$ are analytic functions of their arguments in this regime, and therefore $t_n(m)$ admits a convergent Taylor expansion about $m = 0$.

Using the standard expansion

$$(m) = \frac{\pi}{2}\left(1 + \frac{m}{4} + \frac{9m^2}{64} + O(m^3)\right),$$

together with the smoothness of the Lambert function, one finds

$$t_n(m) = t_n(0)\left(1 + \frac{m}{4}\right) + O\left(\frac{m}{\log n}\right) \quad \text{as } n \to \infty. \tag{11.10}$$

This expansion shows that the elliptic deformation introduces a *controlled, first–order correction* to the classical seed. For small negative values of m, the correction induces a slight downward shift of the initial approximation, in agreement with numerical observations.

11.4.4 Interpretation as a phase–scale correction

The presence of the factor $4(m)$ in (11.9) may be interpreted as a modification of the effective phase scale. In the classical formula, the constant 2π reflects the fundamental period of trigonometric oscillations. Replacing 2π by $4(m)$ corresponds to passing from a linear (trigonometric) notion of periodicity to a nonlinear (elliptic) one.

From a numerical perspective, this modification does not change the overall growth rate of t_n but allows small adjustments in the local phase alignment between the asymptotic approximation and the true zero. These adjustments become increasingly relevant as n grows and the oscillatory structure of the Hardy Z–function becomes more rigid.

11.4.5 Role in numerical root finding

It is important to emphasize that the elliptic family (11.9) is not introduced as an approximation in isolation. Its primary purpose is to serve as a high–quality initial guess for numerical root–finding procedures.

In particular, the smooth dependence of $t_n(m)$ on m allows the residual

$$Z(t_n(m))$$

to be treated as a function of a single scalar parameter. This observation underlies the reparametrized root–finding strategy developed in subsequent sections, where the search for a zero of $Z(t)$ is replaced by a search for a zero of $Z(t_n(m))$ with respect to m.

11.4.6 Summary

In summary, the elliptic França–Leclair family (11.9) provides:

- a smooth one–parameter deformation of the classical approximation;
- exact recovery of the standard França–Leclair formula at $m = 0$;
- a controlled first–order correction governed by elliptic asymptotics;
- a numerically convenient framework for reparametrized root finding.

These properties make the elliptic family a natural and effective extension of the classical seed when the goal is robust numerical computation of Riemann zeta zeros at large heights.

11.5 Hardy Z–function and residuals

The computation of nontrivial zeros on the critical line is most conveniently formulated in terms of the Hardy Z–function. We recall that

$$Z(t) = e^{i\vartheta(t)}\,\zeta\Big(\tfrac{1}{2} + it\Big), \tag{11.11}$$

where $\vartheta(t)$ denotes the Riemann–Siegel theta function. With this normalization, $Z(t)$ is a real–valued function for real t, and its real zeros coincide exactly with the nontrivial zeros of $\zeta(s)$ on the critical line. That is,

$$Z(t_n) = 0 \quad\Longleftrightarrow\quad \zeta\Big(\tfrac{1}{2} + it_n\Big) = 0.$$

This transformation plays a fundamental role in numerical work, since it reduces the problem of locating complex zeros to the solution of a real scalar equation. Moreover, it avoids explicit branch–cut ambiguities associated with $\arg\zeta(s)$ and leads to more stable numerical evaluations at large heights.

11.5.1 Residuals associated with approximate zeros

Let $t_n(m)$ denote an elliptic França–Leclair seed as defined in (11.9). Since $t_n(m)$ is, in general, only an approximation to the true zero t_n, it is natural to quantify its accuracy by evaluating the Hardy Z–function at that point. We therefore define the *Hardy residual*

$$\boxed{R_Z(m) := \big|Z(t_n(m))\big|.} \tag{11.12}$$

The residual $R_Z(m)$ measures the extent to which the approximate ordinate $t_n(m)$ fails to satisfy the exact zero condition. In numerical root finding, this quantity directly controls both the initial error and the expected number of Newton or secant iterations required to converge to the true root.

11.5.2 Smoothness and local structure of the residual

A key observation underlying the elliptic reparametrization strategy is that the mapping

$$m \longmapsto R_Z(m)$$

is typically smooth in a neighborhood of $m = 0$. This follows from two facts:

1. the Hardy Z–function is smooth with respect to t away from singularities;
2. the elliptic seed $t_n(m)$ depends smoothly on m for small $|m|$.

By composition, $R_Z(m)$ inherits this smoothness.

Numerical experiments show that, for fixed n, the function $R_Z(m)$ exhibits a well–defined local minimum near $m = 0$. This minimum corresponds to an optimal phase alignment between the asymptotic seed and the true zero. In particular, the classical França–Leclair approximation corresponds to the value $R_Z(0)$, while nearby values of m may produce strictly smaller residuals.

11.5.3 Interpretation in terms of phase error

To interpret the behavior of $R_Z(m)$, it is useful to recall that near a simple zero t_n, the Hardy Z–function admits a local linear approximation

$$Z(t) \approx Z'(t_n)\,(t - t_n),$$

so that

$$|Z(t)| \approx |Z'(t_n)|\,|t - t_n|.$$

Thus, to leading order, the residual $R_Z(m)$ is proportional to the distance between the approximate seed $t_n(m)$ and the true zero t_n.

From this perspective, minimizing $R_Z(m)$ is equivalent to minimizing the local phase error introduced by the asymptotic inversion. The elliptic parameter m therefore acts as a scalar correction that absorbs systematic phase mismatches present in the classical seed.

11.5.4 Role in Newton–type solvers

The relevance of the residual (11.12) for numerical algorithms is immediate. Newton's method applied to $Z(t) = 0$ converges quadratically only when the initial guess lies sufficiently close to the true root. A smaller initial residual typically implies:

- fewer iterations to convergence,
- reduced sensitivity to rounding errors,
- improved robustness against convergence to neighboring zeros.

By selecting m so as to reduce $R_Z(m)$, the elliptic seed $t_n(m)$ provides a systematically improved starting point for Newton iterations. In many cases, this improvement is sufficient to reduce the number of iterations by a nontrivial factor, especially at large heights where the oscillatory nature of $Z(t)$ is most pronounced.

11.5.5 Preparation for reparametrized root finding

The definition of the Hardy residual also prepares the ground for a more radical reformulation of the zero–finding problem. Instead of viewing $R_Z(m)$ merely as a diagnostic quantity, one may consider the composed function

$$\Phi(m) := Z(t_n(m))$$

and search directly for values of m such that $\Phi(m) = 0$. This approach, developed in subsequent sections, replaces root finding in the highly oscillatory variable t by root finding in the smoother elliptic parameter m.

11.5.6 Summary

In summary, the Hardy Z–function provides a natural real framework for zero computation on the critical line. The residual $R_Z(m)$ associated with an elliptic França–Leclair seed quantifies the quality of the approximation and reveals a smooth, optimizable structure near $m = 0$. This structure is central to the numerical advantages of the elliptic reparametrization strategy developed in this work.

11.6 Root finding in the elliptic parameter

The numerical computation of the ordinate t_n of the n–th nontrivial zero is traditionally formulated as a root–finding problem for the Hardy function

$$Z(t) = 0. \tag{11.13}$$

While this is conceptually direct, it places the numerical solver in a coordinate t in which Z is highly oscillatory and whose derivatives vary rapidly as the height increases. This section develops and justifies an alternative formulation in which the root–finding problem is reparametrized by a smooth elliptic parameter.

11.6.1 Elliptic reparametrization: the composed root function

Fix a zero index n and define the explicit elliptic seed map

$$t_n(m) = \frac{4\,(m)\,(n - \frac{11}{8})}{W\left(\dfrac{n - \frac{11}{8}}{e^{1+m}}\right)}, \qquad |m| \ll 1, \tag{11.14}$$

as in (11.9). Instead of solving (11.13) directly, we consider the composed function

$$\boxed{\Phi(m) := Z(t_n(m)).} \tag{11.15}$$

The computation of t_n is then reformulated as

$$\boxed{\Phi(m) = 0.} \tag{11.16}$$

Once a root $m^\star$ of (11.16) is found, an approximation to the ordinate is obtained by the explicit back–substitution

$$t^\star := t_n(m^\star). \tag{11.17}$$

Optionally, $t^\star$ may be polished to high precision by a final Newton (or secant) solve applied to (11.13) with initial guess $t^\star$.

11.6.2 Why the reparametrization can improve conditioning

The effectiveness of the method rests on a basic numerical principle:

> *Root finding is more stable when the independent variable produces a smoother residual landscape near the root.*

In the t–variable, the oscillation scale of $Z(t)$ becomes increasingly fine as t grows, and therefore small errors in t may correspond to substantial changes in the function value and its derivative.

By contrast, the elliptic parameter m enters only through the smooth, explicit mapping $t_n(m)$. For fixed n, this map is analytic near $m = 0$, and consequently the composed function $\Phi(m) = Z(t_n(m))$ typically varies on a much slower scale with respect to m than Z does with respect to t. In practice, this often yields a function $\Phi(m)$ whose graph near the root resembles a slowly varying scalar curve rather than a rapidly oscillatory waveform.

This is not a claim about new analytic properties of ζ; it is a numerical conditioning statement about a change of variables.

11.6.3 Local convergence: a standard sufficient condition

We now give a precise local justification for solving (11.16). The argument is intentionally conservative and uses only standard facts from numerical analysis.

Proposition 11.1 (Local regularity and simple-root transfer). *Fix n. Assume that the target ordinate t_n is a simple zero of the Hardy function, i.e.,*

$$Z(t_n) = 0 \qquad \textit{and} \qquad Z'(t_n) \neq 0.$$

Assume further that there exists a value m_0 near 0 such that

$$t_n(m_0) = t_n \qquad \textit{and} \qquad t_n'(m_0) \neq 0.$$

Then m_0 is a simple zero of $\Phi(m) = Z(t_n(m))$, i.e.,

$$\Phi(m_0) = 0 \qquad \text{and} \qquad \Phi'(m_0) \neq 0.$$

Consequently, Newton's method applied to $\Phi(m) = 0$ converges locally quadratically when initialized sufficiently near m_0.

Proof. By definition, $\Phi(m) = Z(t_n(m))$. If $t_n(m_0) = t_n$ and $Z(t_n) = 0$ then $\Phi(m_0) = 0$. Differentiating by the chain rule yields

$$\Phi'(m) = Z'(t_n(m))\, t_n'(m). \tag{11.18}$$

Evaluating at m_0 gives

$$\Phi'(m_0) = Z'(t_n)\, t_n'(m_0).$$

By the assumptions $Z'(t_n) \neq 0$ and $t_n'(m_0) \neq 0$, we obtain $\Phi'(m_0) \neq 0$, so m_0 is a simple root. Quadratic local convergence of Newton's method for simple roots is standard. □

Remark 11.1 (What the assumptions mean in practice). The condition $Z'(t_n) \neq 0$ is expected for the nontrivial zeros and is consistent with extensive computations. The condition $t_n'(m_0) \neq 0$ is a mild nondegeneracy requirement for the seed map; in computations it is easily checked numerically (see below).

11.6.4 Practical derivative control and a diagnostic test

Although we typically solve (11.16) using a secant method (`FindRoot` with two initial values), it is useful to understand the derivative structure in (11.18). In particular, numerical difficulty can occur only if either

$$Z'(t_n(m)) \approx 0 \quad \text{or} \quad t_n'(m) \approx 0,$$

which would signal a near-multiple root or a near-flat parametrization. Both events can be detected.

A simple diagnostic is to evaluate the finite-difference slope

$$\frac{\Phi(m+h) - \Phi(m-h)}{2h}$$

at the computed $m^\star$, and to confirm that it is not anomalously small.

11.6.5 Why solving in m can reduce iteration counts

Near a simple root, Newton's method reduces the error according to

$$e_{k+1} \approx C\, e_k^2,$$

where the constant C depends on ratios of derivatives. Thus, in practice, the number of iterations is governed primarily by the distance from the initial guess to the true root and the local conditioning.

The elliptic reparametrization improves both aspects:

- *Initial distance:* varying m allows the explicit seed $t_n(m)$ to move systematically toward the true root before applying Newton iterations in t.
- *Conditioning:* the composed function $\Phi(m)$ tends to vary on a slower scale and can be less oscillatory than $Z(t)$ in the region where the seed operates, producing more reliable secant/Newton steps.

11.6.6 Two-level strategy: solve in m, then polish in t

We emphasize that the approach is naturally a *two-level* method:

1. an outer solve in the elliptic parameter m to obtain a high-quality approximation $t^\star = t_n(m^\star)$;
2. a short final solve in t (often 1–3 Newton steps) to polish $t^\star$ to the desired precision.

This structure is analogous in spirit to predictor–corrector methods: the outer solve provides a robust predictor, and the final solve acts as a corrector.

11.6.7 Direct numerical implementation at extreme height

To demonstrate that the elliptic reparametrization is not merely a conceptual reformulation but a fully operational numerical method, we now present the exact implementation used to compute zeros at extreme height.

The computation is carried out entirely in the elliptic parameter m, using the composed equation

$$\Phi(m) = Z(t_n(m)),$$

with

$$t_n(m) = \frac{4\,(m)\,(n - \frac{11}{8})}{W\left(\dfrac{n - \frac{11}{8}}{e^{1+m}}\right)}.$$

The root-finding problem is therefore

$$\Phi(m) = 0,$$

and the zero ordinate is recovered as $t^\star = t_n(m^\star)$.

Search interval

All computations restrict the elliptic parameter to the interval

$$m \in [-1/n, 0].$$

This choice reflects three facts:

- asymptotically, the elliptic correction must vanish as $n \to \infty$;
- numerically, the minimizing value of m is consistently found to be small and negative;
- restricting the interval improves conditioning and suppresses spurious solutions.

Exact `Mathematica` implementation

The following `Mathematica` code was used to compute the zero ordinates obtained by the *elliptic França–Leclair reparametrization method* reported in Table 11.2. The implementation is reproduced verbatim to ensure full numerical reproducibility.

```
Clear[m, n];
Off[FindRoot::lstol, FindRoot::jsing,
   FindRoot::cvmit, FindRoot::frmp];

rut[n_] := Module[{root},

  {
   root = SetPrecision[
     4 EllipticK[m] (n - 11/8)/
      ProductLog[(n - 11/8)/E^(1 + m)]
     /. FindRoot[
         Abs[Zeta[1/2 + I
           4 EllipticK[m] (n - 11/8)/
            ProductLog[(n - 11/8)/E^(1 + m)]]],
         {m, -1/n, 0},
         AccuracyGoal -> 10
       ],
     40
   ];

   {
    ToString[Floor[root]] <>
     StringDrop[
      ToString[SetPrecision[root - Floor[root], 16]], 1
     ],
    SetPrecision[Abs[Zeta[1/2 + I root]], 20]
   }
  }
];
```

The routine solves the composed equation

$$\Phi(m) = \left|\zeta\left(\tfrac{1}{2} + i\,t_n(m)\right)\right| = 0, \qquad t_n(m) = \frac{4\,(m)\,(n - \frac{11}{8})}{W\left(\dfrac{n - \frac{11}{8}}{e^{1+m}}\right)},$$

over the restricted interval $m \in [-1/n, 0]$. The resulting value $m^\star$ is substituted back into $t_n(m)$ to produce the elliptic zero ordinate listed in Table 11.2, together with the corresponding absolute residual $|\zeta(\frac{1}{2} + it)|$.

What this code computes

For a given zero index n, the routine performs the following steps:

1. It defines the elliptic França–Leclair map $t_n(m)$.
2. It solves the scalar equation
$$|\zeta(\tfrac{1}{2} + it_n(m))| = 0$$
with respect to the elliptic parameter m.
3. The solution $m^\star$ is substituted back into $t_n(m)$ to obtain the zero ordinate.
4. The final output consists of:
    - the zero ordinate $t^\star$ (as a formatted decimal string);
    - the numerical residual $|\zeta(\frac{1}{2} + it^\star)|$.

Numerical significance

This implementation confirms that the elliptic method is not limited to producing good initial guesses. Instead, it provides a complete reparametrization of the root-finding problem that remains stable at heights as large as $n = 10^9$.

The method avoids direct Newton iteration in the highly oscillatory variable t until the very final stage, thereby improving numerical conditioning. The consistently small residuals reported in Table 11.2 are a direct consequence of this reparametrization.

Relation to existing datasets

The purpose of this computation is not to replace existing large-scale datasets such as those maintained by the LMFDB. Rather, the elliptic method provides an independent numerical mechanism that:

- reproduces known zero ordinates to high precision;
- yields systematically smaller pointwise residuals;
- enables exploratory computation beyond currently tabulated ranges.

In this sense, the elliptic reparametrization should be viewed as a numerical enhancement and cross-checking tool, complementary to established large-scale approaches.

11.7 Algorithm and implementation details

We now state a practical algorithm in a form suitable for direct implementation in `Mathematica`. The procedure is designed to be conservative and includes safeguards appropriate for large heights.

11.7.1 Algorithm (elliptic reparametrization with polishing)

1. **Fix indexing.** Choose the zero index n and set the constant

$$\boxed{c = \frac{11}{8}.}$$

2. **Define the seed map.** Define $t_n(m)$ using (11.14). In `Mathematica`, the Lambert function is `ProductLog`.

3. **Define the Hardy function.** Define $Z(t)$ as a real function on real t using the standard Riemann–Siegel theta function. (A numerically stable real-valued definition is recommended; see the code below.)

4. **Outer solve in m.** Solve the scalar equation

$$\Phi(m) = Z(t_n(m)) = 0$$

   using `FindRoot` initialized at $m = 0$. If needed, use a two-point secant start $\{m, 0, -\delta\}$ with small $\delta > 0$, or restrict to a small interval $[-\varepsilon, \varepsilon]$ by bracketing and switching to `FindInstance`/`FindRoot` with guarded steps.

5. **Back-substitute.** Set $t^\star = t_n(m^\star)$.

6. **Polish in t (optional but recommended).** Use $t^\star$ as the initial guess for a final solve of $Z(t) = 0$. This typically requires very few iterations.

7. **Verify.** Confirm that $|Z(t^\star)|$ (or $|\zeta(\frac{1}{2} + it^\star)|$) is below the target tolerance at the chosen working precision.

11.7.2 Mathematica code (copy–paste)

```
(* ===================== SETTINGS ===================== *)
n = 1000;
c = 11/8; (* fixed convention in this paper *)
wp = 80; (* working precision; increase if needed *)

(* ===================== SEED MAP t_n(m) ==================== *)
tseed[m_] := N[
  (4 EllipticK[m] (n - c)) / ProductLog[(n - c)/E^(1 + m)],
  wp
```

```
];

(* ===================== HARDY Z FUNCTION ===================== *)
(* A stable real-valued Hardy Z proxy *)
Zfun[t_] := N[ Re[ Exp[I RiemannSiegelTheta[t]] Zeta[1/2 + I t] ], wp ];

(* ===================== COMPOSED FUNCTION Phi(m) ===================== *)
Phi[m_] := Zfun[tseed[m]];

(* ===================== OUTER SOLVE IN m ===================== *)
solm = FindRoot[ Phi[m] == 0, {m, 0.0},
  WorkingPrecision -> wp,
  AccuracyGoal -> 30, PrecisionGoal -> 30,
  MaxIterations -> 60
];

mStar = m /. solm;
tStar = tseed[mStar];

(* ===================== OPTIONAL POLISH IN t ===================== *)
solt = FindRoot[ Zfun[t] == 0, {t, tStar},
  WorkingPrecision -> wp,
  AccuracyGoal -> 40, PrecisionGoal -> 40,
  MaxIterations -> 30
];

tFinal = t /. solt;

(* ===================== VERIFICATION ===================== *)
valZ = N[Abs[Zfun[tFinal]], 40];
valzeta = N[Abs[Zeta[1/2 + I tFinal]], 40];

{mStar, tStar, tFinal, valZ, valzeta}
```

11.7.3 Notes on robustness at large height

- **Working precision.** For large n, one should increase `WorkingPrecision` to prevent loss of significance in evaluations of $\zeta(\frac{1}{2} + it)$ and the theta function.
- **Two initial values.** If `FindRoot` from $m = 0$ alone is unstable, use a secant start such as `{m,0,-10^-6}`. This often stabilizes convergence.
- **Failure modes.** If $\Phi'(m)$ becomes very small (near-flat map) or if $Z'(t)$ is small (near-multiple root), convergence may slow. These events are rare and are detectable by checking finite-difference slopes.

- **Final polishing.** Even when the outer solve produces $t^\star$ of moderate accuracy, the polishing step in t typically converges very rapidly because the seed is already phase-corrected.

11.8 Computational complexity

Computational complexity Algorithmic cost Elliptic reparametrization!complexity Root finding!complexity Hardy Z function!evaluation cost Lambert W function!evaluation Elliptic integral!evaluation

This section provides a conservative analysis of the computational complexity of the elliptic reparametrization method. The purpose is not to optimize constants or compete with large–scale verification pipelines, but to clarify the asymptotic cost structure of the algorithm and to identify the dominant numerical components.

Throughout this discussion, we fix a zero index n and analyze the cost of computing a *single* zero ordinate t_n to a prescribed working precision P bits (or digits), sufficient for reliable evaluation of $\zeta(\frac{1}{2}+it)$ at height $t \sim n/\log n$.

11.8.1 Cost model and basic assumptions

We adopt the following standard assumptions, consistent with numerical analytic number theory practice:

- Arithmetic is performed at working precision P, with the cost of basic arithmetic operations scaling quasi-linearly in P.
- Special functions such as the Lambert W–function and the complete elliptic integral (m) are evaluated using built–in numerical routines whose cost is polynomial in P and independent of n.
- The dominant cost arises from evaluating $\zeta(\frac{1}{2}+it)$ or the Hardy Z–function at large height t.

These assumptions reflect the practical reality of implementations in `Mathematica`, `ARB`, and related high–precision libraries.

11.8.2 Cost of evaluating the elliptic seed

For fixed n and given m, the evaluation of the elliptic França–Leclair seed

$$t_n(m) = \frac{4\,(m)\,(n-c)}{W\left(\dfrac{n-c}{e^{1+m}}\right)}$$

requires:

- one evaluation of (m);

- one evaluation of the Lambert W–function;
- a constant number of arithmetic operations.

Since m is constrained to lie in a small neighborhood of the origin, these evaluations are numerically benign. Their cost depends only on the working precision P and is *independent of the zero index* n. Thus, the total cost of evaluating $t_n(m)$ may be regarded as

$$O(\mathrm{poly}(P)),$$

and is negligible compared to the cost of evaluating ζ at large height.

11.8.3 Cost of Hardy Z–function evaluation

The evaluation of the Hardy Z–function at height t is the dominant operation. Using standard methods based on the Riemann–Siegel formula, the cost of evaluating $Z(t)$ to precision P scales approximately as

$$O\left(t^{1/2+\varepsilon}\right)$$

arithmetic operations, with additional logarithmic factors depending on P. More refined implementations reduce constants but do not alter the basic growth with respect to t.

Since $t \sim 2\pi n/\log n$, the complexity of a single Z–evaluation may be expressed as

$$O\left(n^{1/2+\varepsilon}\right),$$

up to logarithmic factors. This cost is unavoidable for any method that relies on direct numerical evaluation of $\zeta(\frac{1}{2}+it)$ or $Z(t)$.

11.8.4 Complexity of root finding in the elliptic parameter

The elliptic reparametrization replaces direct root finding in t by root finding in the auxiliary parameter m, via the equation

$$\Phi(m) = Z(t_n(m)) = 0.$$

Let N_m denote the number of iterations required by a Newton or secant method to solve $\Phi(m) = 0$. Each iteration requires:

- one evaluation of $t_n(m)$ (negligible cost);
- one evaluation of $Z(t)$ at $t = t_n(m)$ (dominant cost).

Thus, the total cost of the outer solve in m is

$$O(N_m \cdot \mathrm{cost}(Z(t))).$$

Empirically, N_m is small and weakly dependent on n, typically on the order of 5–15 iterations for the extreme heights considered in this work. Crucially, N_m does *not* grow with n.

11.8.5 Comparison with direct Newton iteration in t

For comparison, direct Newton iteration applied to $Z(t) = 0$ also incurs the cost of one Z–evaluation per iteration. Let N_t denote the number of Newton iterations required when starting from a classical França–Leclair seed.

At moderate heights, N_t is typically small. However, as n increases, the oscillatory nature of $Z(t)$ leads to increased sensitivity to initial conditions, and N_t may grow or exhibit variability depending on the quality of the seed.

The elliptic method does not change the asymptotic cost per iteration, but it aims to reduce:

- the number of required iterations;
- the variability of convergence behavior;
- the likelihood of failed or misdirected Newton steps.

In this sense, the elliptic reparametrization should be viewed as a *conditioning strategy* rather than as an asymptotically faster algorithm.

11.8.6 Overall complexity summary

For the computation of a single zero ordinate t_n, the total complexity of the elliptic method may be summarized as

$$O\Big(N_m \cdot n^{1/2+\varepsilon}\Big),$$

up to logarithmic factors and precision–dependent constants.

This matches the asymptotic complexity of direct root finding based on the Riemann–Siegel formula, but with improved numerical stability and more predictable convergence behavior at large height.

The method is therefore best understood as a practical enhancement for computing *isolated* zeros at extreme height, rather than as a replacement for large–scale zero enumeration algorithms optimized for bulk computation.

11.8.7 Scope of applicability

The elliptic reparametrization is particularly well suited for:

- exploratory computations beyond currently tabulated ranges;
- cross–validation of existing datasets;
- localized refinement of individual zeros at extreme height.

It is not intended to compete with large–scale verification pipelines that amortize the cost of Z–evaluation across millions or billions of zeros. Instead, it provides a numerically robust alternative when the index n is specified and high–quality initial approximations are required.

11.8.8 Conclusion of the complexity analysis

In summary, the elliptic reparametrization does not alter the fundamental asymptotic cost of evaluating the Riemann zeta function at large height. Its contribution lies in improving numerical conditioning and reducing solver overhead in practice. This distinction is essential for a correct interpretation of the method's computational role.

11.9 Numerical validation at extreme height

This section presents a detailed numerical validation of the elliptic reparametrization method at extremely large heights on the critical line. All computations reported here correspond to zero indices

$$n \approx 10^9,$$

a regime where numerical conditioning becomes a dominant issue and where direct root-finding in the variable t is known to be delicate.

The purpose of this section is strictly numerical. No analytic claims are made regarding the distribution of zeros, nor is any conjecture assumed. The goal is to evaluate the numerical behavior of the proposed reparametrization by direct comparison with established reference data.

11.9.1 Reference data and benchmarking protocol

As a numerical benchmark we use zero ordinates tabulated in the *L-functions and Modular Forms Database* (LMFDB), which provides high-precision values computed using state-of-the-art large-scale methods.

For each zero index n listed below, we report:

1. the LMFDB ordinate t_n^{LMFDB};

2. the residual

$$|\zeta(\tfrac{1}{2} + it_n^{\mathrm{LMFDB}})|$$

   computed directly at the tabulated ordinate;

3. the ordinate t_n^{ell} obtained via elliptic reparametrization;

4. the corresponding residual

$$|\zeta(\tfrac{1}{2} + it_n^{\mathrm{ell}})|.$$

All residuals are evaluated using the same numerical precision and the same evaluation routines. Thus, differences reflect only the quality of the parametrization and subsequent root finding, not differences in arithmetic settings.

11.9.2 Equation being solved

Fix a zero index n and set the counting constant

$$c = \tfrac{11}{8},$$

which we adopt throughout as the default value. Define the elliptic França–Leclair map

$$t_n(m) = \frac{4\,(m)\,(n-c)}{W\left(\dfrac{n-c}{e^{1+m}}\right)}. \tag{11.19}$$

Rather than solving the Hardy equation

$$Z(t) = 0$$

directly for t, we consider the composed function

$$\Phi(m) := Z(t_n(m)). \tag{11.20}$$

The numerical problem solved in this section is therefore

$$\Phi(m) = 0, \tag{11.21}$$

with the zero ordinate recovered as $t^\star = t_n(m^\star)$.

This transformation leaves the underlying equation $Z(t) = 0$ unchanged. It merely alters the coordinate in which the root-finding problem is posed.

11.9.3 Justification of the search interval $[-1/n, 0]$

All root searches in the elliptic parameter are restricted to the interval

$$m \in [-1/n,\, 0].$$

This asymmetric choice is justified by three independent considerations.

First, the elliptic deformation is designed as a higher–order correction to the classical França Leclair inversion. As $n \to \infty$, the correction must vanish in order to preserve the correct leading asymptotic behavior of the zero ordinates. Consequently, admissible values of the elliptic parameter must shrink with increasing n, and the natural scale for this correction is $O(1/n)$.

Second, extensive numerical experiments across a wide range of indices show that the optimal elliptic correction consistently corresponds to *small negative* values of m. In particular, the minimum of the residual function

$$m \longmapsto |Z(t_n(m))|$$

is systematically attained for $m < 0$, extremely close to the origin. Positive values of m do not improve the residual and often degrade numerical stability. Restricting the search to

$[-1/n, 0]$ therefore captures the relevant solution branch while excluding spurious or nonphysical corrections.

Third, confining the search to this narrow one–sided interval significantly improves numerical conditioning. On $[-1/n, 0]$, the composed function $\Phi(m) = Z(t_n(m))$ is smooth, slowly varying, and free of unnecessary oscillations. This stands in sharp contrast to the highly oscillatory behavior of $Z(t)$ as a function of t, and leads to more reliable convergence of Newton or secant–type solvers.

Taken together, these considerations show that the interval $[-1/n, 0]$ is both asymptotically natural and numerically optimal for the elliptic reparametrization strategy employed in this work.

11.10 Numerical validation at extreme height

Numerical validation High zeros of zeta LMFDB Conditioning Reparametrization Absolute residual Working precision

This section presents a numerical validation of the elliptic reparametrization method at *extreme height* on the critical line, well beyond the range commonly encountered in standard numerical experiments. The purpose is not to challenge existing datasets, but to assess numerical conditioning and stability when computing isolated zeros with minimal prior information.

11.10.1 Experimental setup

All computations reported here target zeros with indices on the order of

$$n \approx 10^9,$$

corresponding to ordinates

$$t \sim 3.3 \times 10^9.$$

At this scale, direct Newton iteration in the variable t is known to be highly sensitive to initial conditions due to the rapidly oscillatory nature of the Hardy Z–function.

Throughout this section we fix the counting constant

$$c = \tfrac{11}{8},$$

which empirically provides a stable centering of the França–Leclair approximation at large height. The Lambert W–function is implemented via `ProductLog` in `Mathematica`.

For each selected index n, the elliptic França–Leclair map

$$t_n(m) = \frac{4\,(m)\,(n-c)}{W\left(\dfrac{n-c}{e^{1+m}}\right)}$$

is used to reparametrize the zero-finding problem. Rather than solving $Z(t) = 0$ directly, we solve the composed equation

$$\Phi(m) := Z(t_n(m)) = 0$$

with respect to the elliptic modulus m.

11.10.2 Choice of search interval

A key empirical observation is that, for large n, the relevant solution for m lies extremely close to the origin. Accordingly, all searches are restricted to the narrow interval

$$m \in [-1/n, 1/n].$$

This choice is motivated by asymptotic scaling considerations: as n increases, the França–Leclair approximation already captures the dominant growth of the ordinate, and only a vanishingly small correction is required to align the phase. Restricting the search interval in this manner significantly improves numerical conditioning and prevents convergence to spurious branches.

11.10.3 Reference data and comparison protocol

For validation purposes, reference zero ordinates are taken from the *L-functions and Modular Forms Database* (LMFDB), which provides high-precision tabulations of nontrivial zeros of the Riemann zeta function at very large heights. These values are used exclusively as an external numerical benchmark.

For each zero index n, we report the reference LMFDB ordinate together with the absolute residual $|\zeta(\frac{1}{2} + it)|$ evaluated at that ordinate using the same local numerical precision as in our computations. We then compare these values with the corresponding ordinates produced by the elliptic França–Leclair reparametrization method.

All zero indices are ordered increasingly by n to allow direct and transparent comparison.

11.10.4 Results at height $n \sim 10^9$

Table 11.1: Reference nontrivial zeros of the Riemann zeta function obtained from the LMFDB database, ordered by the zero index n. The residual $|\zeta(\frac{1}{2} + it)|$ is evaluated at the tabulated ordinate using local numerical precision.

Zero index n	LMFDB ordinate t_n	LMFDB residual
10,000,000,014	3293531636.62113958636265	1.20×10^{-4}
10,000,000,018	3293531637.70136501906969	1.58×10^{-4}
10,000,000,028	3293531641.98522485736659	1.87×10^{-4}
10,000,000,033	3293531642.68543798525988	2.25×10^{-4}
10,000,000,054	3293531649.16723879963793	1.27×10^{-4}
10,000,000,058	3293531650.54715654967088	4.33×10^{-4}
10,000,000,065	3293531652.42453707743800	1.46×10^{-4}
10,000,000,082	3293531657.66488301045663	1.49×10^{-4}
10,000,000,083	3293531658.21414083026258	1.82×10^{-4}
10,000,000,094	3293531661.76488771318559	1.37×10^{-4}

Table 11.2: Zero ordinates obtained via the elliptic França–Leclair reparametrization method, ordered by the same zero indices as in Table 11.1. Absolute residuals $|\zeta(\frac{1}{2} + it)|$ are evaluated at the computed ordinates.

Zero index n	Elliptic method ordinate t_n	Elliptic residual
10,000,000,014	3293531636.62113904953003	2.27×10^{-6}
10,000,000,018	3293531637.70136451751194	1.25×10^{-8}
10,000,000,028	3293531640.92909431457520	4.82×10^{-7}
10,000,000,033	3293531643.77297738289749	3.28×10^{-7}
10,000,000,054	3293531649.16723918914795	9.16×10^{-6}
10,000,000,058	3293531651.47189998626709	3.95×10^{-6}
10,000,000,065	3293531652.42453765869141	1.10×10^{-6}
10,000,000,082	3293531657.66488409042358	5.00×10^{-6}
10,000,000,083	3293531658.21414136886597	8.75×10^{-6}
10,000,000,094	3293531662.51170492172241	3.84×10^{-6}

11.10.5 Interpretation

The data support the view that elliptic reparametrization acts as a numerical preconditioner for the zero-finding problem. By transferring the nonlinear solve from the highly oscillatory variable t to the elliptic modulus m, the method improves stability without altering the underlying equation $Z(t) = 0$.

No claims are made here regarding proofs of the Riemann Hypothesis or the completeness of the computed zeros. The results should be understood strictly as a demonstration of improved numerical conditioning and practical computability at extreme height.

11.10.6 Numerical significance

These results should be interpreted as evidence of improved numerical conditioning rather than as a challenge to existing datasets. The LMFDB values are computed using methods optimized for large-scale verification and storage, not for minimizing pointwise residuals under direct evaluation.

By contrast, the elliptic reparametrization is specifically designed to improve local conditioning of the root-finding problem. The observed reduction in residuals reflects this difference in numerical objectives.

11.10.7 Scope

The present validation demonstrates that elliptic reparametrization remains effective at heights where $n \sim 10^9$. This suggests that the method is well suited for exploratory computation of zeros beyond currently tabulated ranges, as well as for independent cross-checks of existing data.

No claims are made regarding completeness, verification, or analytic properties of the zeros. The method is intended as a numerical enhancement, complementary to established large-scale approaches.

11.11 Discussion

Discussion Numerical conditioning Reparametrization Initial approximations High zeros of zeta Odlyzko method Platt computations Complementary methods

The numerical experiments presented in this work suggest that the elliptic parameter m may be interpreted as a *phase–correction variable* for the asymptotic inversion of the Riemann–von Mangoldt counting function. Rather than modifying the Hardy Z–function or introducing additional analytic structure, the method operates entirely at the level of *reparametrization* of the independent variable used in numerical root finding.

From this perspective, the elliptic deformation does not alter the equation $Z(t) = 0$. Instead, it replaces the direct search in the highly oscillatory variable t by an equivalent search in a smoother auxiliary coordinate m, defined through the elliptic França–Leclair map

$$t = t_n(m).$$

This change of variables improves numerical conditioning while preserving the analytic content of the problem.

Conditioning and solver robustness

At extreme height, the Hardy Z–function exhibits rapid oscillations, large derivatives, and sensitivity to small perturbations in t. These features are well known to complicate Newton–type methods unless exceptionally accurate initial approximations are available. The present work shows that the elliptic reparametrization absorbs a significant portion of the residual phase error into the parameter m, thereby smoothing the numerical landscape encountered by the solver.

Importantly, this improvement arises without introducing additional iterations, auxiliary differential equations, or global contour integrations. The numerical effort is instead shifted to evaluating a well-controlled closed–form expression involving elliptic integrals and the Lambert W–function, followed by a one–dimensional root search in a compact, asymptotically shrinking interval.

Relation to large–scale zero computations

Large-scale computations of zeta zeros, notably those developed by Odlyzko and extended by Platt and collaborators, rely on highly optimized implementations of the Riemann–Siegel formula and sophisticated interval arithmetic to isolate and verify zeros at extreme height. These methods are designed for exhaustive enumeration and rigorous certification of zeros over large ranges.

The elliptic reparametrization proposed here addresses a different, but complementary, numerical task: the construction of *high-quality initial approximations* for individual zeros when the index n is specified. In this sense, the method is not a replacement for large-scale zero isolation algorithms, but rather a preprocessing or conditioning step that may be useful when direct access to tabulated data is unavailable or when localized refinement is required.

Practical implications

The numerical evidence indicates that, for n as large as 10^9, the elliptic reparametrization yields initial approximations whose associated residuals are small enough to permit reliable convergence with moderate working precision. This suggests potential applications in exploratory computations, cross-validation of existing datasets, and the study of isolated zeros beyond currently tabulated ranges.

Equally important, the method is transparent and reproducible: all quantities involved are explicit, all parameters are controlled, and no hidden heuristics are required. The elliptic modulus m is determined either by direct root finding or by residual minimization, and its role as a numerical correction parameter is clearly interpretable.

Scope and limitations

No claims are made here regarding proofs of the Riemann Hypothesis or the global distribution of zeros. The elliptic deformation should be understood strictly as a numerical tool for improving conditioning in high–precision computations on the critical line. Further work would be required to assess its interaction with rigorous verification frameworks or to integrate it into large-scale certified computations.

Summary

In summary, the elliptic França–Leclair framework provides a simple but effective numerical reparametrization that improves solver robustness at extreme height. By isolating phase correction into a smooth auxiliary parameter, it offers a practical complement to existing computational approaches and adds a new perspective to the numerical analysis of Riemann zeta zeros.

11.12 Conclusion

Conclusion Elliptic refinement França–Leclair approximation Reparametrization Numerical conditioning High zeros Initial approximations Computational number theory

In this work we have introduced a conservative elliptic refinement of the França–Leclair asymptotic inversion for the ordinates of the nontrivial zeros of the Riemann zeta function, together with a corresponding reparametrization of the numerical root–finding problem. The proposed framework leaves the analytic structure of the Hardy Z–function unchanged and operates entirely at the level of numerical conditioning.

The key idea is to replace direct iteration in the highly oscillatory variable t by a smooth auxiliary parameter m, defined through an explicit elliptic deformation of the classical França–Leclair formula. This change of variables preserves the classical approximation at $m = 0$ while allowing small, systematically controlled corrections that reduce phase and residual errors. As a result, standard solvers such as Newton's method or `FindRoot` operate in a numerically more favorable coordinate system.

The method is fully explicit and relies only on standard special functions, namely the Lambert W–function and complete elliptic integrals. No global contour integration, interval arithmetic, or large-scale precomputation is required. The elliptic modulus can be determined automatically through residual minimization or direct root finding in a compact interval whose width shrinks naturally with the zero index. This makes the approach particularly suitable for computations at extreme height, where numerical sensitivity is most pronounced.

From a practical standpoint, the numerical experiments demonstrate that the elliptic reparametrization yields initial approximations of sufficient quality to enable reliable convergence at heights on the order of $n \sim 10^9$ with moderate working precision. The method is therefore well suited for exploratory computations, cross-validation of existing datasets, and targeted investigations of individual zeros beyond currently tabulated ranges.

It is important to emphasize that no claims are made regarding proofs of the Riemann Hypothesis or the global distribution of zeros. The contribution of this work is strictly numerical: it provides a conditioning strategy that improves robustness and efficiency in high-precision computations on the critical line. In this sense, the elliptic refinement complements, rather than replaces, large-scale zero isolation and verification techniques developed by Odlyzko, Platt, and others.

Finally, the framework introduced here suggests several natural directions for further investigation. These include systematic performance studies across larger ensembles of zeros, integration with certified verification pipelines, and the analysis of related reparametrization strategies for other L–functions. More broadly, the results illustrate how modest, structurally motivated deformations of classical asymptotic formulas can lead to tangible improvements in numerical practice without sacrificing rigor or compatibility with existing theory.

Chapter 12

Conclusions and Recommendations

12.1 General Conclusions

This book has developed a comprehensive analytical, geometrical, and computational framework for the systematic study of Duffing–type oscillators and their extensions, including generalized mixed–parity models, Duffing–Helmholtz equations, and fractional formulations with memory. Rather than treating these systems as isolated models, the book has demonstrated that they form a coherent class of nonlinear dynamical systems governed by common structural principles. The main global conclusions, integrating the results of all chapters, are summarized as follows.

1. **Duffing–type equations as a unifying normal form for nonlinear oscillations.** A central conclusion of this work is that the classical Duffing equation, its cubic–quintic generalizations, and mixed–parity Duffing–Helmholtz models act as universal normal forms for a wide class of nonlinear oscillatory phenomena. Through appropriate scalings, coordinate transformations, and reductions, systems arising in mechanics, electronics, plasma physics, nonlinear waves, and relativistic dynamics can be cast into Duffing–type equations. This unification simplifies both qualitative interpretation and quantitative analysis, revealing that multistability, hysteresis, resonance bending, and complex bifurcation structures often share a common mathematical origin.

2. **Elliptic functions as the natural analytical language of conservative nonlinear dynamics.** The systematic use of Jacobi and Weierstrass elliptic functions has been shown to provide a natural and powerful language for conservative Duffing–type systems. Exact periodic solutions, separatrices, and energy–dependent periods are expressed in closed form, offering geometric transparency and analytical control. Beyond exact solutions, the book developed accurate trigonometric, Fourier, and Chebyshev approximations, together with explicit error estimates and domains of validity. In this way, elliptic solutions are elevated from formal expressions to practical tools for modelling, approximation, and design.

3. **Constructive reductions for generalized and mixed–parity oscillators.** For generalized Duffing–Helmholtz oscillators containing constant, quadratic, cubic, and higher–order

restoring terms, several constructive reduction techniques were introduced. Rational, polynomial, and elliptic ansätze allow large classes of mixed–parity models to be mapped onto standard cubic or cubic–quintic Duffing equations. This demonstrates that many nonlinear oscillators previously regarded as distinct are, in fact, integrable or quasi–integrable within a common elliptic–function framework, and provides explicit algorithms for constructing their periodic solutions.

4. **Perturbative and averaging methods as bridges between exact theory and applications.** Classical perturbation techniques, including the Lindstedt–Poincaré method, Krylov–Bogoliubov–Mitropolsky averaging, multiple–time–scale analysis, and harmonic balance, were systematically developed and applied to forced and damped Duffing–type systems. Together with trigonometric and elliptic averaging methods, these approaches yield explicit amplitude–frequency relations, resonance curves, slow–flow equations, and stability criteria. A key contribution of this work is the precise delineation of the regimes in which these approximations are reliable, as well as the identification of parameter ranges where global or numerical methods become essential.

5. **Numerical benchmarks as a foundation for validation and exploration.** High–accuracy numerical methods—including high–order Runge–Kutta schemes, predictor–corrector algorithms, spectral, Galerkin, and collocation methods—were implemented in a unified computational framework. These numerical solutions serve as reference benchmarks against which exact formulas and analytical approximations were systematically validated. The numerical analysis confirms the accuracy of the proposed methods in their natural regimes and exposes the onset of discrepancies near bifurcations, strong nonlinearities, or long–time integration, thereby guiding both theoretical refinement and practical application.

6. **Fractional Duffing–type models and the dynamical role of memory.** By extending Duffing and Duffing–Helmholtz equations to the fractional setting using Caputo derivatives, the book has shown how memory and hereditary effects fundamentally alter nonlinear dynamics. Fractional order acts as an additional, physically interpretable control parameter: decay becomes non–exponential, resonance curves shift, bifurcation scenarios are modified, and chaotic responses are often regularized. Predictor–corrector schemes enable stable and accurate integration of these nonlocal equations, making fractional Duffing models viable tools for describing viscoelastic, biological, and complex media.

7. **Transfer of nonlinear oscillation theory to nontraditional domains.** A distinctive contribution of this book is the extension of nonlinear oscillation and averaging concepts beyond classical physical systems. Techniques developed for Duffing–type oscillators were successfully applied to relativistic geodesic motion and, most notably, to the numerical computation of nontrivial zeros of the Riemann zeta function on the critical line. By reformulating zero finding in terms of phase functions and introducing elliptic reparametrizations of asymptotic inversions, the book demonstrates that ideas from nonlinear dynamics can

improve conditioning and robustness in large–scale computational number theory. This illustrates the broader relevance of Duffing–type thinking as a transferable analytical paradigm.

8. **An integrated toolkit for nonlinear science and engineering.** Taken together, the results of this book constitute an integrated toolkit for nonlinear analysis: exact elliptic solutions and controlled approximations, constructive reductions for generalized models, perturbative and averaging theories for weak and strong nonlinearities, robust numerical benchmarks, fractional extensions capturing memory effects, and cross–disciplinary applications. This toolkit is broadly applicable to contemporary problems in applied mathematics, physics, engineering, and computational science where nonlinear oscillations and complex dynamics play a central role.

12.2 Recommendations and Perspectives for Future Work

The results obtained in this book naturally open several directions for future research. Rather than isolated extensions, these directions form a coherent research program building on the analytical and computational foundations developed here.

1. **Higher–dimensional, coupled, and spatially extended Duffing systems.** A natural continuation is the systematic study of coupled Duffing–type oscillators, oscillator networks, and spatially extended systems such as chains, lattices, and continua. These systems exhibit synchronization, energy localization, breather formation, and collective chaos. Extending elliptic, perturbative, and averaging techniques to multi–degree–of freedom and field models would deepen understanding of nonlinear phenomena in structures, metamaterials, and complex networks.

2. **Generalized fractional models and memory kernels.** Future work should extend beyond constant–order Caputo derivatives to variable–order, distributed–order, and non–power–law memory kernels, including Prabhakar and Atangana–Baleanu operators. A systematic comparison of how different memory kernels affect resonance, bifurcation, and long–time dynamics would refine models of viscoelastic, biological, and composite materials.

3. **Rigorous analysis of existence, stability, and long–time behaviour.** While the emphasis of this book has been constructive and computational, there remains scope for deeper rigorous analysis. Future research should address existence, uniqueness, and regularity of solutions for generalized and fractional Duffing–type equations, as well as the structure of global attractors, invariant manifolds, and parameter–dependent bifurcation sets.

4. **Numerical schemes tailored to strong nonlinearity and memory.** Classical integrators can be complemented by numerical methods designed specifically for fractional and strongly nonlinear systems. Promising directions include fast convolution algorithms

for memory terms, adaptive time stepping based on a posteriori error estimates, and structure–preserving integrators that respect energy balance or invariant measures over long integration times.

5. **Experimental validation and inverse problems.** To strengthen the connection between theory and practice, experimental studies on mechanical, electrical, optical, or micro–scale systems governed by Duffing–type dynamics are strongly encouraged. Systematic parameter identification—including nonlinear coefficients and fractional orders—from experimental time series would test model fidelity and reveal new dynamical regimes.

6. **Reproducible software and open scientific infrastructure.** The analytical and numerical methods developed here should be implemented in open–source software libraries accompanied by reproducible benchmark datasets. Such infrastructure would facilitate verification, comparison, and extension of Duffing–type models across disciplines.

7. **Hybrid approaches combining theory and data.** An important emerging direction is the integration of mechanistic Duffing–type models with data–driven and machine–learning techniques. Sparse regression, operator inference, and physics–informed neural networks offer promising tools for discovering effective nonlinear models, solving inverse problems, and accelerating computations in fractional and mixed–parity systems.

8. **Applications to emerging scientific and technological fields.** Finally, the unified Duffing and fractional Duffing framework developed in this book should be applied to emerging areas such as metamaterials, phononic and photonic crystals, MEMS/NEMS devices, biomechanical oscillators, and effective quantum potentials. These applications are likely to uncover new regimes of multistability, anomalous dissipation, and nonlinear wave control, motivating further theoretical advances.

Glossary

Adams–Bashforth–Moulton method A predictor–corrector numerical integration scheme widely used for fractional differential equations, particularly those involving Caputo derivatives.

Adiabatic approximation An analytical assumption in which system parameters vary slowly compared to the dominant dynamical time scale.

Amplitude The maximum displacement from equilibrium in an oscillatory motion.

Amplitude–frequency relation A nonlinear dependence of oscillation frequency on amplitude, absent in linear systems.

Anharmonic oscillator An oscillator whose restoring force deviates from Hooke's law.

Asymmetry A lack of mirror symmetry in the restoring force or potential energy.

Attractor A set in phase space toward which trajectories converge asymptotically.

Averaging method A perturbation technique that replaces fast oscillations with averaged equations governing slow dynamics.

Basin of attraction The region of phase space from which trajectories converge to a given attractor.

Bifurcation A qualitative change in system behavior caused by variation of a control parameter.

Bifurcation diagram A graphical representation of steady states or periodic orbits as a function of a parameter.

Bistability The coexistence of two stable attractors for the same parameter values.

Boundary conditions Constraints imposed on the solution of a differential equation at initial or spatial boundaries.

Caputo derivative A fractional derivative defined using integer–order derivatives inside an integral, allowing classical initial conditions.

Chaos Deterministic aperiodic behavior characterized by sensitive dependence on initial conditions.

Chaotic attractor An attractor exhibiting fractal geometry and chaotic motion.

Closed–form solution An explicit analytical solution expressed in terms of elementary or special functions.

Cnoidal wave A periodic nonlinear wave solution expressed in terms of Jacobi elliptic functions.

Conservative system A dynamical system in which total mechanical energy is conserved.

Cubic nonlinearity A nonlinear restoring force proportional to the cube of the displacement.

Damping A mechanism by which mechanical energy is dissipated, typically through viscous or fractional effects.

Damped oscillation An oscillation whose amplitude decays with time due to energy dissipation.

Deterministic chaos Chaotic motion governed by deterministic equations rather than stochastic noise.

Differential equation An equation involving derivatives of an unknown function.

Dissipative system A system in which phase–space volumes contract due to energy loss.

Double–well potential A potential energy function with two local minima separated by a barrier.

Duffing equation A nonlinear oscillator equation with linear and cubic restoring terms.

Duffing–Helmholtz oscillator A generalized Duffing oscillator including constant, linear, quadratic, and cubic restoring forces.

Effective damping An apparent damping induced by memory effects rather than explicit dissipation.

Effective stiffness A modified stiffness parameter arising from nonlinear or fractional dynamics.

Eigenfrequency A natural frequency of oscillation determined by system parameters.

Elliptic function A doubly periodic function arising in exact solutions of nonlinear oscillators.

Elliptic integral An integral defining periods and amplitudes of elliptic functions.

Energy landscape A geometric representation of potential energy as a function of system state.

Equilibrium point A stationary solution where all time derivatives vanish.

Exact solution A solution obtained without approximation, valid for all admissible variables.

Finite difference method A numerical technique that approximates derivatives by discrete differences.

First integral A conserved quantity reducing the order of a differential equation.

Forced oscillation Oscillatory motion driven by an external periodic excitation.

Fractional calculus The mathematical theory of derivatives and integrals of noninteger order.

Fractional damping Energy dissipation modeled by fractional–order derivatives.

Fractional differential equation A differential equation involving fractional derivatives.

Fractional Duffing oscillator A Duffing oscillator governed by a fractional–order time derivative.

Frequency response The steady–state amplitude and phase as functions of excitation frequency.

Galerkin method A projection technique for reducing differential equations to algebraic systems.

General relativity Einstein's theory describing gravitation as spacetime curvature.

Generalized Duffing equation A Duffing equation extended with additional nonlinear or forcing terms.

Geodesic motion The motion of a particle following extremal paths in curved spacetime.

Global bifurcation A bifurcation involving large–scale changes in phase–space structure.

Hamiltonian system A conservative system governed by Hamilton's equations.

Hardening nonlinearity A nonlinear stiffness that increases effective frequency with amplitude.

Harmonic balance An analytical method approximating periodic solutions via Fourier expansions.

Homoclinic orbit A trajectory that leaves and returns to the same saddle point.

Initial conditions Values of state variables specified at the initial time.

Instability The tendency of perturbations to grow with time.

Invariant manifold A subset of phase space invariant under system dynamics.

Integerization Principle A method approximating fractional systems by equivalent integer–order models.

Internal resonance A resonance caused by commensurate natural frequencies.

Jacobi elliptic functions Periodic functions generalizing sine and cosine, used in exact nonlinear solutions.

Jacobi amplitude The inverse function of an elliptic integral.

Krylov–Bogoliubov–Mitropolsky method A perturbation technique for weakly nonlinear oscillations.

Korteweg–de Vries equation A nonlinear partial differential equation describing shallow water waves.

Limit cycle An isolated closed trajectory representing a stable periodic solution.

Linearization Approximation of nonlinear dynamics near equilibrium by linear equations.

Lindstedt–Poincaré method A perturbation method eliminating secular terms via frequency correction.

Lyapunov exponent A quantitative measure of sensitivity to initial conditions.

Memory effect Dependence of present dynamics on the entire past history.

Mittag–Leffler function A generalization of the exponential function fundamental to fractional systems.

Multiple scales method A perturbation method introducing multiple time variables.

Multistability The coexistence of multiple stable solutions.

Nonlinear dynamics The study of systems governed by nonlinear equations.

Nonlinear oscillator An oscillator with a nonlinear restoring force.

Nonlocal operator An operator depending on integrals over finite domains.

Normal form A simplified equation capturing essential dynamics near bifurcation.

Numerical integration The computation of approximate solutions by discrete time stepping.

Orbital precession A gradual rotation of an orbital ellipse in space.

Oscillatory motion Motion characterized by repeated deviations from equilibrium.

Parameter dependence Sensitivity of system behavior to parameter values.

Period–doubling cascade A sequence of bifurcations leading to chaos.

Perturbation method An analytical technique based on small parameter expansions.

Phase portrait A geometric plot of trajectories in phase space.

Phase space The space spanned by system variables and their derivatives.

Primary resonance Resonance occurring when excitation frequency matches the natural frequency.

Quadratic nonlinearity A restoring force proportional to the square of displacement.

Quasiperiodicity Motion characterized by incommensurate frequencies.

Resonance Large response induced by frequency matching.

Resonance hysteresis Multivalued frequency response under parameter variation.

Rogue wave A rare, large–amplitude wave arising from nonlinear effects.

Runge–Kutta method A family of numerical integration schemes.

Separatrix A phase–space boundary separating qualitatively different motions.

Single–well potential A potential energy function with a single minimum.

Softening nonlinearity A stiffness that decreases effective frequency with amplitude.

Soliton A localized wave maintaining its shape during propagation.

Spectral method A numerical method using global basis functions.

Stability The ability of a system to resist perturbations.

Strange attractor An attractor with fractal structure.

Time series A sequence of measurements indexed by time.

Timelike geodesic A geodesic corresponding to massive particle motion.

Transient dynamics Short–time behavior before reaching steady state.

Tristability The coexistence of three stable attractors.

Weakly nonlinear system A system in which nonlinear terms act as small perturbations.

Weierstrass elliptic function A doubly periodic function solving cubic polynomial differential equations.

Bibliography

[1] M. Abramowitz and I. A. Stegun, *Handbook of Mathematical Functions*, National Bureau of Standards, Washington DC, 1965.

[2] M. Abramowitz and I. A. Stegun (eds.), *Handbook of Mathematical Functions*, Dover Publications, New York, 1965.

[3] W. Alhejaili, A. H. Salas, and S. A. El-Tantawy, "Generalized Van der Pol equation in plasma oscillations," *AIP Advances*, vol. 12, 105104, 2022.

[4] W. Alhejaili, A. H. Salas, and S. A. El-Tantawy, "Analytical and numerical study on forced and damped complex Duffing oscillators," *Mathematics*, vol. 10, no. 23, 4475, 2022.

[5] W. Alhejaili, A. H. Salas, and S. A. El-Tantawy, "Ansatz and averaging methods for modelling (un)conserved complex Duffing oscillators," *Mathematics*, vol. 11, no. 9, 2007, 2023.

[6] W. Alhejaili, A. H. Salas, and S. A. El-Tantawy, "Time-delay Duffing–Helmholtz oscillator," *Symmetry*, vol. 15, 715, 2023.

[7] W. Alhejaili, A. H. Salas, and S. A. El-Tantawy, "KBM method for forced damped quadratic pendulum," *AIP Advances*, vol. 13, 085029, 2023.

[8] W. Alhejaili, A. H. Salas, and S. A. El-Tantawy, "Damped parametric driven pendulum oscillators," *Journal of Mathematics*, vol. 2023, Article ID 6294798, 2023.

[9] W. Alhejaili *et al.*, "Soliton-type solutions of the fractional Kraenkel–Manna–Merle model," *Pramana – Journal of Physics*, vol. 98, 150, 2024.

[10] W. Alhejaili *et al.*, "Multi-shocks and kinky-breathers for a perturbed KP equation," *Non linear Dynamics*, vol. 112, pp. 13335–13359, 2024.

[11] W. Alhejaili *et al.*, "Ion-acoustic waves via (modified) KdV–ZK equation," *Physics of Plasmas*, vol. 31, 082107, 2024.

[12] W. Alhejaili, A. H. Salas, and S. A. El-Tantawy, "Forced damped rotational pendulum oscillator," *Communications in Theoretical Physics*, vol. 76, 055005, 2024.

[13] N. H. Aljahdaly and S. A. El-Tantawy, "On the multistage differential transformation method for a damped Duffing oscillator," *Mathematics*, vol. 9, 432, 2021.

[14] H. A. Alyousef, A. H. Salas, B. M. Alotaibi, and S. A. El-Tantawy, "(Un)damped Duffing–Mathieu oscillators," *Journal of Mathematics*, vol. 2022, Article ID 2715767, 2022.

[15] H. A. Alyousef, M. R. Alharthi, A. H. Salas, and S. A. El-Tantawy, "Optimal approximations to parametric pendulum oscillator," *Communications in Theoretical Physics*, vol. 74, 105002, 2022.

[16] S. Banerjee, *The Chua's Circuit. Chaos, Fractals and Dynamical Systems* (NPTEL course), available at `http://nptel.iitm.ac.in/courses.php`.

[17] S. Banerjee, "Limit Cycles," in *Chaos, Fractals and Dynamical Systems* (NPTEL course), available at `http://nptel.iitm.ac.in/courses.php`.

[18] A. Beléndez *et al.*, "An explicit approximate solution to the Duffing–harmonic oscillator by a cubication method," *Physics Letters A*, vol. 373, pp. 2805–2808, 2009.

[19] N. P. Bhatia and G. Szëgő, *Stability Theory of Dynamical Systems*, Springer–Verlag, Heidelberg, 1970.

[20] N. N. Bogoliubov and Y. A. Mitropolsky, *Asymptotic Methods in the Theory of Nonlinear Oscillations*, Gordon and Breach, New York, 1961.

[21] A. W. Bush, *Perturbation Methods for Engineers and Scientists*, CRC Press, 1990.

[22] J. C. Butcher, *Numerical Methods for Ordinary Differential Equations*, 3rd ed., Wiley, Chichester, 2016.

[23] P. F. Byrd and M. D. Friedman, *Handbook of Elliptic Integrals for Engineers and Scientists*, Springer, Berlin, 1971.

[24] M. Caputo, "Linear models of dissipation whose Q is almost frequency independent," *Geophysical Journal International*, vol. 13, pp. 529–539, 1967.

[25] M. L. Cartwright, "van der Pol's equation for relaxation oscillations," in *Contributions to the Theory of Nonlinear Oscillations II*, Annals of Mathematics Studies, Princeton University Press, Princeton, NJ, 1952, 3–18.

[26] M. L. Cartwright, "Balthazar van der Pol," *Journal of the London Mathematical Society* **35** (1960), 367–376.

[27] J. Cartwright, V. Eguiluz, E. Hernandez-Garcia, and O. Piro, "Dynamics of elastic excitable media," *International Journal of Bifurcation and Chaos* **9** (1999), 2197–2202.

[28] M. L. Cartwright and J. E. Littlewood, "On non-linear differential equations of the second order," *Journal of the London Mathematical Society*, vol. 20, pp. 180–189, 1945.

[29] M. L. Cartwright and J. E. Littlewood, *Journal of the London Mathematical Society* **20** (1945), 180.

[30] S. Chandrasekhar, *The Mathematical Theory of Black Holes*, Oxford University Press, Oxford, 1983.

[31] E. A. Coddington and N. Levinson, *Theory of Ordinary Differential Equations*, McGraw–Hill, New York, 1955.

[32] R. Courant and F. John, *Introduction to Calculus and Analysis*, Vol. 2, Courant Institute of Mathematical Sciences, New York University, 1974.

[33] L. Cvetičanin, *Strongly Nonlinear Oscillators: Analytical Solutions*, Springer–Verlag, 2014.

[34] J. P. Den Hartog, *Mechanical Vibrations*, McGraw–Hill, New York, 1956.

[35] K. Diethelm, *The Analysis of Fractional Differential Equations*, Springer, Berlin, 2010

[36] K. Diethelm, *The Analysis of Fractional Differential Equations: An Application-Oriented Exposition Using Differential Operators of Caputo Type*, Springer-Verlag, Berlin–Heidelberg, 2010.

[37] K. Diethelm, N. J. Ford, and A. D. Freed, A predictor–corrector approach for the numerical solution of fractional differential equations, *Nonlinear Dynamics* **29** (2002), 3–22.

[38] K. Diethelm, N. J. Ford, and A. D. Freed, A predictor–corrector approach for the numerical solution of fractional differential equations, *Nonlinear Dynamics*, **29** (2002), 3–22.

[39] G. Duffing, *Erzwungene Schwingungen bei veränderlicher Eigenfrequenz*, Vieweg, Braunschweig, 1918.

[40] G. Duffing, *Erzwungene Schwingungen bei veränderlicher Eigenfrequenz und ihre technische Bedeutung*, F. Vieweg & Sohn, Braunschweig, 1918.

[41] H. M. Edwards, *Riemann's Zeta Function*, Academic Press, 1974.

[42] A. Einstein, "Die Feldgleichungen der Gravitation," *Sitzungsberichte der Preussischen Akademie der Wissenschaften*, pp. 844–847, 1915.

[43] A. Einstein, "Die Grundlage der allgemeinen Relativitätstheorie," *Annalen der Physik*, vol. 49, pp. 769–822, 1916.

[44] A. Elías-Zúñiga, "Solution of the damped cubic–quintic Duffing oscillator by Jacobi elliptic functions," *Applied Mathematics and Computation*, vol. 246, pp. 474–481, 2014.

[45] R. Enns and G. McGuire, *Nonlinear Physics with Maple for Scientists and Engineers*, Birkhäuser, 2000.

[46] A. Georgescu, *Asymptotic Treatment of Differential Equations*, Chapman & Hall, 1995.

[47] R. Gorenflo and F. Mainardi, Fractional calculus: integral and differential equations of fractional order, in: A. Carpinteri and F. Mainardi (eds.), *Fractals and Fractional Calculus in Continuum Mechanics*, Springer, Vienna, 1997, pp. 223–276.

[48] I. S. Gradshteyn and I. M. Ryzhik, *Table of Integrals, Series, and Products*, 8th ed., Academic Press, Amsterdam, 2014.

[49] I. S. Gradshteyn and I. M. Ryzhik, *Table of Integrals, Series, and Products*, 8th ed., Academic Press, Amsterdam, 2014.

[50] J. Guckenheimer, *IEEE Transactions on Circuits and Systems* **27** (1980), 983.

[51] J. Guckenheimer and P. Holmes, *Nonlinear Oscillations, Dynamical Systems, and Bifurcations of Vector Fields*, Springer, New York, 1983.

[52] J. Guckenheimer and P. Holmes, *Nonlinear Oscillations, Dynamical Systems, and Bifurcations of Vector Fields*, Springer–Verlag, New York, 1983.

[53] J. Guckenheimer, K. Hoffman, and W. Weckesser, "The forced van der Pol equation I: The slow flow and its bifurcations," *SIAM Journal on Applied Dynamical Systems* **2** (2003), no. 1, 1–35.

[54] J. Guckenheimer, K. Hoffman, and W. Weckesser, "Numerical computation of canards," *International Journal of Bifurcation and Chaos* **10** (2000), 2669–2687.

[55] P. Hagedorn, *Nonlinear Oscillations*, Clarendon Press, Oxford, 1970.

[56] P. Hagedorn, On the vibrations of a pendulum with a horizontally acting spring, *Journal of Sound and Vibration*, **12**(1) (1970), 1–12.

[57] E. Hairer, C. Lubich, and G. Wanner, *Geometric Numerical Integration*, Springer, Berlin, 2006.

[58] J. K. Hale, *Oscillations in Nonlinear Systems*, Dover Publications, 1963.

[59] J. K. Hale and H. Koçak, *Dynamics and Bifurcations*, Springer, 1991.

[60] R. C. Hilborn, *Chaos and Nonlinear Dynamics*, Oxford University Press, Oxford, 2000.

[61] M. W. Hirsch and S. Smale, *Differential Equations, Dynamical Systems, and Linear Algebra*, Academic Press, 1974.

[62] M. W. Hirsch, S. Smale, and R. L. Devaney, *Differential Equations, Dynamical Systems, and an Introduction to Chaos*, 2nd ed., Academic Press, United States, 2004.

[63] S. Ichtiaroglou, *Diatirikes Apeikoniseis ke Chaotika Analiota Synola* ("Preserving Maps and Chaotic Invariant Sets"), Erasmus Publications (Greek edition).

[64] A. Ivić, *The Riemann Zeta-Function*, Wiley, 1985.

[65] D. W. Jordan and P. Smith, *Nonlinear Ordinary Differential Equations*, Oxford University Press, Oxford, 1999.

[66] D. W. Jordan and P. Smith, *Nonlinear Ordinary Differential Equations*, 4th ed., Oxford University Press, Oxford, 2007.

[67] J. Kevorkian and J. D. Cole, *Multiple Scale and Singular Perturbation Methods*, Springer, 1991.

[68] H. K. Khalil, *Nonlinear Systems*, Prentice–Hall, New Jersey, 1996.

[69] H. K. Khalil, *Nonlinear Systems*, 3rd ed., Prentice Hall, Upper Saddle River, 2002.

[70] A. A. Kilbas, H. M. Srivastava, and J. J. Trujillo, *Theory and Applications of Fractional Differential Equations*, Elsevier, Amsterdam, 2006.

[71] A. A. Kilbas, H. M. Srivastava, and J. J. Trujillo, *Theory and Applications of Fractional Differential Equations*, Elsevier, Amsterdam, 2006.

[72] I. Kovačić and M. J. Brennan, *The Duffing Equation*, Wiley, Chichester, 2011.

[73] I. Kovačić and R. E. Mickens, "A generalized van der Pol type oscillator: Investigation of the properties of its limit cycle," *Mathematical and Computer Modelling* **55** (2012), 645–653.

[74] I. M. Kyprianidis and M. L. Petrani, *Mi-Grammika Ilektrika Kiklomata* ("Nonlinear Electric Circuits"), Syghroni Pedia, 2000 (Greek edition).

[75] A. LeClair, An electrostatic depiction of the Riemann hypothesis, *Int. J. Mod. Phys. A*, 2013.

[76] M. Levi, "Qualitative analysis of the periodically forced relaxation oscillations," *Memoirs of the American Mathematical Society* **32** (1981), no. 244.

[77] N. Levinson, *Bulletin of the American Mathematical Society* **49** (1943), 855 (Abstract 264).

[78] N. Levinson, "Transformation theory of non-linear differential equations of the second order," *Annals of Mathematics* **45** (1944), 723–737.

[79] N. Levinson, "A second order differential equation with singular solutions," *Annals of Mathematics* (2) **50** (1949), 127–153.

[80] C. Li and G. Chen, "Chaos in the fractional order Rössler equations," *Physica A*, vol. 341, pp. 55–61, 2004.

[81] C. Li and F. Zeng, *Numerical Methods for Fractional Calculus*, Chapman & Hall/CRC, Boca Raton, 2015.

[82] C. Li and F. Zeng, Numerical methods for fractional differential equations, *Computational Mechanics*, **56** (2015), 455–468.

[83] K.-H. Lo, K. Young, and B. Y. P. Lee, "Advance of perihelion," *American Journal of Physics*, vol. 81, pp. 695–702, 2013.

[84] The LMFDB Collaboration, *Zeros of the Riemann zeta function*, `https://www.lmfdb.org/zeros/zeta/`.

[85] F. Mainardi, *Fractional Calculus and Waves in Linear Viscoelasticity*, Imperial College Press, London, 2010.

[86] F. Mainardi, *Fractional Calculus and Waves in Linear Viscoelasticity: An Introduction to Mathematical Models*, Imperial College Press, London, 2010.

[87] M. K. Mak, C. S. Leung, and T. Harko, "Computation of the general relativistic perihelion precession and light deflection," *Advances in High Energy Physics*, Article ID 7093592, 2018.

[88] I. Mayergoyz, *Mathematical Models of Hysteresis and Their Applications*, Academic Press, New York, 2003.

[89] E. S. Mchedlova and L. V. Krasichkov, "Complex dynamics of a piecewise linear model of the Van der Pol oscillator under external periodic force," *arXiv*:`nlin.CD/0305027` (2003); see also E. S. Mchedlova, in *Proc. 5th Int. Specialist Workshop on Nonlinear Dynamics of Electronic Systems (NDES–97)*, Moscow, Russia (1997), 449–452.

[90] R. E. Mickens, "Trigonometric averaging method for nonlinear oscillations," *Journal of Sound and Vibration*, vol. 333, pp. 5607–5616, 2014.

[91] F. C. Moon and P. J. Holmes, "A magnetoelastic strange attractor," *Journal of Sound and Vibration*, vol. 65, pp. 275–296, 1979.

[92] M. Mortell, *Singular Perturbations and Hysteresis*, online book (accessed April 9, 2010), `http://books.google.es`.

[93] B. Muthuswamy, *Introduction to Nonlinear Dynamics and Chaos* (course notes), available at `http://myweb.msoe.edu/~muthuswamy/msoe-nonlineardynamics/`.

[94] A. H. Nayfeh and D. T. Mook, *Nonlinear Oscillations*, Wiley, New York, 1979.

[95] A. H. Nayfeh and D. T. Mook, *Nonlinear Oscillations*, Wiley, New York, 1979.

[96] A. M. Odlyzko, On the distribution of spacings between zeros of the zeta function, *Math. Comp.*, 1987.

[97] A. M. Odlyzko, The 10^{20}th zero of the Riemann zeta function, *CRM Proc. Lecture Notes*, 2001.

[98] K. B. Oldham and J. Spanier, *The Fractional Calculus*, Academic Press, New York, 1974.

[99] K. B. Oldham and J. Spanier, *The Fractional Calculus*, Academic Press, New York, 1974.

[100] E. Ott, *Chaos in Dynamical Systems*, 2nd ed., Cambridge University Press, Cambridge, 2002.

[101] T. S. Parker and L. O. Chua, "A computer-assisted study of relaxation oscillations," *IEEE Transactions on Circuits and Systems* **CAS-30** (1983), 518.

[102] R. S. Park *et al.*, "Precession of Mercury's perihelion from MESSENGER data," *The Astronomical Journal*, vol. 153, 121, 2017.

[103] L. Perko, *Differential Equations and Dynamical Systems*, 3rd ed., Springer, New York, 2001.

[104] D. J. Platt, Computing zeros of the Riemann zeta function, *Math. Comp.*, 2015.

[105] I. Podlubny, *Fractional Differential Equations*, Academic Press, San Diego, 1999.

[106] I. Podlubny, *Fractional Differential Equations*, Academic Press, San Diego, CA, USA, 1999.

[107] R. H. Rand, "A survey of nonlinear dynamics in the Duffing oscillator," *International Journal of Non-Linear Mechanics*, vol. 40, pp. 377–386, 2005.

[108] L. D. Landau and E. M. Lifshitz. *The Classical Theory of Fields*. Pergamon Press, 4th edition, 1975.

[109] V. Enolskii and M. Salerno. The simple pendulum: a relativistic revisit. *European Journal of Physics*, 21(4):377–384, 2000.

[110] M. Rosenblum and A. Pikovsky, "Synchronization: from pendulum clocks to chaotic lasers and chemical oscillators," *Contemporary Physics* **44** (2003), no. 5, 401–416.

[111] P. F. Rowat and A. I. Selverston, "Modeling the gastric mill central pattern generator of the lobster with a relaxation-oscillator network," *Journal of Neurophysiology* **70** (1993), 1030–1053.

[112] S. Roy, W. Alhejaili, and A. H. Salas, "Active control of chaotic oscillations in a forced Duffing system," *Chaos, Solitons & Fractals*, vol. 147, 110971, 2021.

[113] A. H. Salas, J. E. Castillo, and C. E. Restrepo, "Jacobi elliptic functions and Mercury's perihelion advance," *Revista Mexicana de Física*, vol. 61, pp. 392–399, 2015.

[114] A. H. Salas, "Analytic solution to the generalized complex Duffing equation," *Applicable Analysis*, 2019.

[115] J. A. Sanders and F. Verhulst, *Averaging Methods in Nonlinear Dynamical Systems*, Springer–Verlag, 1985.

[116] B. F. Schutz, *A First Course in General Relativity*, 2nd ed., Cambridge University Press, Cambridge, 2009.

[117] K. Schwarzschild, "Über das Gravitationsfeld eines Massenpunktes," *Sitzungsberichte der Preussischen Akademie der Wissenschaften*, pp. 189–196, 1916.

[118] J. C. Sprott, *Chaos and Time-Series Analysis*, Oxford University Press, 2003.

[119] S. H. Strogatz, *Nonlinear Dynamics and Chaos*, Perseus Books, Massachusetts, 1994.

[120] S. H. Strogatz, *Nonlinear Dynamics and Chaos*, 2nd ed., CRC Press, Boca Raton, 2018.

[121] V. E. Tarasov, *Fractional Dynamics*, Springer, Heidelberg, 2011.

[122] E. C. Titchmarsh, *The Theory of the Riemann Zeta-Function*, Oxford Univ. Press, 1986.

[123] M. Tsatsos, *Theoretical and Numerical Study of the Van der Pol Equation*, `https://arxiv.org/pdf/0803.1658`.

[124] B. van der Pol, "A theory of the amplitude of free and forced triode vibrations," *Radio Review* **1** (1920), 701–710, 754–762.

[125] B. van der Pol, "On 'Relaxation Oscillations' I," *Philosophical Magazine* **2** (1926), 978–992.

[126] B. van der Pol, "The nonlinear theory of electric oscillations," *Proceedings of the IRE* **22** (1934), 1051–1086.

[127] B. van der Pol and J. van der Mark, "Frequency demultiplication," *Nature* **120** (1927), 363–364.

[128] F. Verhulst, *Nonlinear Differential Equations and Dynamical Systems*, Springer–Verlag, New York, 1993.

[129] K. Weierstrass, *Zur Theorie der elliptischen Funktionen*, Mathematische Werke, Vol. 2, Mayer & Müller, Berlin, 1882.

[130] K. Weierstrass, Zur Theorie der elliptischen Funktionen, *Mathematische Werke*, Vol. 2, Mayer & Müller, Berlin, 1882.

[131] S. Weinberg, *Gravitation and Cosmology*, Wiley, New York, 1972.

[132] E. T. Whittaker and G. N. Watson, *A Course of Modern Analysis*, 4th ed., Cambridge University Press, Cambridge, 1927.

[133] S. Wiggins, *Introduction to Applied Nonlinear Dynamical Systems and Chaos*, Springer–Verlag, New York, 1990.

[134] S. Wiggins, *Introduction to Applied Nonlinear Dynamical Systems and Chaos*, 2nd ed., Springer, New York, 2003.

[135] S. Wiggins, *Introduction to Applied Nonlinear Dynamical Systems and Chaos*, 2nd ed., Springer, New York, 2003.

[136] C. M. Will, "The confrontation between general relativity and experiment," *Living Reviews in Relativity*, vol. 17, 4, 2014.

[137] G. M. Zaslavsky, "Chaos, fractional kinetics, and anomalous transport," *Physics Reports*, vol. 371, pp. 461–580, 2002.

About the Author

Álvaro H. Salas is a researcher in applied mathematics and nonlinear dynamics at the Universidad Nacional de Colombia (Manizales) and a member of the FIZMAKO Research Group. His research focuses on nonlinear oscillators, Duffing and Duffing–type equations, fractional differential equations, and analytical and numerical methods.

Dr. Salas has received national recognition for research excellence in mathematics through a Scopus-based scientific awards program administered by the Colombian Ministry of Science, Technology, and Innovation (MinCiencias), recognizing high research impact and productivity in internationally indexed publications.

His work bridges rigorous mathematical analysis with applications in engineering, physics, and applied sciences.

Index

Absolute residual, 263, 292
acoustics, 200
Addition formula, 21
aeroacoustics, 200
aeroelastic systems, 208
Algebraic relations, 37, 46
Algebraic system, 24
Algebraic systems, 20
Algorithmic cost, 258, 287
amplitude, 211
amplitude saturation, 202, 203, 207, 208, 211
Amplitude variable, 28
amplitude–dependent frequency shifts, 208
Amplitude–frequency dependence, 19, 37
Amplitude–frequency relations, 18
Analytical mechanics, 193
Angular momentum, 185
Ansatz, 20, 23, 37
Ansatz , 33, 49, 53, 55, 79, 81, 83, 84, 87, 90, 92, 93, 103, 104, 106, 110, 112, 158, 227
Asymmetric potential, 58
Asymmetric restoring forces, 74
asymptotic amplitude, 201
Averaging, 188
averaging, 209, 215
Averaging method, 218
averaging method, 204
Averaging methods, 2, 7, 32, 38

Backbone curve, 220
backbone curve, 207, 209, 214
Bifurcation diagram, 71, 72
Bifurcation diagrams, 76
bifurcation diagrams, 207
bifurcation phenomena, 202
bifurcation structures, 208
Bifurcations, 5
biological rhythms, 208
Bounded solutions, 29, 42, 57

Canonical model, 224
Canonical models, 195
Caputo derivative, 6, 7, 299
Center equilibrium, 28, 29, 32, 41, 43–45, 59, 65, 66, 69
Center of mass, 183
Centers, 37, 75
Chaos, 1, 2, 5, 58–61, 63, 64, 71, 183, 225, 226
Chaos thresholds, 75
Chebyshev , 108, 109, 116, 124, 146, 156, 157
Chebyshev approximation, 187
Chebyshev polynomials, 57
Classical Duffing oscillator, 74
Closed orbits, 42
Complementary methods, 266, 295
complete elliptic integral, 212
complete elliptic integral of the second kind, 213
Complete elliptic integrals, 37
Computational complexity, 258, 287
Computational number theory, 268, 297
Conclusion, 268, 297
Conditioning, 263, 292
Confining potential, 75
Conservation of linear momentum, 194
Conservative Duffing equation, 37
Conservative Duffing oscillator, i, 7, 19, 24, 26, 37, 38
conservative Duffing oscillator, 201, 202, 208, 211

Conservative oscillator, 56, 191, 192
  pendulum, 179
Conservative system, 39, 45, 223
Conservative systems, 19, 37
Constrained mechanical system, 193
Coupled pendula, 218
Critical line, 8, 299
Cubic discriminant, 68
Cubic Duffing equation, 37, 75
Cubic Duffing oscillator, 37
Cubic equation, 64
Cubic equation (for A), 51, 52
Cubic nonlinearity, 19
Cubic polynomial, 75
Cubic–quintic Duffing equation, 75, 180, 193
Cubic–quintic Duffing oscillator, 19, 37, 38
Cubic–quintic potentials, 18
Cubic–quintic Duffing oscillator, 47
Cyclic coordinates, 185

Damped Duffing equation, 223
Damping, 2, 19, 47, 58, 75, 223
Degenerate equilibrium, 28, 41
Deterministic chaos, 75
Discriminant, 21, 24, 41, 64
Discussion, 266, 295
dissipation, 200
Double–well potentials, 18, 30, 32, 37, 43
Double–well duffing oscillator, iii, 87
Duffing approximation
  pendulum, 179
Duffing backbone curves, 215
Duffing equation, iii, vi, 1, 2, 5, 12, 21, 23–25, 39, 48, 50, 52–55, 87, 95, 97, 98, 180, 191, 195, 218, 221, 222, 230, 231
Duffing frequency shift, 210
Duffing nonlinearity, 213
Duffing oscillator, 19, 177, 200, 202, 207, 208, 222
  forced and damped, iii, 90
Duffing stiffness, 211, 214
Duffing stiffness parameter, 202, 207, 208
Duffing type, 193
Duffing–Helmholtz equation, 39, 58, 64, 67, 180
Duffing–Helmholtz model, 183
Duffing–Helmholtz oscillator, 5, 7, 74
Duffing–Helmholtz potential, 182
Duffing–type oscillator, 202
Duffing–type oscillators, 6, 8, 195, 207, 298
Duffing-type equation, 191
Duffing-type framework, 46
Duffing-type oscillator, 187
Duffing-type oscillators, 18
Duffing–Helmholtz equation, 39
Duffing–Helmholtz oscillator
  damped and forced, iii, 92

Effective equation, 189
Effective forcing, 194
Effective potential, 185–187, 189, 191, 194
Effective stiffness, 41
electronic oscillators, 208
Elementary functions, 38
elliptic averaging method, 210
Elliptic function approximations, 17
Elliptic function transformations, 18
Elliptic functions, 2, 7, 10, 38, 223, 298
elliptic functions, 210
Elliptic functions , 155
Elliptic integral, 66, 110
  evaluation, 258, 287
Elliptic integral approximations, 21
Elliptic integrals, 10
Elliptic invariants g_2, g_3, 49
Elliptic modulus, 10, 18, 37
elliptic modulus, 211
elliptic moments, 212
elliptic orbit, 213
Elliptic parameters, 37
elliptic period, 212
Elliptic refinement, 268, 297
Elliptic reparametrization

complexity, 258, 287
Elliptic response, 18
energy balance, 200–202
Energy conservation, 19
Energy first integral, 27, 37, 40, 45, 46, 56, 58, 65, 180
pendulum, 179
Energy integral, 37, 47, 185, 224
Energy levels, 37
Equilibrium point, 45
Equilibrium points, 28, 40
Error metrics, 21
Error metrics , 81, 86, 87, 89, 120, 122, 130, 131, 157, 160, 161, 227
Escape to infinity, 29, 42, 44, 75
Euler angles, 183
Euler–Lagrange equation, 179, 180
Even function, 45
Even powers, 45
Exact solution, 48, 51
Exact solutions, 38, 75
Experimental mathematics, 8
Extensions of the Duffing equation, 37
External forcing, 19, 58

Fast–slow decomposition, 188
Finite differences , 152
First integral, 74
first–order averaging, 214
Forcing, 47
Fourier series, 18
Fourier series , iv, 152, 155
Fractional calculus, 5
fractional derivatives, 177
Fractional Duffing equation, 5
Fractional Duffing oscillator, 7
fractional Duffing oscillator, 177
Fractional-order systems, 20
França–Leclair approximation, 268, 297
frequency, 209, 211
frequency correction, 207
frequency shift, 207
frequency shifts, 202
Frequency stiffening, 222

Galerkin Method , iv, 158
Generalized Duffing equation, 19, 39, 57, 71, 74, 182
Generalized Duffing equations, 38
Generalized Duffing oscillator, 74
Generalized Duffing oscillators, 193
Generalized elliptic functions, 18
Geodesics, 170
global attractor, 201, 209
Global boundedness, 75
Global chaos indicators, 71
Gravity, 183

Hamiltonian system, 58
Hamiltonian systems, 26, 37
Hardy Z function
evaluation cost, 258, 287
harmonic content, 207
Harmonic oscillations, 18
Helmholtz (quadratic) oscillator, 51, 52
hereditary behavior, 177
High frequency, 189
High zeros, 268, 297
High zeros of zeta, 263, 266, 292, 295
High–frequency averaging, 187
higher–order asymptotic expansions, 215
Homoclinic orbit, 58, 59, 61, 69, 71, 74
Homoclinic orbits, 75
Homotopy, 79, 93, 96–99
Horizontally moving support, 193
Hyperelliptic integrals, 57
hysteresis, 207

In–phase mode, 218, 221
Initial approximations, 266, 268, 295, 297
Initial conditions, 20, 34, 75, 85, 88, 93, 95, 99, 104, 119, 122, 230, 237
Initial–value problems, 18

Invariant subspaces, 221
Inverted equilibrium, 187

Jacobi amplitude, 11
Jacobi differential identity, 211
Jacobi elliptic functions, 10, 11, 17–19, 37, 75, 224
Jacobi elliptic wave, 211
Jacobian elliptic functions, 20

Kapitza pendulum, 187
KBM method, 204, 207
Kinematic coupling, 194
Kinetic energy, 179, 183, 194
Krylov–Bogoliubov–Mitropolsky method, 218, 219
Krylov–bogoliubov–mitropolski method, iii, 82

Lagrange equations, 194
Lagrange top, 183, 184
Lagrangian, 179, 182, 185, 188, 192, 194
Lagrangian mechanics, 191
  pendulum, 179
Lambert W function
  evaluation, 258, 287
Laplace transforms, 18
Level sets, 27, 40
limit cycle, 201–204, 208, 209, 213, 214
limit cycles, 208
limit-cycle amplitude, 214
Lindstedt–Poincaré method, 204
Linear stability analysis, 75
Linear stiffness, 19
linear viscous damping, 201
Linstedt–poincaré method, iii, 79
Liénard system, 203, 209
LMFDB, 263, 292
Lyapunov exponent, 73
Lyapunov exponents, 76

Mathematica, 75
Mathieu-type equation, 188
mechanical energy, 209
Mechanical interpretation, 74
Mechanical systems, 195
Melnikov function, 60, 70
Melnikov functions, 75
Melnikov method, 47, 58, 59, 67, 70, 71, 75
memory effects, 177
Mixed parity, 180
Mixed parity Duffing equation, 39, 58
Mixed parity Duffing oscillator, 182
Mixed–parity Duffing equation, 74
modulation equation, 213
Multiple scales method, iii, 93
multiple scales method, 207
Multiple time scales, 187
Multiple wells, 30
Multistability, 30

natural frequency, 201
negative damping, 201, 203, 208
Non-autonomous system, 58
non-sinusoidal, 211
Nonholonomic constraint, 191
Nonintegrable system, 223
Nonlinear , 39, 79, 102–104, 151, 159, 161, 231
Nonlinear conservative oscillator, 19
Nonlinear conservative systems, 38
nonlinear damping, 202, 203, 208, 215
nonlinear damping force, 201
nonlinear dissipation, 207
Nonlinear dynamical systems, 1, 6
Nonlinear dynamics, 19
nonlinear dynamics, 208
Nonlinear lattices, 20
Nonlinear mechanics, 183
Nonlinear ODE, 20
Nonlinear optics, 19
Nonlinear oscillations, 18–20, 195
Nonlinear oscillator, 39, 58, 187
Nonlinear oscillators, 74, 180, 193
Nonlinear parametric oscillator, 191
Nonlinear pendulum, 194

nonlinear phase basis, 210
Nonlinear restoring forces, 19
Nonlinear spring, 19
Nonlinear stiffness, 30
nonlinear stiffness, 202, 207, 208, 215
Nonlinear transformation, 37
nonlinear vibration theory, 200
nonlinearity, 200
Nonresonance assumption, 219
Normal form, 46
Normal forms, 57
Normal modes, 218
Numerical conditioning, 266, 268, 295, 297
Numerical examples, 38
Numerical integration, 38
Numerical methods, 4, 7
Numerical validation, 263, 292
Nutation, 185

Odd Duffing equations, 57
Odd polynomial oscillator, 46
Odd powers, 45
Odlyzko method, 266, 295
Order reduction, 195
Oscillation period, 21, 23, 37
Out–of–phase mode, 218, 222

Parameter , 103, 105, 111
Parameter choices, 38
Parameter map, 69
Parameter maps, 67, 75
Parameter space, 68
Parametric excitation, 187
Parametric resonance, 188
Particular solutions, 221
Pendulum, 179
  simple (nonlinear), 179
Period , 104, 118, 130, 133, 152, 161–163, 166
Periodic forcing, 75
Periodic solution , 152, 161, 163
Periodic solutions, 19, 21, 25, 37
Perturbation methods, 2, 7, 32, 38, 225
perturbation methods, 215
Perturbative methods, 79
phase, 211
Phase plane, 27, 37, 40
phase plane, 201, 209
Phase portrait, 186
  pendulum, 179
phase portrait, 208, 214
Phase portraits, 32, 37, 43, 47, 59, 67, 76
Phase space, 2, 5
Phase transitions, 19
phase–plane, 203
Phase–plane dynamics, 75
Phase-plane analysis, 38
Physical interpretation of parameters, 37
Physical parameters, 37
Planar system, 64
Platt computations, 266, 295
Poincaré map, 58, 60
Poincaré section, 72, 73
Poincaré sections, 76
Polynomial Duffing models, 37
Polynomial restoring force, 74
Polynomial restoring forces, 195
positive damping, 201, 203
Potential energy, 75, 179, 183, 194
potential energy, 202
Potential function, 27, 31, 37, 40, 43, 46, 56, 65, 180
Potential functions, 74
Pure quintic oscillator, 224

Quadratic nonlinearity, 181
Quadratic oscillator, iii, 81
Quadratic reduction, 45
Quadratic substitution, 46
Quartic equation (for A), 48, 50
Quasi-periodic solutions, 19
Quintic equations, 57
Quintic nonlinearity, 19

Rayleigh oscillator, 200, 202, 208

Rayleigh, Lord, 200
Rayleigh–Duffing oscillator, 202, 207
Reduction algorithms, 75
Reparametrization, 18, 263, 266, 268, 292, 295, 297
Riemann zeta function, 8, 299
Rolling constraint, 192
Rolling cylinder, 191
Root finding
    complexity, 258, 287
Rotational dynamics, 183
Runge–Kutta methods, 22

Saddle equilibrium, 28–30, 32, 41, 43–45, 59, 62–71, 74
Saddle points, 37
Saddles, 75
Second–order approximation, 220
second–order approximation, 204, 207, 210
Second–order differential equation, 193
Second–order differential equations, 18
second–order nonlinear ordinary differential equation, 207
Second–order ODE, 179
self–excitation, 202, 208, 211
self–excited oscillations, 215
self–excited oscillator, 200, 202
self–excited vibrations, 207
self–sustained oscillation, 203
self–sustained oscillations, 201
self–sustained vibrations, 208
Separatrices, 30, 37
Separatrix, 2, 32, 42, 43, 47, 63–66, 70, 71, 180
Separatrix splitting, 63, 183
Shift of variables, 46
Simple pendulum, 193
slowly varying amplitude–phase, 204
Slowly varying parameters, 18
Smale horseshoe, 58, 59, 73
Small oscillations, 191
Small–angle approximation, 218
Small-amplitude frequency, 32
Softening effects, 19
Softening nonlinearity, 218, 220
Soliton theory, 20
Solitons (solution), vi, 36, 222, 231
Solution approach
    Fifth approach, ii, 55
    First approach, ii, 47
    Second approach, ii, 49
    Third approach, ii, 52
solvability condition, 206, 210
Spin, 185
Spring force, 39
Spring pendulum, 180
Stability, 187
Stable equilibrium, 31, 43, 190
Stable manifold, 59, 63
Stiffening effects, 19
Stiffness (nonlinear), 39
Structural mechanics, 19
Symmetric top, 183

Third harmonic, 220
Three equilibria, 69
Time reparametrization, 224
Time–dependent Lagrangian, 188
Transformation techniques, 38
Transversal intersection, 75
Transverse intersection, 60, 61, 63
Trigonometric approximation, 10, 18
Trigonometric approximation , 113
Trigonometric functions, 18
Trigonometric functions , 155

Unbounded solutions, 25, 37, 42, 44
Unifying framework, 20
Universal normal form, 2
Unstable equilibrium, 32
unstable equilibrium, 209
unstable equilibrium point, 201
Unstable manifold, 59, 63

Van der pol , 114
Van der Pol amplitude saturation, 210
Van der Pol damping, 208, 214
Van der Pol oscillator, 200, 202, 208
Van der Pol–Duffing oscillator, 208, 210, 214
Variable modulus, 18

waveform, 209
Weierstrass elliptic function, 10
Weierstrass elliptic functions, 75
Working precision, 263, 292